Transitions towards sustainable agriculture and food chains in peri-urban areas

Transitions towards sustainable agriculture and food chains in peri-urban areas

Edited by:
Krijn J. Poppe
Catherine Termeer
Maja Slingerland

Wageningen Academic
Publishers

ISBN 978-90-8686-117-0

First published, 2009

Table of contents

Preface

The Netherlands is a highly urbanised country, situated in a fertile delta with a major export industry in food and flowers. The resulting negative environmental externalities and new consumer concerns like those about animal welfare as well as new demands from the general public for quality recreational areas, health care and local food products are all forcing agriculture to change. Today, the Dutch agri-food industry is facing the threefold challenge to ensure continued profit for the producers; to offer quality products for the people; to minimise pollution of the planet. To meet this challenge, the government has been calling for a 'transition towards sustainable agriculture'.

Producers (farmers), processors (the agri-food industry), the retail sector and governments are struggling to deal with this challenge. In the Netherlands, the government supports this transition towards sustainable agriculture through its main policy instrument – research and education. This book presents the expertise from at least five years of Dutch research by scientists who have been actively working to promote the desired transition. Our aim has been to collate the results of our experiments, to learn from them, to confront them with existing theory, and to share them with a larger audience in order to foster learning about transition. In 2007 the contributors to this book started a series of monthly meetings, called 'professional conversations' as a community of practice to learn from each others' work and to gain new insights by confronting practice with theory. This was the basis for an international conference in 2008 in Wageningen (www.agricultureintransition.eu) and for this book.

Nearly all the papers have been extensively discussed by the group of contributors gathered in these professional conversations. The discussions have also been used to improve the papers. Most of the papers were presented in a joint poster session at the 2008 conference. We chose the poster format to promote interaction and to balance out the contribution from Wageningen UR.

We would like to thank Linda van Mosel (LEI Wageningen UR) for her management assistance in organising the professional conversations and putting together this volume. The production of the book would not have been possible without the skill and efficiency of Mary Montanus, who corrected the English and subedited the manuscripts, thereby helping the editors and authors to maintain the common format of the chapters. Thanks are also due to the Dutch Ministry of Agriculture, Nature and Food Quality for their financial support for this project through their Knowledge Base Fund.

We hope that the book not only furthers understanding about the transition towards sustainability, but that it contributes towards international collaboration on transitions in agriculture. To this end, we look forward to receiving feedback from our readers.

Krijn J. Poppe, Catherine J.A.M. Termeer and Maja Slingerland
Summer 2009, Wageningen UR

Chapter 1

Introduction

Maja Slingerland and Rudy Rabbinge

1.1 Characteristics of Dutch agriculture: continous change

In the second half of the 19th century produce prices in Europe plummeted, causing an agricultural crisis. This was partly a result of the import of cheap small grains from the New World (USA). The reaction to this crisis differed in each European country (Koning, 2004). The United Kingdom decided to liberalise and many farmers went bankrupt but found employment in new industries created after the industrial revolution. The Germans and French chose protection and closed their borders. The Dutch, already a trade nation, chose neither complete protection, nor complete *laissez faire*. They found a third option consisting of strengthening their competitive ability through land reform, stimulating cooperatives with market power and very importantly stimulating knowledge and innovation through private-public investment. During the last century this became the typical attitude and behaviour of the various countries in Europe when crisis occurred. You could characterise Dutch agriculture as being a sector that successfully exploits the constant dynamics of change to improve its competitive power and in doing so, it makes the best possible use of knowledge and innovation.

The focus on science, technology and innovation in the Netherlands has been and still is a critical factor in the development and viability of agriculture. This reflects the Dutch vision that dynamics and change should be seen as an opportunity and a challenge rather than as a threat or an unneeded and undesired liability. This book builds further on this tradition. It describes the changes that are possible and necessary to maintain agriculture as a powerful and prosperous branch of the Dutch economy. In this book, we look at agriculture in its broadest sense including the entire agro-food complex consisting of production, processing, logistics and distribution, retail and consumers. The swift changes of the last few years require an appropriate answer in developments that have taken place and need to take place in Dutch agribusiness related activities. During the last three decades, this process of change, dynamics and renewal has continued and a number of mega-trends (Rabbinge, 2001) can been discerned which have characterised successful developments within the agro-food complex.

1. Economies of scale and technological innovations led to high efficiency in production per hectare, per man hour and per unit of input (fertiliser, water) in crop production, and also per animal and per unit of feed.
2. Agriculture changed from being a craft into an industry which was possible because of technological developments and knowledge-intensive changes. Several forms of production did not require much land anymore; intensive pig and poultry production

takes place entirely in the barn, and flower and vegetable production largely takes place in greenhouses. This intensive form of production is performed under highly controlled environmental circumstances with high-tech self-regulating feedback systems for temperature, humidity, light, disease and predator control, nutrient and water provision, etc.

3. Chain organisation and management replaced individual buyer-seller transactions. At chain level, the whole process from primary production to international trade is covered, including quality control, processing, packaging, transport logistics, improving efficiency, increasing end-users satisfaction and the scale allows for large investments in the private sector.
4. The current globalisation fits nicely in the Dutch tradition of international exploration and trade (related to international transport) and can therefore be seen as another opportunity. Perhaps not an opportunity to sell primary produce (which can be produced more cheaply where labour and land are less expensive) but an opportunity to sell Dutch technology, logistics systems, seeds or value- added products worldwide.
5. Multiple goals now have to be served. Production and productivity alone are no longer sufficient. In today's world, agriculture also has to contribute to landscape, nature conservation, environmental issues and an attractive rural area for the urbanised population.
6. Food and feed have to promote health and have to respond to changing lifestyles.
7. The traditional linear knowledge model that functioned very well in the research-extension-education triptych is now replaced by a more participatory, iterative knowledge model with the buzzword 'co-innovations'.

These trends have considerable effect on the way farmers and other stakeholders in the agro-food complex operate.

One important characteristic of Dutch farming has been its organisation and management as family farms. This concept led to the conservation of a certain degree of diversity among Dutch farming systems and also prevented extreme industrialisation with professional managers and wage labour. However that characteristic is currently much less prominent in most branches of Dutch agriculture. For example, professionalisation of management and a high degree of technology can be found in glasshouse cultivation, flower bulb production, dairy farming, etc. explaining the competitive power of these branches (Jacobs *et al.*, 2005). The lack of uniformity has been an asset as it has allowed for a diversity of adaptation strategies to cope with new and complex issues that have required Dutch farming to change.

Until now we have discussed agriculture's role as that of food and non-food producer yet occupation (land use), planning, management and government of the rural area are other possible entry points for involvement by agriculture. Rural land is not just there for agricultural purposes. There are many other possible designations such as housing, industry, conservation and infrastructure. In the past, the designation for rural areas in the

Netherlands was mainly planned by national policy-makers. Clear examples are polders like the 'Flevopolder', the dikes and dams in the 'delta works' and, more recently, railway transport infrastructure such as the High Speed Train line (TGV) and the 'Betuwe line'. On the other hand, in the domain of nature conservation, initiators are mainly private individuals, NGOs or agencies that either buy a property or manage a property on behalf of the state. The national government, being responsible for compliance to international regulations, plays a role as well and is responsible in particular for infrastructural needs such as the creation of networks of natural conservation areas (Natura 2000 – EC, 2005). The national government regulates environmental pollution and controls compliance to these regulations. Space designated for housing, industry or nature is decided through negotiations between different levels of formal government: national, provincial and local municipalities. The distribution of tasks and decision-making powers between levels of government is dynamic and tends to change over time. Civil organisations may take part in negotiations and discussions or even challenge a government decision in court.

1.2 Challenges to agriculture: current demand for change

The Netherlands is a small country with a high population density. The level of urbanisation is high. Cities are rigidly demarcated from the countryside, which traditionally belonged to the farming communities and to a lesser extent served public functions (e.g. natural dunes function as catchment areas for clean drinking water and as a security system against the sea; natural conservation areas have a recreational function and sustain biodiversity). Continuous population growth and urbanisation, increased standard of living and demand for housing, all put pressure on the rural space. The global climate change discussions and international treaties on biodiversity and reduction of emissions also drive an increased demand for space. Add to that the claims of social welfare and luxury housing projects, industry, infrastructure, leisure, water catchment, conservation, preservation of the national landscape heritage, etc. and you have huge demands on the (limited) rural space. These functions compete for resources (land, water) and it is not a given that agriculture will remain the major activity outside the city. In view of the number of new functions and associated new players in the rural areas, the role of agriculture in shaping and managing the landscape is once again similar to what it was in the centuries before these changes began. Multifunctional land use demands a combination of functions in and outside the agricultural production system and requires new ways of governance at every level affecting rural areas.

The position of agriculture has also changed with respect to its relationship to society as a whole. A shift took place from appreciation to criticism. Side-effects of over-intensive production methods, such as pollution of ground water, residues of pesticides in the food chain and epidemics within animal populations, have discredited agriculture despite its success in food production and its contribution to the Dutch GDP. More recent debates about animal welfare and about greenhouse gas emissions added to this negative picture. Agriculture lost its license to produce and license to sell. To regain the license to operate agriculture needs to shift to more environmentally friendly production methods and to restore consumers' trust in the quality of its produce, for instance, by increased transparency and communication and by providing the consumer with additional functions that are perceived as being positive.

Farmers do not only have to deal with Dutch laws and regulations but they also have to conform to European directives that may impact the boundaries within which they have to operate and that also tend to impact rural areas at large. A whole system of agricultural incentives and levies exists and farmers need specific expertise to find the most profitable way to deal with them. Another factor they have to consider is the international context currently shaped by liberalisation and globalisation. Agricultural trade is increasingly becoming a global activity which increases competition in consumer markets. Maintaining and strengthening the agribusiness' international position, both in a changing international context and in the context of competing claims and social debate, requires new strategies and approaches. The very successful, efficient production and trade of bulk foods by the Dutch agro-food complex is currently being taken over in part by producers with lower labour costs and land prices and with less strict environmental and social laws and regulations (e.g. China) especially as transport to the market tends to become cheaper. To survive it might be a relevant strategy for Dutch agriculture to adopt an approach aimed at added value targeting specific international consumer groups. Agriculture must also accept that corporate social responsibility has become a pre-condition for the license to produce. Sustainable production methods, taking into account the planet and people as well as profit, have become a must for the food industry (Cescau, 2007).

In summary, agriculture is challenged to:

1. find a good balance between economic, ecological and socio-cultural factors;
2. cope with social resistance to the dominant technical and economical rationality;
3. perform well within the complex governance circumstances reigning in the rural area because of imposition of national, European and international laws, and of rules, regulations and treaties for the different sectors which use the rural space;
4. position itself in a competitive way in the global market.

All these issues imply that there are many different players in the field in the agro-food complex. When different functions and stakeholders meet in the rural area each brings its own set of rules and regulations to comply with, its own visions for the future and its own vested interests. Negotiation among these players for the scarce resources (land, water) and harmonisation of the different rules and regulations into a feasible and legal system of governance of the rural area is therefore a highly complex matter.

These challenges are not new, but they require more than ever before, that agriculture adjusts to its new role and position in the rural area and the global market. Agriculture is challenged to intertwine spatial, environmental and socio-economic values to deliver new services, to contribute to spatial quality, vitality and liveability of the rural areas, and to develop new ways to compete through added value and intelligent logistics for agribusiness in the global market. Such a transition requires not only technological innovation but also a change in the knowledge infrastructure and adaptations within governance processes. It is a combination of the appropriate hardware (technologies), software(socio-cultural traits) and organisational-ware (institutions) that is required.

1.3 Change and transition

In 2001 the 4th National Environmental Policy Plan (NMP4) was released (VROM, 2001). In this document the Dutch government formulated its objective that the Netherlands should be a sustainable society within 30 years. The problems that need to be overcome, especially in the environment, were perceived to be persistent and pernicious and not solvable simply by incremental improvements or changes. The policy document (NMP4) mentions seven such persistent problems: loss of biodiversity, climate change, overexploitation of natural resources, threats to health, danger of external security, decrease of quality of living environment and uncontrollable risks. To solve these problems and to reach the objective of a sustainable society, (inter)national structural societal changes are needed, which are called transitions. Based on scenarios for desired sustainable futures, the Dutch government proposed to support and implement technological, economic and institutional changes which would lead to transitions which promote sustainability. They formulated four specific transition areas:

- sustainable energy;
- sustainable use of biodiversity and natural resources;
- sustainable agriculture;
- sustainable mobility.

Transitions themselves are not new. Society has always been subject to constant change. In the past, Dutch society successfully made the change from an agrarian society to an industrial society and the importance of information technology in our current Dutch society can be seen as a result of another transition process. These changes were only called transitions after the fact and were not planned or managed by policy to become a transition. In fact the objectives of these transitions were not determined beforehand, but the transitions and their directions emerged as the result of ongoing developmental processes. What is new is that since 2001, the Dutch government, has decided to develop policy deliberately aiming to promote transitions, and they want to include ordinary citizens, private sector parties, knowledge institutions and civic organisations in the process. From 2001 onwards transitions were not only planned in advance but also had a predetermined direction: a sustainable society. One of these 'planned' transitions is the transition to sustainable agriculture (VROM, 2001).

1.4 Characteristics of transition

Development or change in time can be seen as linear and be defined separately for each domain: technology development, knowledge development, economic growth, increase in wellbeing, etc. Development can also be seen as discontinuous or non-linear and as a product of interaction between domains: co-evolution. It is theoretically possible to predict the future in linear development models by extrapolating trends in each domain – assuming business will go on as usual. However the interplay of factors leads to co-evolution between domains and occasionally to non-linearity or disruption of trends. An example of linear development is the development of higher and stronger dikes, and a system of building more and more dams and making them increasingly efficient to prevent water from flooding the land. An example of co-evolution can be seen in the agricultural landscape: people and nature together have been continuously shaping the rural area. Because of this, you can see large-scale mechanisation on flat fertile soils where it is possible and viable to grow crops, which leads to monotonous large fields with monocultures, while peat areas are more likely to be grazing land and marginal areas and areas with steep slopes tend to end up as nature reserves. Non-linearity can take place in a variety of domains separately or simultaneously. When non-linearity becomes so important that it reshapes our society, we call it a transition. One example of a series of changes is the energy transition: a change from wind energy (sailing, windmills), via wood and coal for the steam engine, to fossil fuel to feed the combustion engine for transport and to provide electricity for many functions in our society. Today, another transition appears to be taking place from the fossil fuel-based to the bio-based economy, using biomass for fuel, but also replacing other oil-based products such as chemical components in plastics, paint, etc. Another transition is currently taking place in water management. As a consequence of global climate change and because of human intervention in the countries where Dutch rivers originate or flow through, the water levels in the rivers and in the sea will increase so much that the Dutch won't be able to keep the water out simply by building progressively higher dikes. A transition is needed from measures to exclude the water to water management designs whereby we can safely live with the water. Instead of a sharp delineation between water and land functions such as housing, grazing animals, etc., more emphasis has to be on combining these two elements, e.g. floating houses. Similarly one can argue that Dutch agriculture itself is going through a transition, providing biomass for the bio-based economy, providing services to health care, nature conservation, etc.

Why do we call these relatively abrupt changes a transition? What are their characteristics? Can transitions be understood? And, more importantly, can transitions be managed? There is clearly a need for research on change processes – which can, potentially, be called transitions.

1.5 Research on transitions and change

Not all activities designed to promote a more sustainable society are dependent on transition policy by the Dutch government. Different parties in society, including industry, have started their own experiments and achieved their private innovations without being promoted or supported by transition policy.

The issues facing agriculture cover multiple domains and involve a multitude of formal and informal institutions. In addition they combine processes that play in the short and long term and at the interface of local, national and global levels. Coping with these issues or, more importantly, trying to provide solutions to problems that arise in all these domains, requires new paradigms, new technologies and new social structures. It also requires that certain trends be disrupted and new avenues be explored. The dominating trend in Dutch agriculture towards ever increasing efficiency in producing bulk products at low cost may be complemented and perhaps even replaced in the future with the production of specialty products for premium prices aimed at local urban consumers. The current trend to intensive crop and animal production creating externalities such as pollution of the environment could be changed in two possible directions: high technology-based agro-production parks or low external input-based organic agriculture, both addressing a specific market. Yet modifying farmers' behaviour and reinforcing this new behaviour is not a solution in itself. Society needs to change as well: consumer behaviour (price consciousness, product choices), retail (visibility and appreciation of attributes of food items), laws and regulations (certification; environmental impact assessment methods for enterprises combining industrial and agricultural functions), logistics (organic separated from conventional), etc. This whole complex of interrelated changes is called a transition and requires new technologies, new policies but also new governance structures and institutional arrangements. Transitions involve new stakeholders and require them to reassess their vision of their own role. Traditional system boundaries creating clear distinctions between urban and rural areas, between water and land, between industry and agricultural production, between policy makers and citizens, between scientists as knowledge-creator and farmer as knowledge-applier will disappear. A combination of functions and a combination of partners involved is needed to create a new society.

Given the complexity of the task ahead, the issues require expertise that not only looks at technology but also at governance, acknowledging that technology is developed and used by people. Furthermore, complex issues cannot be solved by one discipline alone. The nature of training and research should therefore increasingly be *inter*disciplinary and *trans*-disciplinary.

Interdisciplinary activities aim to bridge the gap between disciplines and to go beyond disciplinary boundaries. The object is to create new concepts and to advance scientific knowledge, to increase insight into complex problems and to being able to provide methodologies and options that better address complex issues. New combinations of functions require new bridges to be built between scientific domains, e.g. agriculture and chemistry (bio-based economy), agriculture and medical sciences (care and health farms) and agriculture and pharmacy (neutrigenomics, new plant-based products for use in health care, such as neutriceuticals) to name but a few.

Trans-disciplinary activities aim to bridge the gap between science and society. It is based on the explicit acknowledgement that knowledge is not exclusive owned by scientists but also by other parties in society. A trans-disciplinary approach claims that knowledge is generally co-created in interaction between different knowledge-holders. Instead of science looking at the context of complex problems as a given and addressing just one aspect of this complexity (the reductionist approach), science takes the complexity itself as a topic for research (holistic approach) and/or participates in the context as one of the stakeholders.

Whereas new combinations between theory and practice lead to new insights, new combinations between science and society lead to renewal of the knowledge infrastructure. The emphasis on the knowledge system can be seen in the project plan launched in February 2003 called 'Knowledge Network Transition Sustainable Agriculture' (Stichting Innovatie Netwerk Transitie Duurzame Landbouw, 2003).

Wageningen University and Research centre (Wageningen UR or WUR) is an essential part of the knowledge structure that has traditionally provided training, research and policy advice in the domains of agriculture, food and rural development. In view of the required change in the domains that WUR covers, it has a role to play both in the content and process of change. Realising that the traditional linear knowledge system – from fundamental to applied research to knowledge transfer in education and extension – no longer applies, it has taken up the challenge to develop a new role in the light of the transition process. To participate in the complexity of the problems in agriculture and rural development and to contribute to the transition process WUR will focus on demand-driven interdisciplinary and trans-disciplinary approaches and engagement with stakeholders. Researchers somehow need to become part of the knowledge creation and change process, and scientists can also research the learning and change processes and reflect on them. To this end, Wageningen University has been transformed into a so-called third generation University – a development that will be further discussed in Chapter 3.

1.6 The rationale behind this book

Given the complexity of problems in the agro-food complex, and the rural sector as a whole, and their importance for the future, they cannot be ignored by government, nor anyone else involved. Not all activities which foster a more sustainable society are dependent on transition policy by the Dutch government. Different parties in society, including industry, have launched their own experiments and achieved commercial innovations without being promoted or supported by transition policy. As we write, there are many projects struggling with experiments and innovations that might contribute to this transition and at the same time, a large number of scientists is trying to develop theories to explicate the complex problems in transition in the above-mentioned domains. It is time to collate the results of the experiments, to learn from them, to confront them with existing theory, and to share them with a larger audience in order to foster learning about transition. This learning will be based on past experiences, and progress in different disciplines particularly those disciplines explicitly relating to transition theory. This book is an attempt to make a contribution to that learning process. It will refer to existing literature on theories on transitions, learning, change processes, institutional economy, (complex) system theory, innovation, governance, monitoring and evaluation, etc. In Chapter 2 there will be an overview of these theories in as far as they appear to contribute towards answering the question whether transitions can be managed or not.

Theory and practice are both constantly evolving but not necessarily in interaction with each other. This book aims to juxtapose the two, not only does it enrich and sharpen the current theory, but it also looks at transitions in the field, offering theoretical insights which may help to analyse the situation in practice, in order to jointly discern general principles that would allow for more *adequate performance* in future situations. The confrontation of theory and practice aims at mutual learning; to make an inventory of what is being achieved and to provide a methodological underpinning for these achievements.

The main part of the book (from Chapter 4 onwards) particularly addresses the interaction between practice and theory. Each chapter consists of 6 sections. A case is briefly introduced in the first section, using a real life example. Then the case is abstracted to its main question/problem or its transition task. In the following section a theory is introduced for two reasons: either to allow analysis and deeper understanding of the question/problem or to provide the legitimacy of an intervention(s) in the case. In the fourth section the case is revisited. Depending on the choice in section three: it will be either reinterpreted based on the presented theory which will then serve as a basis for future interventions, or else the theory-based intervention will be evaluated. In the fifth section the lessons that were learned from the theory for the case are highlighted. In the sixth section lessons gleaned from the theory will be considered in their relevance to, and applicability in practice.

The authors of the individual chapters are either participants in a case in practice or proponents of the theories. The authors first came together in evening meetings called 'professional conversations'. During these 'conversations' the practitioner presented a case with emphasis on what the practitioner thought to be the transition task in his particular project (15 minutes). After the presentation, a theory was presented (30 minutes), followed by a discussion (45 minutes) in which the focus was not on analysing or solving the problems raised by the case but on the contribution of the case to the theory. These 'conversations' have motivated the practitioners to think of their cases as examples of transition experiments and, as such, provided them with additional insights. At the same time, the theories needed to be explicit about their usefulness for the transition processes in practice. The contributions stimulated high quality discussions that were much appreciated by the participants. An additional effect was that participants remained in contact with each other afterwards so that the interaction went far beyond the organised evening 'conversations'. This positive experience motivated the group to invest in writing this book, aimed at integrating the lessons learned in a larger framework and sharing their insights with potential readers. Researchers in transition issues in other domains as well as in other countries are the potential audience for this book.

The chapters of this book have been been loosely ordered from micro to macro. Chapter 2, written by Art DeWulf and colleagues provides theoretical notions on transition management and compares transition management to other approaches (economics, change management, etc.). It is followed by a chapter written by Rudy Rabbinge and Maja Slingerland, on Wageningen UR in transition towards becoming a third generation university. This describes how the changes in public management and in agriculture have changed the governance of the Dutch knowledge system. Our aim is to promote understanding of the background of the changes going on in Dutch agriculture (as an extension of this chapter) and to understand the working environment and rationale behind the approach of the authors of the different papers in this book.

The first paper that confronts empirical cases with theory is presented in Chapter 4. Irini Salverda and colleagues describe how individuals in rural development look to history to organise themselves to cope with new challenges. The case is tested against the theory of new institutional economics. Also very much at the micro level Rosalie van Dam and colleagues describe the self-organisation of squatters in Amsterdam and home owners in an enclosed community, using among others Castell's network theory as a reference.

Next, are three papers dealing with learning processes as manifested in the farming community. José Vogelezang and colleagues present the network programs in animal husbandry and test them up against theories about knowledge and learning in innovations. Eelke Wielinga and Floor Geerling-Eiff reflect on the same networks from an ecological point of view. Barbara van Mierlo and Marlen Arkesteijn investigate the usefulness of participatory systems analysis to promote learning.

Chapter 9, by Roel During and colleagues, takes the reader abroad, to the world of the European Union's Interreg programs. This is a fertile area for studying aspects of culture and governance, using Luhman's social system theory. In a multinational, but otherwise quite different environment, Frances Fortuin and Onno Omta present a case study in the European food industry. Using innovation theory they explore how the potential successfulness of transition projects in such a setting can be improved.

With these chapters the focus has shifted from learning to planning. Frank Wijnands and José Vogelezang describe in Chapter 11 two transition pathways for sustainable technology development based on forecasting and backcasting. Bram Bos and Peter Groot-Koerkamp describe how needs for improved animal welfare could be synthesised using methodological design together with reflexive interactive design. Animal housing is a complex issue, but developing agriculture in urban city design for a new city is probably even more complex. Andries Visser and colleagues investigate whether the DEED framework is useful in addressing this challenge. Another complex multi-stakeholder environment concerns the control of animal diseases. Catherine Termeer and Geert van der Peet have choosen this topic to reflect on change management and transition.

The last five papers reflect on a more theoretical level on transitions and have, therefore, not always the standard format of the previous chapters. Jim Woodhill presents the experiences in developing countries with institutional learning. Jan Douwe van der Ploeg questions in his essay the well known S-curve in transition theory. Martijn Duineveld and colleagues discuss the neglect of power issues in transition research and therefore warn scientists to refrain from prescriptive advice. Historian Pim Kooij digs into the past to show that transitions are not unique to our time and that diverse transitions can be prevalent at the same time, together influencing the lives of the people in a small village in a peri-urban area. Krijn Poppe focuses on one of those types of transitions, the major technologies in society as reflected in industrial waves, to reflect on the current economic credit crisis and questioning the usefulness of the productivist/post-productivist dichotomy.

The last chapter of the book rounds off with a summary of what can be learned from the papers in this book in terms of the transition towards sustainable agriculture in the Netherlands and the contribution offered by the research system to that development. This includes an explanation of the rationale behind the various transition activities that form the basis for this book. The final section provides suggestions for further research.

References

Cescau, P., 2007 Beyond corporate responsibility: Social innovation and sustainable development as drivers of business growth. Speech by Patrick Cescau, Group Chief Executive of Unilever, at the 2007 INDEVOR Alumni Forum in INSEAD, Fontainebleau, France (25 May 2007).

European Commission (EC), 2005. Natura 2000, Europe's nature for you. European commission, Luxembourg, ISBN 92-894-9214-7.

Jacobs, D., Lankhuizen, M., Van Ark, B. and Stuivenwold, E., 2005. De sterke Nederlandse clusters volgens de Porter-methodiek anno 2003. Onderzoek uitgevoerd in opdracht van het Ministerie van Landbouw, Natuur en Voedselkwaliteit, Rijksuniversiteit Groningen, Groningen.

Koning, N., 1994. The failure of agrarian capitalism: agrarian politics in the United Kingdom, Germany, the Netherlands and the USA, 1846-1919. Routledge: London, UK and New York, USA.

Rabbinge, R., 2001. Premissen, taboes, mythen, paradoxen en dilemma's: megatrends in landbouwontwikkeling en ruimtelijk beleid [grond en ruimte in Nederland]. Spil 173-174: 18-21.

Stichting Innovatie Netwerk Transitie Duurzame Landbouw, 2003. Knowledge network transition sustainable agriculture. Bsik Project Plan, Renswoude, the Netherlands.

VROM, 2001. Een wereld en een wil: werken aan duurzaamheid, vierde Nationaal Milieubeleidsplan (NMP4). nota 01.0433 14548/176. Ministerie van Volkshuisvesting, Ruimtelijke Ordening en Milieubeheer (VROM), Den Haag, the Netherlands.

Chapter 2

Transition management for sustainability: towards a multiple theory approach

Art E. Dewulf, Catherine J.A.M. Termeer, Renate A. Werkman, Gerard R.P.J. Breeman and Krijn J. Poppe

Abstract

Transition management, as a theory for directing structural societal changes towards sustainability, has become a major topic for scientific research over the last years. In the Netherlands, the concept of transitions was adopted by several governmental agencies as one of the leading principles for 'steering' sustainable development. In this paper we focus on the question of how transitions can be influenced or managed, in particular by governmental actors. We will address this question by theoretically comparing transition management theory to a number of related theories on change and intervention, from the fields of economics, organisational change management, multi-actor collaboration, network governance, policy agenda setting, social learning and adaptive management. From this selective comparison, we argue that (1) these related theories put the managerial assumptions of transition management into perspective, by adding other steering roles and leadership mechanisms to the picture; and (2) transition management tries to overarch a lot of diversities in one theory, while we suggest a multiple theory approach could be more useful for dealing with the enormous challenge of sustainability.

Keywords: change theories, transition management, transition theory

2.1 Introduction

Transition and transition management (Kemp *et al.*, 2007; Loorbach and Rotmans, 2006; Rotmans *et al.*, 2001) have become major topics in scientific research and policy practice over the last years. Transitions are defined as a gradual process of change which transforms the structural character of a societal domain (Rotmans *et al.*, 2001). Transition management aims at influencing the direction and pace of transitions towards a more sustainable society (Loorbach and Rotmans, 2006). For more than 30 years environmental issues had been discussed in different policy arenas, resulting in policy plans and many policy measures. Progress under these earlier plans was considered insufficient to prevent environmental degeneration let alone that they would promote sustainability. The growing recognition of the inter-related nature of contemporary societal problems and the call for fresh approaches and forms of governance has contributed to the rise of the concept of transition management

(Shove and Walker, 2007). The recently burgeoning literature on transitions and the adoption of transition management by government agencies, especially in the Netherlands, testifies to the influence of this concept. In the Netherlands, where the concept has been developed, transition management was used as one of the leading principles for 'steering' sustainable development during the formulation of the Fourth National Environmental Policy Plan in 2001 (Smith and Kern, 2007; Termeer, 2004). The National Environmental Policy Plan project team initiated discussions about the desired steering philosophy, because the tools to implement ambitious policies to foster sustainable development were lacking. Public servants and scientists engaged in extensive discussions on current theoretical concepts. Transition management (Rotmans *et al.*, 2001) emerged as a promising concept and became adopted as a guiding principle for public policy. As a result, transition teams were created in four departments, including a transition team for sustainable agriculture at the ministry of Agriculture, Nature and Food Quality (LNV). Although this transition team was disbanded in 2005 (Aalders and Selnes, 2005), the concept of transition is still alive. The concept is no longer restricted to the domain of policy making. In different places in society transition proves to be an attractive concept for inducing sustainable development. Scholars and practitioners have been developing different frameworks to steer or to facilitate transitions towards a more sustainable future.

In this chapter we address the question if and how transitions can be managed or steered. We analyse this question theoretically by discussing a range of related theories on social change and intervention. In doing so, we avoid the trap of considering transition management to be 'the only model in town, and for exploring other social scientific, but also systemic theories of change' (Shove and Walker, 2007: 768). To clarify our point of reference about transition management we start with summarising important aspects of transitions and transition management, based on three core publications (Kemp *et al.*, 2007; Loorbach and Rotmans, 2006; Rotmans *et al.*, 2001). In the second section we will discuss theories from the fields of economics, organisational change management, multi-actor collaboration, network governance, policy agenda setting and social learning, and the relation of these theories to transition management. In the third section we systematically compare these theories by presenting and discussing a table where key features of all treated theories are assessed. Finally we formulate our conclusions regarding transition management as a theory for social change and intervention.

2.1.1 Transitions

Basic assumption underlying the transition model is the diagnosis that environmental problems are not caused by clearly identifiable actors or factors but by failures of a systemic nature. As most policy strategies are not able to tackle system failures they will lead to suboptimal solutions (Kemp *et al.*, 2007). 'Sustainable development requires structural changes in social-technical systems and wider societal change, in beliefs, values and governance that co-evolve with technology changes' (Kemp *et al.*, 2007: 78). Transitions are linked to system innovations

(Loorbach and Rotmans, 2006), which are much broader than just technological innovations (because the current societal *regime* is supposed to change, cf. infra), and much more radical than incremental system improvements (because the change involved is a transformation of the system).

Transitions are defined as a gradual continuous process of change where the structural character of a society (or complex subsystems of society) is transformed (Rotmans *et al.*, 2001). The co-evolution of a set of slow changes forms the undercurrent for a fundamental change. Transition processes involve multiple actors within a societal subsystem and fundamentally change both the structure of the system and the relations between the actors. Historical examples include transport transitions from sail to steam ships or from horse to car, and the energy transition from coal to gas.

The transition concept was inspired by the dynamics of demographic transitions (Rotmans *et al.*, 2001). These occur, for example, when improvements in hygiene and health care lead to a falling death rate, while the birth rate only starts falling in a later phase. This results in a significant population growth until the system stabilises at a low birth and death rate, yielding the typical S-shaped curve.

Transitions are not linear processes, but involve a shift in the system from one dynamic equilibrium to another equilibrium, over four consecutive *phases* (Loorbach and Rotmans, 2006; Rotmans *et al.*, 2001), as shown in Figure 2.1:

- A *pre-development* phase of dynamic equilibrium, where there is very little visible change at the systems level but a great deal of experimentation at the individual level.

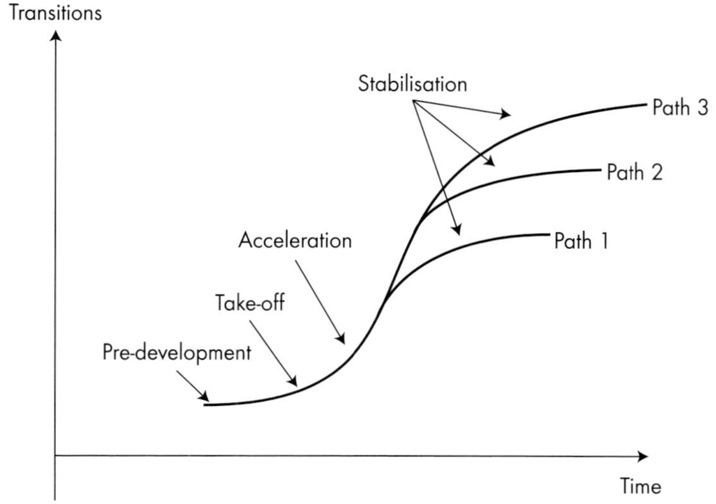

Figure 2.1. The four phases in transitions.

- A *take-off* phase, where the process of change gets under way because the state of the system begins to shift because of different reinforcing innovations or surprises.
- An *acceleration* phase, where visible structural changes take place through an accumulation of socio-cultural, economic, ecological, and institutional changes.
- A *stabilisation* phase, where the speed of social change decreases and a new dynamic equilibrium is reached.

Borrowed from socio-technical systems literature, a distinction is made in transition literature between micro, meso and macro *levels*, respectively referred to as niches, regimes and landscapes (Loorbach and Rotmans, 2006; Rotmans *et al.*, 2001):

- At the macro-level the *landscape* is determined by slow changes in material infrastructure, political culture and coalitions, social values, worldviews and paradigms, the macro economy, demography and the natural environment.
- Operating at the meso-level are the social norms, interests, rules and belief systems that underlie companies', organisations' and institutions' strategies and political institutions' policies. This level is called the *regime* level.
- Acting on the micro-level are individual actors, technologies and local practices – the *niche* level.

At the niche level, variations and deviations from the existing regime can occur (e.g. new technologies or social practices). The socio-technical regime often acts as an inhibiting factor in the early stages of transition, reducing variety and deviations, but it can transform into an enabling factor when a new socio-technical regime is forming and gaining momentum. The landscape factors guide the flow of transitions but remain relatively unaffected themselves (Rotmans *et al.*, 2001). Figure 2.2 (Geels and Schot, 2007: 401) represents this process in more detail.

Figure 2.2 illustrates the view that transitions involve the disintegration of an existing ongoing socio-technical regime, seen as a configuration of industry, markets, policy, technology, science and culture. Niche-innovations break into the existing regime, whose disintegration allows for the emergence of a new configuration of elements, which is then stabilised into a new regime.

2.1.2 Transition management

A distinction can be made between two types of transitions (Loorbach and Rotmans, 2006): evolutionary transitions, in which the outcome is not planned in a significant way; and goal-oriented transitions, in which goals or visions or the end state guide the process. Although most of the studied historical examples look like evolutionary transitions, the promise of this whole transition approach lies in goal-oriented transitions or in the idea that transitions can somehow be steered or managed.

The question is then: how can transitions be managed? Here's a brief summary:

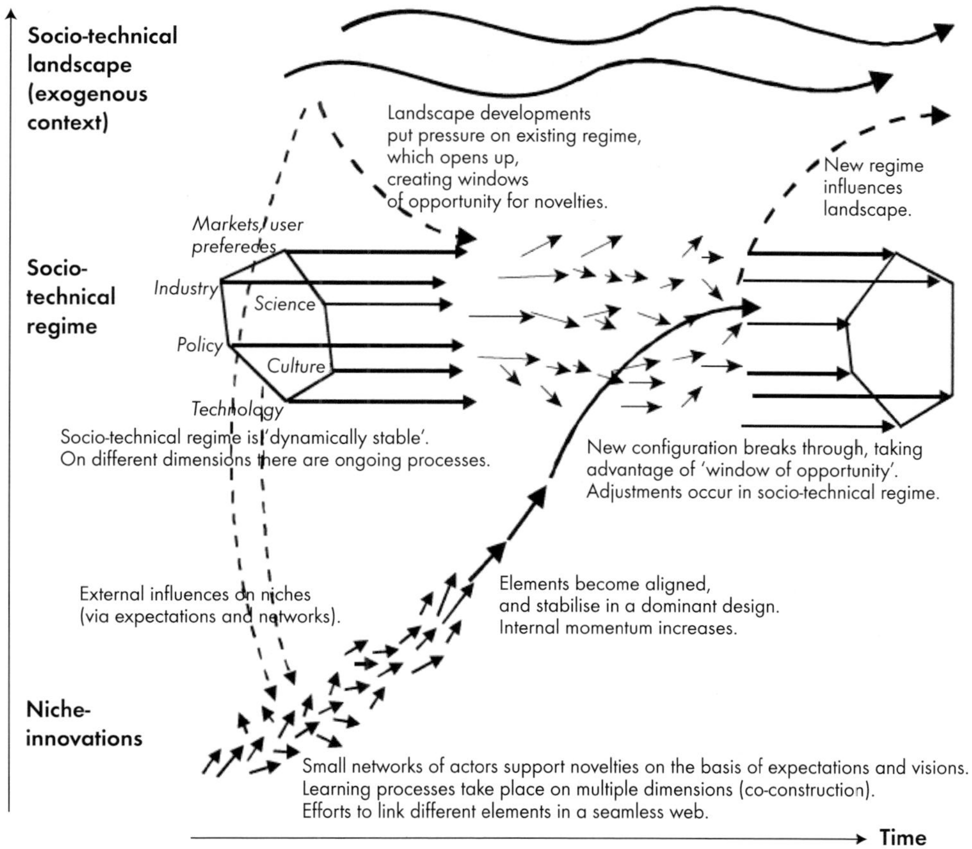

Figure 2.2. Multi-level perspective on transitions (Geels and Schot, 2007).

> *'Although transitions cannot be managed in terms of command and control, they can be managed in terms of influencing and adjusting: a more subtle, evolutionary way of steering. In other words, the direction and pace of transitions can be influenced, even if not controlled directly. Transition management therefore aims to better organise and coordinate transition processes at a societal level, and tries to steer them into a sustainable direction'* (Loorbach and Rotmans, 2006: 5).

Transition management thus deliberately attempts to steer transitions towards a more sustainable future. Three coordination mechanisms can be used to steer transitions (Loorbach and Rotmans, 2006):

- markets: price mechanisms, individual product and service choices;
- plans: setting transition goals and policy strategies;
- institutions: development of transitions arenas, agendas and goals, fostering of networks and learning processes.

A core element of organising transition processes is the *transition arena*. Transition arenas are 'networks of innovators and visionaries that develop long-term visions and images that, in turn, are the basis for the development of transition-agendas and transition-experiments, involving growing numbers of actors' (Loorbach and Rotmans, 2006). Facilitation at both the process (learning, communication) and content level (feeding new information) is needed. A transition manager is expected to bring the parties together, retain an overview and mediate where necessary. Governmental actors can fulfil the function of transition manager, with different roles in different transition phases. In the pre-development stage, for example, there is a need to foster social experimentation and create support for a transition programme (Loorbach and Rotmans, 2006), while in the acceleration phase there is a special need to control the side-effects of large-scale applications of new technologies. The government has both a content role (setting sustainability objectives) and a process role ('stimulating and organising the transition process, mobilising the social actors concerned, creating opportunities and challenges for transition participants, and creating boundary conditions within which the transition process can operate', Rotmans *et al.*, 2001: 12). The government's role is thus manifold. On the one hand, state actors are called upon to steer the transition whilst, on the other hand, they need to facilitate and evaluate procedures that mobilise and engage actors.

With sustainability as the overall guiding principle, multiple transition visions are developed in the transition arena. Multiple visions or *transition images* allow for diversity in the short term, while convergence is sought on long-term ambitions. A selection of the most innovative, promising and feasible multiple transition visions and images is only made later in the transition process. Interim objectives (including content, process and learning objectives) are set through back-casting from the long-term objectives. A concrete and joint transition agenda should result from this, laying out problems, goals, actions and instruments (Loorbach and Rotmans, 2006).

Transition-experiments with new technologies or social arrangements at the niche-level fulfil a central role. These experiments are supposed to contribute to sustainability at the system level and can be linked up with existing innovation efforts and other transition-experiments in complementary ways. Arena participants and their networks play a crucial role in the execution of experiments.

Learning is essential in transition management, in the form of learning-by-doing (developing theoretical knowledge from practice) and doing-by-learning (development of practical knowledge from theory). The expectations for transition management are high: 'In our view, transition management not only makes good sense but is also the only possible (and do-able) way of achieving true sustainability benefits in the long term while maintaining short-term diversity' (Loorbach and Rotmans, 2006: 18).

2.1.3 Transition management as a steering philosophy

Transition management is presented as a multilevel model of governance (Kemp *et al.*, 2007: 78). It can be understood as *goal-oriented modulation* (Kemp *et al.*, 2007): the modulation of ongoing societal developments against a set of collectively chosen goals. When compared to incrementalism on the one hand and long-term planning on the other hand, transition management is presented as a combination of advantages of both types of approaches (Kemp *et al.*, 2007). It inserts a strategic element in incrementalism and makes planning more adaptive (open with regard to outcomes) and participatory (open to stakeholders). Table 2.1 (from Kemp *et al.*, 2007) gives an overview of the distinguishing characteristics of goal oriented modulation as compared to incrementalism and planning.

The resulting steering philosophy combines adaptation and anticipation. Anticipation (the strength of planning) takes place by specifying desired futures and collectively choosing goals and setting up strategic experiments. Adaptation (the strength of incrementalism) takes place though the modulation of ongoing developments, learning, portfolios and re-evaluation of goals. This overview positions transition management, as an example of goal-oriented modulation, in the middle ground between incrementalism and planning. Although this is a very useful comparison, we argue that this middle ground is actually a broad and varied playing field. A range of other theories and approaches can be found between short-sighted incrementalism and traditional blue-print planning. In the next sections we discuss a number of theories of social change and intervention and compare them in order to create a more detailed map of possibilities.

2.2 Transition management compared to other approaches

In this section we assess the claims of transition management about it being able to steer societal change by contrasting it with other approaches. In this way we enter into the much wider debate about systemic social change, which is only partially considered in transition management theories (Shove and Walker, 2007). In each of the subsections, we will first briefly summarise these theories. Attention will be paid to the analytical framework which is used to conceptualise change and interventions, and the role of governmental actors or change agents. Finally, we point out important differences and similarities with reference to transition management.

We will discuss more general economics and organisational change management theories as well as more specific theories on multi-actor collaboration, network governance, policy agenda setting, social learning and adaptive management. In this overview, we are necessarily selective. In the choice of which approaches would be discussed we opted for theories that share some basic features with transition management, especially the focus on change and innovation and their relevance for sustainability issues, but which diverge from transition management in some other relevant respects. We will rely on a limited number of

Table 2.1. Goal oriented modulation: between planning and incrementalism (Kemp et al., 2007).

	Incrementalism	Goal-oriented modulation – of which transition management is an example	Planning
Key actors	Private and public actors	Private and public actors, experts	Bureaucrats and experts
Steering philosophy	Partisan mutual adaptation, learning-by-doing	Modulation of developments to collectively chosen goals, government is facilitator and mediator	Hierarchy
Role for anticipation	Limited (no long-term goals)	Dynamic, adaptive anticipation of desired futures as basis for interaction	Future is analysed and implemented through blueprint plans
Type of learning	First order: learning about quick fixes for remedying immediate ills	Second-order and first-order (rethink following problem structuring)	First order (instrumental)
Mechanism for coordination	Markets and emergent institutionalisation	Markets, network management, institutionalisation (both designed and emergent)	Hierarchy (top-down)
Degree of adaptivity	Adaptive	Highly adaptive thanks to institutionalised evaluation and (policy) learning, portfolios and re-evaluation of goals	Hardly adaptive
Role for strategy and plans	Limited role	Important role for goals and strategic experiments for exploring social trajectories, undertaken as part of adaptive programmes for system innovation	Plans with steps
Things against which policies are evaluated	Individual goals and short-term gains	Policy goals and learning goals, helping to determine what to do next	Predefined outcomes
Interest mediation/conflict resolution	Individual gains for everyone	Rewards for innovators, phase out of non-sustainable practices through markets and politics (collective choice)	Little mediation (implementation and enforcement)
Policy integration	Minimal	Important but typically evolving	Narrow if present
Type of change that is sought	Incremental, non-disruptive change (system improvement)	System innovation (renewal) and system improvement	Predetermined outcomes which could be an improved old system or a new one

representative publications, because we are trying to capture the core ideas of these theories. These are not always the most recent publications, because recent publications often involve attempts to incorporate elements of different theories and combine them, thus providing a less clear picture of the distinctive contributions of each theory.

2.2.1 Economics

Economics is the science which studies human behaviour as a relationship between ends and scarce means that might have alternative uses (Robbins, 1935). In other words, economics is about choice. It studies how people choose between alternative futures, but it is not about the future itself[1]. This science can, however, sometimes help people to map economic aspects of the future: how the inflation rate might develop, the exchange rate, the price of cereals?

Central in the economists' thinking is that society's pursuit of the highest maximum welfare does not demand central planning, but is generally guaranteed automatically if individuals pursue their own interests and follow their given preferences. This is the idea of the invisible hand of the market that leads to optimal allocation of resources and wealth, as stated by Adam Smith (1776). It also applies to innovations. In the so called induced innovation theory, Hayami and Ruttan (1970) developed the idea that innovation focuses on bottlenecks. Take the example of 19th century US agriculture. In those days agricultural land was abundant and cheap and labour was scarce, innovation delivered machines to reduce the input of labour to keep extensive agriculture going. At the same time, in Japan land was scarce and labour abundant. Technology was developed to maximise intensive agriculture (irrigation, etc.).

The way people organise themselves is also a choice between alternatives that involve scarce resources. Sometimes the market is the cheapest solution, in other cases it is the hierarchy of an organisation (including a government). In a market solution it costs time to (self) organise a group, to gather information, to negotiate a deal, to monitor its execution. Organisations help to decrease transaction costs in cases where markets could be even more inefficient. They are especially useful if contracts are incomplete or markets are imperfect (having external effects). This is the field of economic organisation theory or institutional economics, proposed by Coase (1937). Essential to economists' thinking is that institutions change over time, e.g. as transaction costs change due to technical or structural change (see Chapters 4 and 19 for more details).

The assumption that scarcity induces innovation and that institutions matter is shared by both transition management and economic theory. Nevertheless, economists tend to be sceptical about transition *management*. Economics is not meant as a tool for predicting the future; its basic idea is that people know how to pursue their own preferences and interests, and that the world progressively improves by catering for these preferences.

[1] See Chapter 19 for long term business cycles.

2.2.2 Organisational change management

Three schools of thought can be distinguished within the realm of organisational change managment: planned change, organisational development and continuous change (Boonstra, 2004; Weick and Quinn, 1999). *Planned change* is a programmatic and top-down approach to change. Dramatic and painful changes are often considered necessary to fulfil the (changing) requirements of the organisation's surroundings and to increasing economic value. Managers and consultants take a rational approach: they analyse the surroundings, formulate goals, develop a strategy, and then implement the change (Boonstra, 2004). Planned change may be useful in stable and predictable situations where the problems are unambiguous. It is a form of episodic change (Weick and Quinn, 1999) steered by top-down planning.

Organisational development (OD), in which members are involved in all phases of the change process, is thought to be a more successful approach (Boonstra, 2004). OD has emerged from the psychology field (Lewin, 1951) and criticises the command and control systems in which organisations' members are seen as lazy, self-centred and resistant to change. OD focuses on helping organisations to create conditions that promote people's learning capacity to address specific problems or, more radically, to learn how to continuously transform and renew themselves. It aims at a joint optimisation of organisational effectiveness and the quality of working life (Cummings, 2004). In OD, many advanced techniques were developed like: individual coaching and counselling, team development, conflict management, search conferences, visioning, problem-solving in networks, multi-stakeholder dialogues, group model building, methods to uncover underlying value conflicts, action learning etc. OD can take the form of episodic change (e.g. a large-scale appreciative inquiry summit) or continuous change (e.g. aimed at improving the continuous learning capacity of an organisation) (Seo *et al.*, 2004; Weick and Quinn, 1999).

The concept of *continuous change* can be understood as a third school of thought within change management, developed as a possible new approach to change (Weick and Quinn, 1999). A description of continuous change is that of an ongoing process of adjustments to, or experiments with, everyday contingencies, exceptions, opportunities, or unintended consequences. The basic assumption is that all organisations have people somewhere who are adjusting to changing environments. Small wins are important as each shift in practice creates the conditions for further breakdowns and institutional change. Multiple change agents become important for their ability to make sense of change dynamics that are already under way. They recognise emergent change, make it more salient and reframe or redirect it (Weick and Quinn, 1999).

It's surprising that transition management literature does not explicitly refer to literature on change management, as they share many dilemmas and concepts. The distinctions between planned change, organisational development and continuous change has similarities with the search to bridge the gap between top-down planning and bottom-up incrementalism in transition management undertaken by Kemp *et al.* (2007). Transition management literature combines elements of planned change, organisational development and continuous change insights. We recognise features from planned change like: the content as main driver, the focus on goals, the phased structure and the importance of a transition manager. Compared to planned change approaches, transition management does not pay much attention to the more painful interventions and the ambition to realise change in a scheduled timeframe.

We also recognise that OD's fundamental belief that all members should be treated as equals by management and be actively involved in change. Apart from techniques aimed at the individual – like individual coaching and counselling – a lot of OD techniques are being used and elaborated on in transition management. Transition management shares OD's continuous change view of the world, which sees it as continually in motion, and it also shares its focus on people in the margin who are already adapting to new trends. As change management mainly focuses on organisational change it confines itself to the organisations' members instead of the huge array of actors that transition management has to cope with. It also deals with a shorter time-horizon than transition management.

2.2.3 Multi-actor collaboration

Multi-actor collaboration theory (Gray, 1989; Huxham and Vangen, 2005) addresses cooperation and negotiation between multiple interdependent actors in the context of a 'wicked' problem domain in which they all have a stake, like e.g. environmental pollution, city regeneration or water management. Ouchi (1980) differentiates this concept from other kinds of relationships like market or hierarchically regulated relationships. Gray (1989: 5) defines collaboration as 'a process through which parties who see different aspects of a problem can constructively explore their differences and search for solutions that go beyond their own limited vision of what is possible'. Stakeholders include all individuals, groups or organisations that are directly affected by or interested in actions taken by others to solve the problem. A step-by-step process is proposed for collaboration initiatives, which includes: (1) problem setting (culminating in a shared appreciation of the complex problem domain); (2) direction setting (culminating in a negotiated agreement); and (3) implementation (culminating in tangible actions and changes) as main phases.

Getting the necessary actors together and creating awareness of their interdependencies is considered crucial for obtaining leverage to effectively deal with wicked problem domains. One of the involved persons or organisations usually functions as a *convener* who brings the parties together. Given that wicked problem domains usually defy unilateral intervention, the convenor is very much dependent on other actors to bring about any change in the collaboration or problem domain.

With multi-actor collaboration (Gray, 1989; Huxham and Vangen, 2005), transition management shares a focus on multiple actors and on crossing the boundaries of different policy domains or sectors. Both theories also coincide in their search for innovative solutions based on the variety of knowledge and perspectives that the involved actors bring to the table.

Transition management takes a more selective approach to involving actors by focusing on innovators (niche players) and visionaries, whereas multi-actor collaboration theory would deliberately try to get the 'whole system in the room', involving those actors who oppose the innovations championed by the visionaries as well. This reflects a different relative emphasis on innovation versus negotiation in transition management and multi-actor collaboration respectively. What gets more attention in multi-actor collaboration theory is the negotiated nature of the framing of the problem domain itself (Dewulf *et al.*, 2004). Defining the initial idea that transition rather than stability should be strived for, setting sustainability as a broad ultimate goal and defining the boundaries of the system to be changed are all potentially contentious issues. Transition management also takes a longer term perspective (25 to 50 years) as the relevant time-frame for social change, while collaboration theory pragmatically focuses on reaching an agreement and effectuating a change in a few years time.

Whereas multi-actor collaboration theory speaks of *conveners* with a precarious leadership position, transition management assumes a presumably external (governmental) *transition manager* who steers the transition from an overview position. Leadership in multi-actor situations comes in different shapes: through leadership *media* such as structures, processes and participants (Huxham and Vangen, 2000). Open versus closed membership *structures*, for example, influence the development of the collaborative agenda. Specific work *processes* (e.g. workshops versus formal meetings) influence the kind of people who are likely to take prominent roles. Finally, any *participant* associated with a collaboration who has the power and know-how to influence others may enact a leadership role. None of these leadership media is wholly within the control of the members of a collaboration, echoing one of the principles of the theory of collaborative advantage: assume that no one is in control and that partners and environment are continually changing (Huxham and Vangen, 2005).

2.2.4 Network governance

Network governance refers to theories that take into account the interdependencies of public, private and semi-private actors in self-organising networks (Klijn and Koppenjan, 2000). The theoretical shift from 'government' to 'governance' indicates a shift from hierarchical and well institutionalised forms of government towards less formalised, interactive forms of governance in which state authority makes way for an appreciation of mutual interdependence with different stakeholders. Core concepts are network management, self-governance, deliberative policy making or stakeholder dialogues.

Rooted in the network approach to policy (Klijn and Koppenjan, 2000), network management focuses on mediating and co-ordinating policy making in policy networks. Two types of network management strategies can be distinguished: *process management*, focused on improving the interaction between actors by seeking convergence of perceptions, creating temporary organisational arrangements and managing conflict; and *network constitution*, focused on changing the institutional characteristics of the network, by changing the actor constellation, changing the network rules or reframing ideas about the functioning and the substantive problems of the network.

Governmental actors have multiple options when confronted with network-like situations (Klijn and Koppenjan, 2000): (1) not joining in network games and trying to unilaterally impose their ideas and goals; (2) co-operating as a partner in networks with other public, semi-public and private actors; (3) taking up the role of process manager and facilitating interaction processes in networks; or (4) taking up the role of network builder, for which governments, with their special resources, are well suited.

Network governance shares with transition management a focus on less formalised, interactive forms of governance. Network governance focuses primarily on actors who are already included in policy networks, while transition management focuses on actors in innovative niches and tries to use this marginal position to foster radical innovation. The network manager bears some resemblance to the transition manager, but steering takes more varied forms in network governance, ranging from network building over process management (Klijn and Koppenjan, 2000) to unilateral interventions in networks (De Bruijn, 2005), rather than the external overarching position assumed by transition management. Network theory also pays more attention to multiple, partially overlapping or disconnected, decision-making arenas and to the different roles that governmental actors can take up (Klijn and Koppenjan, 2000; Termeer, 2007). As network theory has been developed within the domains of public management, political aspects receive special attention such as the position of politicians in networks, the use of power and the tensions between network governance and representative democracy.

2.2.5 Policy agenda setting

Theories of agenda-setting focus on the politics of attention for policy issues and the concomitant punctuated-equilibrium dynamics of policy change. Punctuated-equilibrium theory tries to explain the pattern of policy stability (or small incremental changes), which are occasionally interrupted by abrupt major policy changes (True *et al.*, 2007).

In times of stability this involves prioritising the ongoing policy. When priorities shift, big policy changes may result. When a certain issue rises to the top of the policy agenda (e.g. climate change), pressure on politicians to take action increases, new actors may be mobilised and current policies attacked. Policy is executed by a small number of officials, experts and stakeholders, working together in a small network of various (public) organisations; sometimes referred to as a policy community. Usually, this community does not make major policy changes and operates without much political interference. Agenda-setting theory suggests that many policy domains attract minimal political attention because politicians have a limited amount of time and resources and must set priorities. There can only be a few issues that flow from 'low politics' to 'high politics'.

Exogenous events, such as a crisis or a natural disaster, or an endogenous event, such as an accounting scandal or a big organisational failure, can cause heightened media attention for a specific issue. These events are focussing events which create windows of opportunities for policy entrepreneurs to change policies. If certain issues rise to the top of the media agenda, as climate change did when Al Gore presented his movie '*An inconvenient truth*,' pressure on politics increases to take action, and new actors may be mobilised to attack current policies. The punctuation of the stable period is complete when the media reacts to politicians, and politicians react to each other in an escalation of the issue, causing an overkill of attention. Once the policy is changed, or other issues have captured the attention of media and politics, the policy is likely to been drawn back into a new period of stability and incremental adjustments.

The general outcome of much agenda-setting research is that the size of yearly policy changes is not distributed normally, as one would expect when only incremental change takes place. The changes conform rather to a leptokurtic distribution, i.e. with a relatively large number of small changes, a relatively large number of big changes, and a relatively low number of moderate changes. Punctuated equilibriums apply both to policy inputs (e.g. media coverage), processes (e.g. congressional hearings) and outputs (e.g. laws and regulations), but outputs deviate more from a normal distribution than inputs – e.g. media coverage adapts more rapidly to a changing policy agenda than laws (Jones and Baumgartner, 2005; Jones *et al.*, 2003).

Transition management and agenda-setting theory have some important features in common. In agenda-setting theory, agendas are created in different places, such as in the parliamentary arena, in expert meetings, through lobby activities, in formal and informal meetings, and in the media; agendas also develop on various levels: on the national level, the international or the local level. All the different meeting places in which agendas are moulded are referred to as policy venues. They resemble the transition arenas, as described above, albeit that the transitional arena focuses on the entire transition process, whereas the policy venue is somewhat limited in its scope, focusing on the agenda stage only. However, both theories are multi-level and multi-actor. Another important feature in agenda setting theory is the concept of policy image, which bears some resemblance with the transition image. A policy image refers to the way a problem is framed. It connects different concepts and re-formulates the problem in such a way that the media and politicians take up the issue. The policy entrepreneur, who connects a policy frame to his desired outcome and is able to insert it into the right policy venue, is likely to turn matters to his advantage. Furthermore, both approaches share the systemic ideas about non-linear changes in the form of alternating punctuated equilibriums, although agenda setting theory suggests shorter time frames for policy punctuations than transition management. Agenda setting theory also stresses the largely unpredictable nature of punctuations, while transition management somehow assumes that attention can remain focused on the transition issue for the duration of an entire generation. However, both theories suggest an s-curve development of the actual change. Finally, agenda setting theory focuses on policy entrepreneurs who take advantage of windows of opportunity to achieve their individual goals, rather than on transition managers who keep a more distant stance from the actual transition, supervising and guiding the entire process for a longer period of time.

2.2.6 Social learning

Social learning can be defined as learning through participatory systems such as groups, networks, organisations and communities, in conditions which are new, unexpected, uncertain, conflictive and hard to predict (Wildemeersch *et al.*, 1998). In this perspective, four processes are central to the concept of social learning: action (involving a tension between deficit and competence, allowing for pro-activity), reflection (involving a tension between distance and identification, allowing for double loop learning), communication (involving a tension between unilateral and multilateral control) and cooperation (involving a tension between consensus and dissent).

Social learning has to do with collective learning, whereby different community members or stakeholders generate new knowledge, skills, confidence, resources, insights and perspectives on which action can be based (Leeuwis, 2000). The emphasis is on joint situation and problem analysis; on effective and open communication of stakeholder ideas and perspectives inspired by Habermas' theory of communicative action; and on the necessity of higher order learning (e.g. double loop learning, cf. Argyris, 2004) to overcome conflicts of interest. This process of social learning requires critical self-reflection, the development of participatory democratic processes, and reflexive capabilities of individuals and societies (Pahl-Wostl, 2002).

Concretely, the following elements are deemed important in processes of social learning (Pahl-Wostl, 2002: 400):
- the development of a shared problem perception in a group of actors, in particular when the problem is largely ill defined;
- the building of trust as base for a critical self-reflection, which implies recognition of individual mental frames and images and how they pertain to decision making;
- the recognition of mutual dependencies and interactions in the actor network;
- reflection on assumptions about the dynamics and cause-effect relationships in the system to be managed;
- reflection on subjective valuation schemes;
- engagement in collective decision making and learning processes.

The facilitator plays a central role in processes of social learning. The implicit assumption is that people do not communicate openly and effectively mainly because there is no one to organise and facilitate such a process (Leeuwis, 2000). The facilitator is then a neutral outsider whose prime concern is to enhance communication and learning, without steering the content of the dialogue.

The importance of social learning has been stressed in a variety of fields, including rural development, natural resources management, adult education and policy change. With respect to this last field social learning is understood, in contrast to government learning among state officials, as learning in policy communities with the outcome of a policy paradigm shift (Bennett and Howlett, 1992).

From a social learning perspective and, similarly, in transition management, change depends to a large degree on what people learn, on the way this learning is the result of exchange between a diverse group of societal actors and on the nature of that learning (double loop rather than single loop learning). While social learning theories stress the primary importance of open communication and mutual understanding, transition management assumes a more conflictive setting in which niche players 'learn against the regime'. While social learning stresses the involvement of laymen and often disadvantaged groups in society, in transition management learning appears to be more restricted to a highly educated avant-garde.

2.2.7 Adaptive management

Adaptive management can be defined as 'a systematic process for improving management policies and practices by learning from the outcomes of management strategies that have already been implemented' (Pahl-Wostl *et al.*, 2007: 4). Originally developed as a management approach for ecological systems, adaptive management has evolved into an interdisciplinary field of research and action, often referred to as 'adaptive governance of social-ecological systems' (Folke *et al.*, 2005). Adaptive management assumes a world that changes continuously in unpredictable directions. These changes can be gradual, but abrupt or turbulent changes tend to become more prominent. In turbulent change episodes, available experience and expertise often proves to be incomplete, consequences of action are unclear and the future of the system is uncertain. Vulnerable ecosystems, for example, can rapidly shift into undesired states and stop providing ecosystem services (like food or scenery) to society. Similarly, social-ecological systems can lose their resilience to keep fulfilling basic functions in conditions of change or disturbance. In this sense, adaptive management pays attention to both 'change as growth' and 'change as destruction'. The panarchy concept (Gunderson and Holling, 2002) connects both aspects into one ¥-shaped iterative cycle, consisting of exploitation, conservation, release and reorganisation phases. The S-shaped curve of transition management can be read into this cycle as the exploitation and conservation phases (the growth aspect of change).

Attempts at managing or steering have to take into account uncertainties and both gradual and abrupt changes. Therefore, learning plays a central role in adaptive management, as a way of keeping knowledge up to date with continuously changing conditions. Social networks and social memory are considered important bases for building and maintaining the capacity to learn (Folke *et al.*, 2005). Combining different types of knowledge (scientific, professional, experiential, indigenous, etc.) is an important feature of this learning. Learning is not a goal in itself but serves to adapt management strategies and policies as changing conditions require. As not all uncertainties can be 'learned away', another focus in adaptive management is on devising measures or strategies that are robust (stay functional under a range of different scenarios) or flexible (can be adjusted as needed or applied only when necessary). This often requires some redundancy in the system, e.g. in the form of polycentric institutional arrangements, with nested quasi-autonomous decision-making units operating at multiple scales (Folke *et al.*, 2005). To be able to mobilise and link the necessary actors and knowledge quickly and effectively, bridging organisations (between scientists and policy makers, between actors, between networks, between levels, ...) are suggested, as well as the creation of conditions for self-organisation through enabling legislation and policies. Self-organised local responses can foster active adaptation, even in crisis situations. In short, critical factors for adaptive management include learning to live with change and uncertainty, combining different types of knowledge for learning, creating the opportunity for self-organisation and nurturing sources of resilience for renewal and reorganisation (Folke *et al.*, 2005).

Leadership plays an important role in adaptive governance networks by providing key functions, such as 'building trust, making sense, managing conflict, linking actors, initiating partnership among actor groups, compiling and generating knowledge, and mobilising broad support for change' (Folke *et al.*, 2005: 451). Important as this leadership may be in steering adaptive management, it does not involve a position like 'adaptive manager'. Apart from leaders, bridging organisations fulfil an important role in directing adaptive management efforts.

2.3 Comparing multiple approaches to social change and intervention for sustainability

A search for theories implicitly or explicitly related to transition management results in a whole range of paradigms, concepts, methods and tools. Of course, the overview is not complete. The list of relevant theories could easily be expanded with concepts such as soft systems theory, innovation literature, cultural theory or evolutionary economics amongst others. Because transition management is both a field of research and a field of practice, it will continue to grow and to develop as well.

Although at first glance the theories appear to share many insights they differ in underlying assumptions, values and rationalities. As a result the practical applications of these approaches will vary widely. We have adopted the ideas of De Caluwé and Vermaak (2004), who stress the necessity of clarifying conceptual differences between approaches to change because (1) it facilitates clearer communication between people involved, for example between stakeholders, public actors and (action) researchers; (2) it helps create a more complete and complex picture of transitions; (3) it provides a map of possible strategies to intervene; and (4) it offers a tool for reflection to the people engaged in transition.

In Table 2.2 we compare the theories on a number of aspects (the columns of the table) to summarise differences and similarities. The theories make different assumptions about the *nature of change*: what is it that changes and in what direction. They variously focus on

Table 2.2. Transition management compared to other change theories.

	Nature of change	Change trajectory	Main actors
Transition management	Long term structural change of a societal domain	S-shaped curve, with pre-development, take-off, acceleration and stabilisation phases, over the course of an entire generation	Regime players and niche players (innovators) Public authorities
Economics	Change is a reflection of new scarcities	Continuous, path dependent with shocks from new technologies	Innovators, entrepreneurs
Planned organisational change	Changing technology, structure and culture of organisation	Short term programmed change, from one stable situation to a new stable situation	Top managers, experts, employees
Organisational development	Developing organisational human capital (learning and change capacity)	Episodic (a focused change episode) or continuous (aimed continuous learning and change capacity)	Managers and organisation members, using their creativity and experience
Continuous organisational change	Change is always already happening, is continuous and emergent	Numerous small adaptations cumulate and amplify	All people in the centre and in the margin of change processes
Multi-actor collaboration	Negotiated structuration of an under-organised problem domain	Stepwise exploration negotiation and implementation over a number of years	Representatives of organisations having a stake in the problem domain
Network governance	Change in policy and/or change in policy networks	Policy games in successive rounds in policy networks	Public and private actors linked in networks, supporting or hindering policy strategies

Relationship between actors	Steering/ influencing	Leading figures	Role for government	Success
Conflictive and competitive in the short term, shared long term goal of sustainability	Creating transition arenas, starting transition experiments Niche management	Transition manager Visionary innovators	Transition manager, creating support and conditions for a transition programme	More sustainable societal domain
Market exchange (or hierarchy within an organisation)	Prices that reflect scarcities (invisible hand)	Entrepreneurs with creative destruction changing the markets	Correct market failures (externalities), set property rights	Highest maximum welfare, improved income
Hierarchical relations	Using power, planning and control, external knowledge and financial incentives	Change manager, assisted by change experts	n.a.	Organisation functions as was designed in the plan
Open to learning and constructive confrontation	Using advanced tools for mobilising knowledge and insights of members	Process driven facilitators	n.a.	Improving effectiveness, organisation and quality of working life
Loosely coupled relations	Recognising and redirecting change, reducing blocks	Multiple distributed change agents	n.a.	Preventing exclusion and blocked adaptation
Interdependent, conflictive/ collaborative	Leadership through participants, processes and structures. No one is in control	Convener	None, convener or participant	Negotiated agreement (win-win) on the future direction of a problem domain
Sustainable interdependencies between actors, engaged in overlapping policy networks	Providing incentives for co-operation, process management, network constitution	Network manager or process manager	Partner, process manager, network builder or staying out	Win-win situations Enriched chance of policy implementation Democracy

Table 2.2. Continued.

	Nature of change	Change trajectory	Main actors
Policy agenda setting theory	Change in policy input, agenda and output	Incremental changes punctuated by abrupt and large policy change	Politicians, administrators, media, interest organisations
Social learning	Developing new understanding in interaction with other societal actors	Gradual increase of understanding, followed by change in actions	Societal actors
Adaptive governance	Adaptation to the changing conditions in social-ecological systems	Dealing with gradual and abrupt changes through close monitoring and learning	Scientists, policy-makers, ngo's

change in a societal domain, change in policy, or the relation between both. Differences are also apparent in assessing when change has occurred. Is it about changes in understandings, networks, structures, technologies, policies, markets, problem domains or entire societal domains? The change can be directed towards structuring an under-organised domain or on changing existing structures. Transition management is among the more ambitious theories, focusing on structural changes in an entire societal domain. Conceptualisations of the *change trajectories* vary in their focus from short term to long term changes, and assumptions about the continuous (change happens all the time) versus episodic (change comes in big shocks) unfolding of change. Transition management focuses on long term changes (one or more generations), with gradual or continuous changes in the early phases, episodic change in the acceleration phase, and once again more gradual changes in the stabilisation phase.

Theories focus on different *main actors,* playing different *roles,* and standing in different *relations* to each other. The actors that are portrayed as the crucial ones in the different theories can be roughly classified as policy actors, business actors, societal actors, science actors or a combination of these. While some theories focus mainly on one kind of actor – economics focuses mainly on business actors while social learning focuses mainly on societal

Relationship between actors	Steering/ influencing	Leading figures	Role for government	Success
Competitors and allies for attention on the policy agenda	Connecting problems and policies during windows of opportunity, framing of policy images, inserting these in policy venues	Policy entrepreneur	Policy entrepreneur, responsive to societal or political demands	New agenda for policy
Co-creators of knowledge, partners in dialogue	Facilitating open communication	Facilitator	None, organising or participating	Concertedly designed future
In need of each other's knowledge, jointly adapting to changing circumstances	Bridging science and policy, bridging networks of actors	Adaptive network leaders, bridging organisations	One of the multiple decision units	Social-ecological system keeps fulfilling basic functions

actors – transition management distinguishes itself by addressing all these actors. The range of roles that actors play includes innovator, manager, entrepreneur, user/employee, policy maker, politician, gatekeeper, facilitator, expert. The relationship between actors is also conceptualised in different ways, in terms of hierarchy (e.g. planned organisational change), competition (e.g. technology firms in economic theories, or multiple actors competing for attention in agenda setting theory), or mutual interdependency (most of the other theories) (Powell, 1990). Transition management shares the idea of bypassing with multi-actor collaboration. Change trajectories are organised independently of existing organisations, drawing a distinction between actors in the centre of change (the innovators or negotiators) and actors in the margin. Policy and governance theories explicitly focus on actors hindering change, like gatekeepers or the 'usual suspects' who resist change by using power. Transition management mentions the dominant regime as resisting change, but simultaneously assumes that important governmental actors (supposedly part of the current regime) have to pave the way for the transition to a new regime.

Steering or influencing concepts are based on assumptions about how people or things change. In planned-change thinking it is assumed that people change if clearly specified results are laid down beforehand and incentives (or punishments) prove to be sufficient to motivate them. Other assumptions are that people will change when their interests are integrated in win-win situations, when they are encouraged and motivated, when they interact and learn or when space exists for spontaneous adaptation (De Caluwé and Vermaak, 2004). Transition management is not explicit in this respect, but we infer the assumption that people can really innovate and induce system innovations when actions are coordinated in the right settings. Transition management also reflects confidence in the potential of new technologies to reach a sustainable society, provided that they become part of new dominant technological regimes.

The prominent figures in steering change range from facilitators who limit their influence to the process (e.g. in organisational development or social learning), through to entrepreneurs (e.g. the convener as a social entrepreneur in multi-actor collaboration, the policy entrepreneur in punctuated policy change, or the technical entrepreneur with an innovative idea in transition management or economics), and including managerial figures (e.g. transition manager, network manager or the planned change manager). Most approaches consider governmental actors as possible incumbents of these leading roles, though in different degrees. The approaches from organisational change literature, of course, do not consider the role of government actors, but their role can be compared to that of the change manager in planned organisational change (as in the blueprint approach to policy), the facilitator in organisational development (as in social learning), and to one of the multiple change agents in continuous organisational change (as in adaptive management). In some of the approaches the government simply takes part as one actor amongst others (Termeer, 2007), especially in network governance, multi-actor collaboration and social learning. In the last two approaches, and also in the economic approach, societal change can very well occur without the involvement of governmental actors.

2.4 Conclusion

The increased interdependencies in our technology and information driven globalised world turn sustainability into an enormous societal challenge. Sustainable agriculture in the context of climate change, pollution and threats to biodiversity requires the combination of knowledge and possible avenues of action for a wide variety of agents, who position themselves very differently. Governments, local and international businesses, non-governmental organisations, experts and scientists, and well organised or badly organised social pressure groups of all kinds influence the direction of activities that lead to either more or less sustainability. The sustainability question is also an enormous challenge for the social sciences; to try and understand how societal change positively or negatively affects sustainability and to develop intervention concepts and methods. We argue that transition management has made an important and distinctive contribution to this field, without making other theories superfluous.

When comparing transition management with other theoretical approaches to societal change and intervention, what strikes us is the attempt to incorporate a very wide range of aspects into a single theory. Where other approaches to complex societal problems are more inclined to make a choice when faced with the recurring dilemma's that complicate every attempt at societal steering, transition management's answer often embraces both sides of the dilemma. Transition management addresses multiple actors, multiple sectors, multiple levels, multiple time scales, multiple objectives and multiple options:

- multi-actor: multiple actors are needed and often maintain a conflictive relation;
- multi-sector: system innovations affect multiple sectors;
- multi-level: co-evolution of developments at niche, regime and landscape level;
- multi-time scale: both short and long term orientation;
- multi-objective: maintain multiple images of the future;
- multi-option: keeping options open by developing multiple innovative niches.

This attempt at overarching is also reflected in the way transition management is presented as a 'third way' (hinting at the third way in social economic policies which combined features from both socialism and liberalism). The central dilemma in change and innovation management is that both planning as well as incrementalism are thought to be insufficient to tackle system failures and, in the end, to reach sustainable development (cf. Table 2.1. taken from Kemp *et al.*, 2007). The planning obstacles mentioned are (1) the impossibility of handling dissent and uncertainties (2) the lack of a coordinating leadership with shared goals in our pluralist society and (3) the danger of lock-ins (Kemp *et al.*, 2007). However the more modest strategies of bottom up incrementalism are also insufficient as they cannot meet the huge challenges of sustainability. The model of transition management is put forward as combining the best of both worlds into 'goal oriented modulation', but it is, for example, not very clear how a transition management process can survive multiple short term political changes, or how long-term structural changes may be achieved through short term steps (Kemp *et al.*, 2007).

Integrating all this multiplicity into one theory is an important achievement, but it is also a paradoxical one in two important respects. Firstly, the core S-shaped change curve is based on the assumption of a single outcome parameter – the pace and the end result of the change may not be clear, but the desired direction of the change (upwards) is supposed to be clear – which contrasts sharply with the huge range of factors the theory wants to integrate. A single outcome parameter assumes a pretty clear and one-dimensional definition of what will count as sustainability for a certain transition development path, while the direction of change or even the framing of the problem domain is usually a very contentious issue. Some argue 'there is politics to the very processes of abstraction involved in defining something to manage (the 'it', or system) and to the implication that there are managers of the 'it' who sit outside 'its' boundaries' (Shove and Walker, 2007: 765). Sustainability is in itself a multidimensional concept and examples like the bio-fuel debate show that a curve indicating the use of biofuels would hide thorny issues and big discussions about the most relevant dimensions of sustainability.

Secondly, transition management tries to integrate a broad range of varieties into a single theory, thereby drawing upon concepts and methods from the other theories we discussed. It is not very clear within transition management theory how all this variety should be handled. A lot more theory and practice seems to be needed to face the enormous challenge to overarch this multiplicity of factors. Therefore a second paradoxical aspect lies in the very attempt to integrate everything into one theory. Another option, which we would like to put forward, is an approach that rests on the multiplicity of theories. The basic argument is that multiple theories (the ones we discussed here and others) will continue to be needed simultaneously for dealing with the complex societal sustainability issues. Only variety beats variety. This is true for theories as well. A variety of available theories serves as a box of conceptual tools to analyse situations and to design interventions. This does not mean that each of the theories should proceed as if the others didn't exist. In fact, by writing this chapter we're assuming that it is worthwhile to compare theories and to look for points of similarity and difference. This approach can be understood as a meta-paradigmatic approach (Gioia and Pitre, 1990), which recognises the value of the distinctiveness of each individual theory and the value of exploring zones where theories overlap or can benefit from each other, but does not try to integrate everything into one paradigm.

Both issues have implications for the question we started this chapter with, namely, if and how transitions can be managed or steered. Firstly, a distinctive trait of transition management appears to be the assumption of an overarching position of (governmental) transition managers who can apply management tools, niche-building machinery and engineering devices from a privileged, knowledgeable and external position (Shove and Walker, 2007), towards a clear and one-dimensional target. As we have shown, quite different assumptions on this issue appear in related theories. Any transition management arena, however, is likely to be only one of the arenas where sustainability-relevant issues are discussed or sustainability-related decisions are taken, and the sustainability issue at hand may get framed quite differently in those other arenas. Also in transition management, 'steering for sustainability typically surfaces as isolated moments of reflexivity amid a sea of everyday politics' (Hendriks and Grin, 2007: 334). Secondly, steering societal developments in areas as complex as sustainability is unlikely to be successful when only one theory is used, especially when the relevant time frame extends over one or more generations. A broad base of different (and partially overlapping) theories is probably needed to deal with the multiple challenges that present themselves at any point in time – especially because these challenges (and available theories) will also change over time. Sustainability issues are complex and important enough to warrant the generation of insights and interventions about them from the variety of conceptualisations of change that the human mind has developed.

Acknowledgements

The authors thank the participants in the 'Professional Conversations on Transitions' seminar series at Wageningen University for their feedback on earlier versions of this chapter.

References

Aalders, M. and Selnes, T., 2005). Landbouwtransitie op dood spoor: onderzoekers hekelen vastgeroeste beleidsstructuur. Milieutijdschrift ArenA 11(3): 10-11.

Argyris, C., 2004. Double loop learning and organizational change. Facilitating transformational change. In: J.J. Boonstra (Ed.) Dynamics of organizational change and learning. Wiley: Chichester, UK.

Bennett, C.J. and Howlett, M., 1992. The lessons of learning: reconciling theories of policy learning and policy change. Policy Sciences 25(3): 275-294.

Boonstra, J.J. (Ed.), 2004. Dynamics of organizational change and learning. John Wiley and Sons Ltd.: West Sussex, UK.

Coase, R.H., 1937. The nature of the firm. Economica 4.

Cummings, T., 2004. Organization development and change: foundations and applications. In: Dynamics of organizational change and learning. Wiley: Chichester, UK.

De Bruijn, H., 2005. Roles for unilateral action in networks. International Journal of Public Sector Management 18(4): 318-329.

De Caluwé, L. and Vermaak, H., 2004. Change paradigms: on overview. Organization Development Journal 22(4): 9-18.

Dewulf, A., Craps, M. and Dercon, G., 2004. How issues get framed and reframed when different communities meet: A multi-level analysis of a collaborative soil conservation initiative in the Ecuadorian Andes. Journal of Community and Applied Social Psychology 14: 177-192.

Folke, C., Hahn, T., Olsson, P. and Norberg, J., 2005. Adaptive governance of social-ecological systems. Annual Review of Environment and Resources 30(1): 441-473.

Geels, F.W. and Schot, P., 2007. Typology of sociotechnical transition pathways. Research Policy 36(3): 399-417.

Gioia, D.A. and Pitre, E., 1990. Multiparadigm perspectives on theory building. Academy of Management Review 15(4): 584-602.

Gray, B., 1989. Collaborating. Finding common ground for multiparty problems. Jossey-Bass: San Francisco, USA.

Gunderson, L.H. and Holling, C.S., 2002. Panarchy: understanding transformations in human and natural systems. Island Press: Washington, USA.

Hayami, Y. and Ruttan, V.W., 1970. Agricultural productivity differences between countries. American Economic Review 60: 895-911.

Hendriks, C.M. and Grin, J., 2007. Contextualizing reflexive governance: the politics of Dutch transitions to sustainability. Journal of Environmental Policy and Planning 9(3): 333-350.

Huxham, C. and Vangen, S., 2000. Leadership in the shaping and implementation of collaboration agendas. Academy of Management Journal 43(6): 1159-1175.

Huxham, C. and Vangen, S., 2005. Managing to collaborate. Routledge: London, UK.

Jones, B.D. and Baumgartner, F.R., 2005. A model of choice for public policy. Journal of Public Administration Research and Theory 15(3): 325-351.

Jones, B.D., Sulkin, T. and Larsen, H.A., 2003. Policy punctuations in American political institutions. The American Political Science Review 97(1): 151-169.

Kemp, R., Loorbach, D. and Rotmans, J., 2007. Transition management as a model for managing processes of co-evolution towards sustainable development. International Journal of Sustainable Development and World Ecology 14(1): 78-91.

Klijn, E. and Koppenjan, J., 2000. Public management and policy networks: foundations of a network appraoch to governance. Public Management 2(2): 135-158.

Leeuwis, C., 2000. Reconceptualizing participation for sustainable rural development: towards a negotiation approach. Development and Change 31(5): 931-959.

Lewin, K., 1951. Field theory in social science: Selected theoretical papers. Harpers: New York, USA.

Loorbach, D. and Rotmans, J., 2006. Managing transitions for sustainable development. In: A.J. Wiczorak and X. Olshoorn (Eds.) Industrial transformation – disciplinary approaches towards transformation research. Kluwer Academic Publishers: Dordrecht, the Netherlands.

Ouchi, W.B., 1980. Markets, bureaucracies, and clans. Administrative Science Quarterly 25: 129-141.

Pahl-Wostl, C., 2002. Towards sustainability in the water sector. The importance of human actors and processes of social learning. Aquatic Sciences, 64(4), 394-411.

Pahl-Wostl, C., Sendzimir, J., Jeffrey, P., Aerts, J., Berkamp, G. and Cross, K., 2007. Managing change toward adaptive water management through social learning. Ecology and Society 12(2): 30.

Powell, W.W., 1990. Neither market nor hierarchy: network forms of organization. Research in Organizational Behavior 12: 295-336.

Robbins, L.C., 1935. An essay on the nature and significance of economic science. MacMillan Ltd.: London, UK.

Rotmans, J., Kemp, R. and Van Asselt, M.B.A., 2001. More evolution than revolution. Transition management in public policy. Foresight 3(1): 15-31.

Seo, M., Putnam, L.L. and Bartunek, J.M., 2004. Dualities and tensions of planned organizational change. In: M.S. Poole and A.H. Van de Ven (Eds.) Handbook of organizational change and innovation. University Press: Oxford, UK.

Shove, E. and Walker, G., 2007. Caution! Transitions ahead: politics, practice, and sustainable transition management. Environment and Planning A, 39: 763-770.

Smith, A., 1776. An inquiry into the nature and causes of the wealth of nations. Penguin Classics: London, UK.

Smith, A. and Kern, F., 2007. The transitions discourse in the ecological modernisation of the Netherlands. Universiyt of Essex, SPRU Electronic Working Paper Series, Paper No. 160.

Termeer, C.J.A.M., 2004. Duurzame transities. Een beknopte methodologie voor trajectmanagement. In: P. Van der Knaap, A. Korsten, C.J.A.M. Termeer and M.J.W. Van Twist (Eds.) Trajectmanagement, beschouwingen over beleidsdynamiek en organisatieverandering. Lemma: Utrecht, the Netherlands.

Termeer, C.J.A.M., 2007. Vital differences. On public leadership and societal innovation. Inaugural Speech, Wageningen University, the Netherlands.

True, J.L., Jones, B.D. and Baumgartner, F.R., 2007. Punctuated equilibrium theory: explaining stability and change in policymaking. In: P.A. Sabatier (Ed.) Theories of the policy process. Westview Press: Colorado, USA, pp. 155-187.

Weick, K.E. and Quinn, R.E., 1999. Organizational change and development. Annual Review of Psychology 50: 361-386.

Wildemeersch, D., Jansen, T., Vandenabbeele, J. and Jans, M., 1998. Social learning. A new perspective on learning in participatory systems. Studies in Continuing Education 20(2): 251-265.

Chapter 3

Change in knowledge infrastructure: the third generation university

Rudy Rabbinge and Maja Slingerland

Abstract

During the last 150 years one dominating aspect of Dutch agriculture has been it's dynamism and ability to innovate. Both the government and the private sector have believed that science and technology have a lot to offer in improving the agricultural sector's competitive ability. Investments in research, education and valorisation of knowledge were promoted, even in periods of crisis. In the 1990s, when the diversification of goals for agriculture and a negative public attitude towards it caused enormous challenges for the sector, the government pursued a similar policy. Wageningen UR was founded as one response to those challenges, signalling a renewed major investment in the knowledge infrastructure. The ambition to pursue its new course is strongly present in Wageningen, where the emphasis in the renewed university structure lies in extended public-private cooperation to promote and fund the desired dynamism and innovation. The reformulated mission, the role that the university plays in education, research and, last but not least, utilisation of that knowledge, was made possible by merging the university, research institutes and experimental stations. The character of this new, third generation university is different from that of its predecessors. It is a university where the hierarchical structure, present in the medieval and Humboldt University is replaced by a network structure where excellence and authority prevail. Direct government influence is limited, the orientation on science for impact determines the attitude of all scientists and there is a willingness to accept that graduate schools have the right to exclude non-performing professors. The strength of a third generation university is determined by three widely accepted characteristics, the first is the internal coherence which comes as a result of the generally accepted vision and mission of the university. The second is the flexibility, but nonetheless stability of the institution's finances and, finally, the willingness and ability to work with partners from very varied backgrounds. Wageningen University can be seen as one of the first universities that has made the change and reflects these characteristics. Wageningen University and Research Centre adopted this strategy about 10 years ago, and the results are already visible in scientific output, growing graduate students (MSc and PhD) numbers, its sound financial structure and close ties with partners within the Netherlands and a growing network outside the Netherlands. The history and background of that development is described in this paper.

Keywords: third generation university, research, education, collaboration

3.1 Introduction

The dynamism and innovative capacity within the various branches of the agricultural sector in the Netherlands has been a major cause of its strength during the last century. It explains why the sector is still vital and continues to be relevant and competitive, as opposed to developments elsewhere in Europe. This process began in the second half of the 19th century and has continued right into the 21st century. The timely adaptation and innovation in promising activities and branches is illustrated by the creation of Wageningen University and Research Centre at the end of the 20th century as a third generation university. The implementation of this strategy in the 21st century is a typical example of second mode science and development.

3.2 Historical sketch in a birds-eye view

3.2.1 Agricultural policies in times of crisis

When steamships made bulk transport of agricultural produce possible, cheap imports from the new world (USA) to the Netherlands in the second half of the 19th century caused grain prices to collapse. The reaction of each European country was very different. In the UK, a *laissez faire* policy was adopted and consistently followed causing many farmers to go bankrupt and forcing them to find alternative employment in the new industries founded in the UK's industrial revolution, which had begun in the early part of that century. In France and Germany the governments, still influenced by feudal system in the rural areas, decided to protect the agricultural sector through import levies and other regulations and measures. In Germany the emerging industries needed protection against free trade competition. These protective measures resulted in the consolidation of not very dynamic production structures. In the Netherlands both the protecting and the liberalisation attitude were impossible. The Netherlands had been a trade nation since the 16th century and could not close its borders. At the same time, more than 50% of the population was still dependent on agriculture and other employment opportunities didn't exist, so the government had to look for another survival technique. In the end, they found a third option; neither liberalisation nor protection, but strengthening competitive ability was seen to be the best way to deal with the situation. Measures to achieve this end included: new structural land improvement through efficient infrastructure; improved layout of agricultural lands in the landscape and increased farm sizes; strengthening the position on the market by, for example, stimulating cooperatives; and finally, knowledge intensification through research, education and extension. All these measures proved to be successful.

For more than 100 years each European country has reacted consistently with their chosen policy in times of crisis. It is not surprising that, in the Netherlands, the policy of strengthening competitive power has remained and continued during the last decades. The creation of Wageningen University and Research centre is a natural continuation of that policy. Its creation in the '90s was a government response to a crisis caused by decreasing student numbers, the degraded funding for agricultural research and the implosion of the powerful and influential negotiation structures of the agricultural sector of the past. To address all these problems, a third generation university was created with very different structures than were characteristic of first and second generation universities.

3.2.2 Development towards the third generation university

The first generation university, or medieval university, was founded in Bologna in 1158, this was followed by the establishment of the university of Paris in 1200. By 1300 there were some 20 European universities. They were set up and run by national governments and had faculties in theology, medicine, law and arts. They were loosely organised around individual teachers and had a rector, chancellor and General Assembly as governing bodies. Research, as we now know it, was absent and the university's objectives can be characterised as enlightenment of the world and stimulation of student obedience. Mobility and intellectual exchange were integral parts of academic 'training'.

In the 19th century we see first in Germany and then in other European countries the rise of the second generation university or Humboldt university. This type of intellectually elite university explicitly stimulated research. Education was called 'Bildung': not power but rationality and authority based on knowledge, experience, insight and expertise was the leitmotif. Excellence was embedded in the institution in the form of centres of excellence for the elite and, as a result, mobility was no longer a factor of importance.

This type of university had two major characteristics: firstly, the objective of science was to search for the truth in an environment free of all social pressure, and outside the influence of religion and politics. Research was organised into disciplines and all disciplines were seen as branches of the tree of reason, although closely connected to ethics and art. Scientific research could only be judged by reason, hence by science itself. However 'Bildung' was not restricted to scientific training aimed at producing clever students, but also encompassed moral and ethical education in order to promote the development of wisdom in the broadest sense. Education was seen as an integral part of research and 'Bildung' was seen as a life long learning process.

Secondly, embedded in this university concept, was the idea that increased knowledge and the process of uncovering the truth, would increase people's insights and confidence and promote a true intellectual freedom based on reason. In this view scientists are seen as an elite that can further society's development towards freedom by increasing intellectual insights. Societal development is, in fact, seen as the same as intellectual development. The metaphor of the Ivory Tower, rising above society, housing the elite, whose exclusive purpose is the acquisition of truth, fits this idea perfectly. Applied research or seeking applications for knowledge are seen as mundane skills and are certainly not regarded as the domain of science at second generation universities.

In the change from first generation universities, Latin disappeared as *lingua franca* and because second generation universities are nationally oriented, inter-university mobility, widely promoted in the medieval university, decreased. Governance of the second generation university is by a senate, fulltime professors with a chair and an executive board comprising chancellor and vice-chancellor or rector.

This type of university prevailed until the end of the 20th century although some important developments had been putting this type of university under pressure for some time. For instance a number of philosophers and social science theories deny the existence of a single truth which can be found and known. In their extreme forms they claim that the world as we see it is mainly a construct of individual and collective mindsets and does not 'exist', except in the minds of the observers (constructivism). Another force is the strong societal desire for science to provide answers to problems instead of merely increasing knowledge. Research topics are less curiosity-driven and have become more dependent on potential clients or end-users for scientific findings. Judging the value of scientific research is not only dependent on its contribution to 'the truth' or to 'the body of knowledge' but increasingly on its societal relevance. Science can no longer function as superior to, or independent of society (in its ivory tower) but is becoming an integral part of it. Scientific disciplines have differentiated, creating new branches, some of which limit themselves to studying empirical reality, and avoid involving ethics and art. Mono-disciplinary research aiming to deepen insights is gradually being complemented by multidisciplinary and interdisciplinary research seen as more suited to solving complex societal problems. Within universities themselves, research and education are drifting apart. Transfer of knowledge has become increasingly more important than the art of 'Bildung' – creating scientists to be part of an elite with high moral standards and superior insights, able to guide society towards true freedom, based on reason. Graduate students are increasingly required to use knowledge and scientific methods (know-how) to contribute to society. This is reflected in the variety of professions that graduate students occupy after their studies. They have become 'experts' instead of 'academics', but they are still 'free' from societal influence as they are trained to be objective, able to observe or measure facts in the empirical world. PhD students are also still trained to develop knowledge and to become career academics, aiming to replace their professors in due time, in the tradition of 'Bildung'. Yet, as there are only a limited number of vacancies for professors at universities many of these students end up outside academia – and this requires them to learn additional skills.

Wissema (2005) pointed to a number of additional developments which had forced the Humboldt University to change. Research tasks were no longer the exclusive domain of universities but a number of public non-university research centres have emerged. University studies were not only open for the elite but university training was democratised (in the Netherlands especially from the 1960s onwards) becoming available for masses of students. The increase in numbers led to the need for professional managers from non-university backgrounds. The large number of new scientific fields (many of which are applied sciences) and the increase in collaborations with societal partners have promoted a drift away from the clear linear faculty structure of the Humboldt University, and a mix of working structures is replacing it. Governmental steering has become increasingly problematic and at least a partial shift to a more demand-driven orientation has become a more attractive option. Internationalisation, globalisation, computers and internet, and demand for English fluency, have led to increased contacts and competition between universities, complicating the steering issue and putting the concept of 'a centre of excellence, developing itself in isolation' under threat.

All of these processes and developments indicate clearly that a new type of university is needed. The know-how carousel concept is becoming more dominant. The entire knowledge chain is becoming more and more integrated and the typical distinction between fundamental, curiosity-driven research, strategic research, applied research and application of research outcomes is becoming obsolete. In the know-how carousel, research at the university and outside of it, in techno-parks, by techno-starters, by private companies but also in close contact with civil society partners in land designation, planning and conservation projects, becomes integrated and includes both academic and professional education. This know-how carousel forms the core of the third generation university. The third generation university is a network university with many partners and many research locations. Education, research and utilisation/valorisation of knowledge are all objectives in this model. This integration of know-how exploitation and education and research also requires an academic response in the form of a clear analysis and description to guarantee academic freedom and integrity.

Direct government influence has been reduced and funding comes in the form of grants, contracts and facilitative structures. To raise these funds the third generation university needs to be more client-oriented, aiming to provide answers or advice to end-users of research. International and national science foundations play a major role in the funding of research and education. The third generation university needs to be competitive in terms of scientific excellence to become eligible for these funds. The international character of the university is growing, and this combined with its collaboration in various partnerships has increased the mobility of its staff and programmes.

3.2.3 Transition within the knowledge infrastructure to support dynamism and change

The need to change has been felt in all the different European countries as well as in the Netherlands and this has resulted in a variety of responses to the same challenge. In Denmark, they took another road than the Netherlands. The Royal University for Veterinary and Agricultural Sciences (KLV) was not merged with research institutions to form one new entity, it was itself split up. One part formed a university with a research organisation in Arhus and another part became a more academic teaching and research institute, the

University of Copenhagen. Agricultural research was placed under the umbrella of more general universities. The aim was to optimise the influx of students from other disciplines like biology, mathematics, etc. and for that reason it was considered a better option to be associated with universities with studies in these science domains rather than creating a separate agricultural centre.

In Germany there was considerable resistance to fundamental change in all the different States, and that only allowed for smaller incremental changes in existing institutions. Merging research institutes was not considered to be an option and that has meant that the dispersed character of expertise throughout the Federal Republic of Germany has been maintained. In France, the INRA (Institut National de la Recherche Agronomique) extended its area of responsibility and made major structural changes, yet the far-reaching agreements to collaborate with '*Haute Ecoles*' or regional universities did not materialise. From 1998 onwards, thematic collaborations in joint research units (Unités Mixtes de Recherche) were formed instead as a way to promote more structural inter-institutional collaboration. The staff members of INRA, CIRAD (Centre de Coopération Internationale en Recherche Agronomique pour le Développement), IRD (Institut de Recherche pour le Développement) and a diversity of universities and institutions for higher agricultural and veterinary education participate in these entities. In 2005 more than 50% of all INRA research units and about 40% of the CIRAD research units were *unités mixtes* (INRA, undated; CIRAD, undated). Merging the institutions themselves is not foreseen. Other types of collaborations (*recherche contrats*) tend to include powerful regionally-based industries and remain regional initiatives, without penetrating to national level.

It remains to be seen whether the Wageningen model, described in more detail below, is better then the Danish, French or German models. The preliminary results in the Netherlands indicate that Wageningen University and Research Centre (WUR) is functioning rather well, but proper evaluation of its scientific and societal impact can only be done about 10 years after full implementation (2012).

3.3 The creation of Wageningen UR as an example of a third generation university

To understand why Wageningen UR was established in 1997, we need to look at the driving forces behind it from the perspective of the agricultural research organisations and the Ministry of Agriculture. During the 1950s and 1960s a large proliferation of agricultural research institutes and stations occurred. Because of their large numbers and diverse locations, there were a lot of overlapping mandates and activities. This was followed by a period of centralising activities around two of the major agricultural research centres, Wageningen and Lelystad, during the 1980s, when funding became tighter. The responsibility for coordinating agricultural research in the research stations was assumed by the National Agricultural Research Council, whereas the Directorate of Agricultural Research assumed responsibility for all ministerial research organisations. The parallel structures led to duplication in policy formulation and coordination. Driven by the desire to increase the research's accountability to its clients, the Dutch government decided in 1986 to privatise public agricultural research, leading to a new client-provider relationship between the Ministry and the agricultural research department (DLO) and output funding in 1991 (Van den Berg, 2001). Over time, the ministry became not only accountable for research to increase productivity by agricultural producers but also for environment and landscape issues expressed by consumers and society at large. In 1995, to decrease overlap and to take account of a broadening in the policy domain, the responsibility for all research, extension and education facilities (Wageningen Agricultural University included) were brought together under one roof, the Directorate of Science and Knowledge Transfer (Roseboom and Rutten, 1998). In 1995 a study on the future of the Dutch agricultural knowledge system revealed that there was also substantial overlap and competition between activities within the research components (Peper, 1996) and a merger between Wageningen Agricultural University (WAU) and DLO was recommended. The Ministry of Agriculture, which commissioned the study, adopted this recommendation and started the integration by merging the boards of the two entities and appointing one new chairman in 1997. The new organisation was called Wageningen University and Research Centre (WUR). The new status provided DLO with more flexibility to attract foreign clients and investors. In the mean time the applied research stations were also changing in status.

They used to collaborate with experimental farms and gardens and were governed by representatives of the agricultural and commodity board. The ministry stopped subsidising experimental farms and gardens in 1996 to force farmers and their representatives to take more responsibility for the research by co-funding it themselves (Van den Berg, 2001). The many applied research organisations responded by merging into fewer, larger units. Applied Plant Research and Applied Livestock Research became part of WUR in 2001, in line with the focus of WUR on a more participatory model of knowledge creation.

The creation of Wageningen University and Research Center (WUR) under the Dutch Ministry of Agriculture should be seen as part of the Ministry's greater struggle to maintain its independent position. Prior to the creation of WUR the Ministry of Agriculture had progressively lost prestige, partly because of the negative image that agriculture had accrued in the previous few years and its subsequent diminished political clout as agriculture lost popularity. However, agriculture still remains one of the major economic sectors in the Netherlands and this explains why the Ministry of Economic affairs has supported the Ministry of Agriculture in its struggle for survival. During budget negotiations in 1994 there was a strong lobby to shift agricultural education and research, accounting for one third of the ministry's budget and staff, to the ministry of education (OCW). This would have severely affected the legitimacy of the Ministry of Agriculture. Seen in this light, the idea to create a strong, competitive research and education centre such as WUR was a timely and brilliant idea, justifying the Ministry's existence by giving it a prominent guiding/steering role and thereby, strengthening its political position.

From the university's perspective, the creation of WUR has been the result of a decision to go the route of the third generation university. Wageningen University has not been defeated by the negative image crisis, low student numbers and the lack of political power from the constituency in the agricultural domain. It has not accepted its extinction as an independent university, allowing itself to be absorbed within mainstream Dutch universities, surviving merely in the form of a limited number of chair groups. Instead it has taken the initiative to invest in increasing its competitive power by combining different levels of research and education: university, research centers (DLO), experimental stations and professional education (Van Hall Larenstein), complemented by special centres for knowledge valorisation, business schools, professional mid-career training and capacity building, in the traditional Wageningen domains: 'healthy food and a living environment' (Kropff and Kalwij, 2008).

To express its strength and unity Wageningen UR has formulated a visionary mission statement which is: 'to explore the potential of nature to improve the quality of life'. The WUR model has lead to the creation of a coherent package of research and education activities with sufficient critical mass to develop excellence. Its focus makes it clear to stakeholders what they can expect from Wageningen UR, and its additional statement 'Science for Impact' underlines its relevance to issues that affect the general public.

In *sensu stricto,* the third generation aspect can be seen by looking at the curricula that show a balance between purely discipline-oriented and integrated courses; between focus on development of knowledge and skills and competencies needed to use knowledge in society. The prominence given to the concept of T-shaped skills means that disciplinary depth (adding to disciplinary knowledge development) is combined with sufficient interdisciplinary width (adding to discipline-overarching knowledge). Another major aspect is the attention given to beta-gamma interaction, on the assumption that synergy between natural and social sciences should increase societal relevance both in improving the articulation of relevant research questions and in developing adequate technologies. In each of the master study programmes all students have to do individual master thesis research projects but all students also have to participate in a multidisciplinary research project commissioned by a societal party (academic master cluster).

The financial structure at Wageningen UR also reveals its third generation university character. The funding structure is flexible and consists of different components. One is the successful acquisition of research fellowships for excellent research from Dutch *veni-vidi-vici* programmes and from international programmes sponsored by organisations like the European Union, the *Marie Curie Fund.* Playing a leading role, or simply taking part in a large number of extremely large and complex interdisciplinary European Union funded integrated programmes underline the institution's scientific and organisational competence in functioning in such complex arenas. Its comparative scientific excellence in the European research and development arena helps it secure operational funding. The university and research institute components function side by side in these programmes, complementing and reinforcing one another's distinctive attributes. WUR takes part in many international development oriented research programmes such as the Alliance for the Green Revolution in Africa (AGRA) and the challenge programmes of the Consultative Group on International Agricultural Research (CGIAR). It also plays a leading role or participates in large, privately funded programmes, for instance those set up by the Bill Gates or Ford Foundations. Their participation in world-wide studies such as IPCC, IAASTD, Hunger Task Force and the Inter-Academy Council study on Africa show an international recognition of their academic qualities. Wageningen University also uses core funding to support development-oriented research by its Interdisciplinary Research and Education Fund (INREF) aimed at contributing to the solution of important societal problems and at the same time investing in building up the skills and competence of partners and partner institutions in developing countries. These programs aim at the same time to develop the WU staff's own capacity to initiate, coordinate and implement interdisciplinary research programs, building on T-shaped skills and beta-gamma interactions, which are typical components of a third generation university. One further aspect which underlines Wageningen's international character is the fact that 30% of MSc students and 50% of PhD students that graduate from Wageningen are not Dutch.

Last, but not least, the WUR's relationship with the Dutch government has become one of output funding. Joint decisions about the research agenda have lead to an annual allocation of research budgets not only to 'provide recommendations to the ministry' but also to invest in knowledge development. Dutch politicians are aware of the crucial role that knowledge has to play in our society. The profits from sales of Dutch natural resources such as gas are invested in the intensification of the knowledge system. These funds support larger and longer term government-funded research programmes (ICES-KIS). They are competitive and Wageningen UR is present in most of them not only due to excellence in its 'own' domain, but also because of its effective partnering with other disciplines, research organisations and, in particular, private sector parties – to create added value through synergy.

A characteristic of third generation universities is their collaboration with private sector parties. During the last 10 years, the research institutions and activities of more than 20 national, international and multinational companies have been concentrated in *Food Valley,* which was set up in Wageningen at WUR's initiative. In this 'valley' structural assistance has been created to foster techno starters by facilitating housing, providing support services, etc. This vibrant business environment stimulates research at Wageningen UR as well. Mutual benefit lies in the fact that partners easily find each other to create and exchange knowledge and to work on innovations. The international character leads to spin-offs in the European market and beyond. In his inaugural address on techno-starters Wissema (2005) explicitly mentioned WUR and its *Food Valley* initiative as evidence that this third generation university is probably the most developed in the Netherlands and an example to be followed by the Technical University (TU) Delft.

It is also important to note that a third generation university does not mean a drop in academic quality compared to a second generation university. In fact, a system of internationally peer-reviewed graduate schools guarantees the quality of PhD research and PhD teaching, and also of university staff performance. The graduate schools have developed procedures so that students are offered tailor-made courses both in scientific depth and personal development aimed at delivering well-balanced academics. At the same time research and training proposals are critically evaluated by international peer review to ensure quality standards. In addition the relative independence of the students in deciding their research themes, selecting their supervisors and developing their own research proposals, choosing methodologies, research locations and the degree of involvement of potential end users, provides all the ingredients for students to develop into critical, independent, reflexive researchers that have a contribution to make to both science and society. Wageningen Business School also caters for 'life long learning' by providing short, in-depth courses whereby many alumni come back to the university, business school or Wageningen International to provide input in research and education from their professional life. In this way society interacts with the university enriching it with practical experience from the field.

Although the innovations in the knowledge infrastructure at Wageningen are full of promise, all is not rosy. Many paradoxes, dilemmas and debates still take place. The steering management model adopted by WUR, for instance, remains a highly debatable issue within the organisation. Management by control and procedures focussed on economic indicators appears difficult to reconcile with the characteristics of an organisation of professionals where scientific quality and peer-reviewed outputs have always dominated. Similarly the debate between contracted, commissioned, market-driven research as opposed to curiosity-driven research and scientific independency is still ongoing. However, these debates are only to be expected in an organisation undergoing radical transformation and, as such, should be seen as a challenge rather than a problem.

In summary one can say that the preliminary results are impressive. The university's funding has improved, collaboration with private firms has increased, its international character has become more prominent, it's graduate schools have a leading position in the world and the name 'Wageningen' carries more prestige than ever before. This has big advantages for Dutch agribusiness and the agricultural sector as a whole, which can be demonstrated by the increase in added value in agricultural exports. In 1980 the Netherlands ranked 27th and in 2005 it ranked 6th of all countries adding value to agriculture per capita. The added value in agriculture (expressed in constant 2000 US$) increased from 337,366 US$/capita in 1980 to 579,336 US$/capita in 2005 (Nationmaster, undated). The development of this third generation university confirms the historical tradition whereby, in times of crisis, the Dutch opt neither for protection nor liberalisation but choose to strengthen their competitive capability by investment in the knowledge infra-structure.

Acknowledgements

The authors would like to thank Krijn Poppe for his constructive comments, and Wiebe Aans for putting some of his internal notes at their disposal to augment their insights into this subject matter.

References

CIRAD, undated. Available at: www.cirad.fr/en/pg_recherche.php. Last consulted 10 June 2008.

INRA, undated. Available at: www.inra.fr/l.institut. Last consulted 10 June 2008.

Kropff, M.J. and Kalwy, J.J. (eds.), 2008. Science for impact on science, society and business, 90 years Wageningen University. Wageningen University: Wageningen, the Netherlands. 103 pp.

Nationmaster, undated. Agriculture, value added > constant 2000 US$ per capita by country. Available at: www.nationmaster.com/red/Agriculture Statistics. Last consulted 22 August 2008.

Peper, B., 1996. Duurzame kennis, duurzame landbouw; Een advies aan de Minister van LNV over de Kennisinfrastructuur van de Landbouw in 2010. Ministerie van Landbouw, Natuurbeheer en Visserij, The Hague, the Netherlands.

Roseboom, J. and Rutten, H., 1998. The transformation of the Dutch Agricultural Research System: an unfinished agenda. World Development 26 (6): 1113-1126.

Van den Berg, M.H., 2001. The agricultural knowledge infrastructuur: public or private? An explorative study about evolving public and private goods and roles. FAO SD Knowledge. FAO Research, Extension and Training Division.

Wissema, J.G., 2005. Technostarters en de derde generatie universiteit (Technostarters and the third generation university), Rede uitgesproken bij de aanvaarding van het ambt van hoogleraar aan de Faculteit Techniek, Bestuur en Management in het vak Ondernemerschap op 22 april 2005 (Inaugural speech for the Faculty of Technology, Policy and Management, 22 April 2005), Aula TU Delft. DUP Satellite: Delft, the Netherlands. 26 pp.

WUR, undated. Available at: www.wur.nl. Last consulted 10 June 2008.

Chapter 4

History is alluring: self-organisation and the significance of history in the search for a new local sense of collectivity

Irini Salverda, Louis Slangen, Jeroen Kruit, Titus Weijschedé and John Mulder

Abstract

This research reveals the significance of the reference back to historical organisational forms in present-day self-organisation initiatives in the Dutch rural landscape. Three self-organisations which, in name at least, refer back to the historical *marke* [bounded common land], *meent* [common grazing grounds] and *naoberschap* [the obligation within a small social community to help each other] have been analysed in this chapter. This analysis was done by means of a theoretical framework, based on the four levels of institutions from the New Institutional Economics (NIE), and an historic analysis of the *boermarke* [bounded common land] in the province of Drenthe. It reveals that the initiatives don't necessarily wish to revive the historical organisations. The historical connection appears merely to refer to the modern-day desire for collectivity on a local scale for the organisation of the landscape. This desire for a small-scale organisation with local commitment, for communal use and management of land as well as the local exchange of rights and responsibilities, is timeless but, nowadays, it is not a common phenomenon in the planning and management of the landscape.

Keywords: self-organisation, local collectivity, New Institutional Economics, landscape, historical organisations

4.1 Introduction

In the sleepy, picturesque little village of Dwarsgracht in the east of the Netherlands, a small group of villagers gets together because they have grave concerns about the future of their community. In the heart of a natural marshland, this centuries old, small-scale farming landscape is in danger of disappearing. The type of farming that took place in the past is best described as 'water farming'. Small strips of land were surrounded by wide ditches full of water. Transport of materials and livestock was traditionally done with flat boats. The farms had gradually disappeared, and the strips of farmland were reverting into a wilderness. In 2004 these determined people formed a group called the '*Naobers van Zudert*' to preserve the unique qualities of their local landscape. This name hearkens back to the times when rural communities in parts of the north-east of the Netherlands – isolated from the rest of the world – were dependent on one another for help and for their daily needs.

4.1.1 Nostalgia rules?

If one is to believe the media, we have become slaves to nostalgia. In a world that is increasingly globalised we feel threatened and, to preserve a sense of our own identity, we hold onto an idealised vision of our past, thereby ignoring the challenges of the future. This is expressed, for instance, in the way we build our thematic holiday resorts, the revival of the architecture of the 30's of the 20th century and the move to conserve our agricultural heritage. Besides these material expressions of nostalgia – if it is nostalgia – one can also detect an immaterial manifestation. For example, the revival of old traditions and the reanimation of historical types of organisation. In this study we look at types of organisation that apparently model themselves on forms that effectively died out 200 years ago – like the *marke*, the *meent* and the *naoberschap*. It could appear that these organisational forms have been called out of obsolescence in order to promote a modern collective responsibility for the development and management of the landscape at a local level. The *'Naobers van Zudert'* is one example of a group of rural neighbours that is cooperating to preserve the characteristics of their village landscape. The *'Nieuwe Marke in de Rijssenervallei'* and the *'Meent in Hof van Twello'* are two new concepts for a local and collective way of organising the development, use and management of the rural landscape. But is it really nostalgia that rules these self-organisation initiatives in our rural landscape?

4.1.2 The issue

This research aims to reveal the significance of the historical correlation to present-day self-organisation initiatives in the Dutch rural landscape. Why do these initiatives hearken back to historical organisations? What is the significance of the historical correlation for the way they want to organise the planning, use and management of the landscape? Is it their aim to revive an extinct way of organising or are they making use of historical elements for their present day objectives? Why are these initiatives taking off now? What difficulties do they face from within the existing institutional environment?

In order to be able to answer these questions we first developed a theoretical framework (paragraph 4.2). Secondly, we take a trip through time to visit past and present *boermarke* organisations in the province of Drenthe (paragraph 4.3). In the light of these two focal points we will analyse three present-day self-organisation initiatives that refer back to the traditional *marke*, *meent* and *naoberschap*. These possible forerunners of institutional transition are described and analysed (paragraph 4.4). Conclusions are drawn on the cases and the theory (paragraph 4.5 and 4.6). Since the study has both an exploratory and a programming character, we conclude with suggestions for future research (paragraph 4.7).

4.2 Theoretical framework

4.2.1 Four levels of institutions

As mentioned above, we used the four levels of institutions as defined by New Institutional Economics (NIE) as the basis for the theoretical framework. These four levels are (1) social embeddedness, (2) institutional environment, (3) institutions of governance and (4) incentives and motives (see Table 4.1). With this model, NIE aims to explain the origin of institutions of governance or modes of organisation (level 3). The central thought is that the institutional environment (the formal rules of the game in level 2) and the social embeddedness (the informal rules of the game in level 1) have a great influence on the origination of and changes in the institutions of governance (how the game is played). In the social environment, changes in social preferences and societal trends influence the institutions of governance, along with the aforementioned informal rules. The existing organisations reflect the possibilities provided by the social and institutional environment (North, 1993, 1994). Nevertheless, as a result of changes and shortcomings in society, new types of organisations originate that don't fit into the social and institutional environment (yet). Incentives and motives at level 4 – which could be called the stakes or objectives of the game – determine people's choice whether or not to take part in organisations.

At the level of social embeddedness, institutions change very slowly (in approximately 50 to 100 years). The formal rules of the game (institutional environment) are partly the product of evolutionary processes and partly the result of political actions. Institutions at this level change over a period of 10 to 50 years. The period for decisions at the third level (institutions of governance) is 1 to 10 years. The self-organisation initiatives in this research can be seen

Table 4.1. Four levels of institutions (based on Williamson, 2000: 597).

	Level		Period of change (years)	Core element
↓	Level 1 Social embeddedness	↑	50 to 100	Informal rules of the game – e.g. customs, traditions, norms, prevailing notions, preferences en ideologies
↓	Level 2 Institutional environment	↑	10 to 50	Formal rules of the game – e.g. policy, laws, regulations and the issuing of rules on ownership
	Level 3 Institutions of Governance		1 to 10	How the game is played – e.g. government, firm and club/self-organisation
↑	Level 4 Incentives and motives	↓	Continous	Stakes of the game – self-interest and shared ideals

as new institutions of governance. Their reference back to historical types of organisations suggests a yearning for a type of organisation that doesn't exist anymore. A framework can be deduced from the NIE model to help gain a better understanding of how these initiatives have come about.

The four levels influence each other. Government fiscal laws (level 2) for instance, or a social norm that puts pressure either to do or not to do something (level 1) influences the stakes of the game (level 4). In this chapter we focus on the influence of these three levels on level 3 – the origination and development of institutions of governance (like that of self-organisation initiatives that refer back to traditional organisations). This is indicated by the coloured arrows in Table 4.1.

Which developments in the social and institutional environment promote the formation of the new self-organisation initiatives in Level 3 (see the red and the blue arrow in Table 4.1)? What motivates and stimulates people to join these initiatives (see the green arrow)? Which existing institutional rules prevent the establishment and success of these initiatives (see the blue arrow)? We must bear in mind that the formal rules in the institutional environment generally change at a slower rate than the institutions of governance.

In this instance we have chosen an institutional view – that explicitly takes the social embeddedness into account – to gain a better understanding of the origination of and the motivation for self-organisation. Naturally, we will consider other developments that influence self-organisation, such as economic, cultural and political developments, because these always have consequences for the social and institutional environment of these organisations. Further, we will use the framework of the four levels of institutions to introduce other theories that can explain or interpret the origination and functioning of self-organisation initiatives.

4.2.2 The social embeddedness and the institutional environment: theories that explain the origination of self-organisation

Self-organisation is often a reaction to changes in or failure of the current social and institutional environment. The thesis of the institutional gap (Bargeman, 1996) states that the origination of associations is a reaction to shortcomings in existing institutions like the government, but also within the market, church, company or family. These institutions form part of the establishment and have largely determined the rules of the game in the institutional and social environment. New associations form an alternative that has a correcting and complementary effect on the existing structures.

The problem situation thesis is an addition to the thesis of the institutional gap (Bargeman, 1996). According to this thesis, problems that arise because of changing situations stimulate the origination of associations. These problems bring about a disruption to the social system. People form groups in order to balance the system or to solve the experienced problematic situations.

4.2.3 Institutions of governance: the Club-theory

Landscape is a quasi public good (Table 4.2). There are a number of options for the provision of a quasi public good: (1) the organisation is provided for by the *government,* (2) by private businesses contracted by the government (*contracts*) or (3) by *clubs*. The *market* is only an option for individual goods. In the past 150 years the government was mostly responsible for the planning and management of the landscape.

According to the Club-theory however, a 'club' is in many cases a more efficient type of organisation for the provision of a quasi public good than a government. An important reason for this is the high level of involvement of the club members, as a result of the generally exclusive nature of the club and the short relation between 'paying – deciding – enjoying'. According to Cornes and Sandler (1996) a club is a small-scale and voluntary group of individuals that collectively benefits from the sharing of: (1) the production costs of goods or services; (2) the characteristics of the members; and (3) a good that is defined by excludable (internal) benefits.

The most important feature of a club is the presence of a mechanism that can be used to (partly) exclude non-members and/or non-payers (for example a membership fee). Without such a mechanism there wouldn't be an incentive for members to join a club. This means excluding non-members from the internal benefits, such as control over the activities of the

Table 4.2. Characteristics of goods in the rural area based on excludability and rivalry (based on Van Huylenbroeck en Slangen, 2003).

	Rivalry	Non-rivalry goods and services (indivisible)	Rivalry goods and services (divisible)
Excludability		100%	100%
Impossibility of exclusion or rejection	100%	(1) Pure public goods open space/rest/biodiversity/natural habitat/cultural heritage	(2) Common goods ground and surface water/fish in the ocean, rivers and canals/wildlife
Possibility of exclusion or rejection	100%	(3) Quasi public goods nature/landscape	(4) Pure individual goods agricultural products/agricultural tourism/health care farms

club. The exclusion of non-members from external benefits like an attractive recreational landscape is difficult or even impossible because of the nature of the good (partly non-exclusive). Formal types of club are for instance foundations, associations and co-operatives. Nevertheless a club can also be an informal form of organisation. There are many different kinds of clubs, such as a card-club, a sports-club, a Rotary-club or a squatters' movement. The self-organisation initiatives in this research concern clubs in which the jointly produced quasi public good is not restricted to the members themselves (like in a card-club), but is partly available to non-members as well (see also Bargeman, 1996). In other words, it also fulfils a public need, because non-members can enjoy the external benefits too (for instance a nicer landscape for recreation).

4.2.4 Incentives and motives: theories on motives for co-operation

As said before, the founding of a club or an initiative for self-organisation is a reaction to the environment, such as the failure of government, a change in the social system, a financial incentive or a practical problem. However, the motives for people to co-operate to achieve a quasi public good come from within. According to Komter *et al.* (2000), prominent social theories distinguish two types of co-operation:

- co-operation whereby shared norms, values and emotions are the basis for solidarity and unity;
- co-operation with an instrumental motive, whereby the rational self-interest of individuals drives solidary behaviour.

Komter *et al.*, (2000) claim that these two notions have been incorrectly put opposite of each other. In reality, the two types of motives don't exclude each other and can therefore coexist. The intrinsic motivation for self-organisation can thus come from idealistic motives *and* motives of self-interest.

4.2.5 Coordination mechanisms in self-organisation

Depending on the motives of an organisation a manner is chosen in which the joint activities are coordinated. Four groups of coordination mechanisms can be distinguished that don't exclude each other and therefore can be simultaneously applied in a certain mix (see Figure 4.1)[2].

The 'Handshake' is the coordination mechanism which describes a form of co-operation based on shared norms, values and emotions where there is a presence of mutual trust. Important elements are mutual adjustment and reciprocity. This mechanism is often applied

[2] Only the Handshake and the Handbook are relevant mechanisms for cooperation within self-organisation initiatives. The Invisible hand is the mechanism for the market and the Visible hand in hierarchical organisations. Elements of these mechanisms could play a role in some initiatives, but when it has the upper hand it isn't a club anymore.

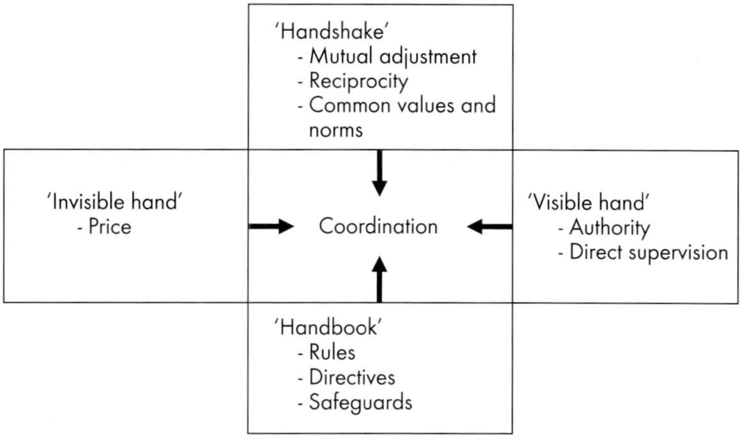

Figure 4.1. Four groups of coordination mechanisms (based on Borgen and Hegrenes, 2005).

in non-hierarchical organisations with highly motivated people where there is a strong sense of mission, esprit de corps or ideology (Douma and Schreuder, 2002). The 'Handbook' is the coordination mechanism for co-operation based on an instrumental motive (driven by the rational self-interest of individuals. Rules, agreements and (detailed) contracts are a features of this mechanism and can be seen as expressions of distrust. In practice the coordination of an initiative of self-organisation is a mix of both coordination mechanisms.

4.3 Past and present of the boermarke in the province of Drenthe

This paragraph goes into the origin, functioning and development through time of the *boermarke,* one of the few examples of an historical type of organisation that is still functioning in some way today. The confrontation of this historic analysis and the present-day use of the historic *marke, meent* and *naoberschap*[3], enables us to gain a better understanding of the correlation between the historical form and the present-day initiative. The theory described in paragraph 2 can be applied to both present-day and historical organisations. This paragraph shows the influence of changes in the social and institutional environment on the origination and development of the historical *(boer)marke.* Furthermore, it illustrates that the *(boer)marke* function as a club, with the corresponding coordination mechanisms 'Handshake' and 'Handbook'.

4.3.1 The origin of the marke

The *marken* originated when the Germanic tribes, a nomadic people, settled for permanent farming. All members of the tribe could use the area taken into possession by the tribe.

[3] In this research we see the *marke, meent* and *naoberschap* as more or less comparable concepts.

'*Marke*' originally meant 'border'. The communal lands formed a bordered area that belonged to a settlement. Due to an increase in population in the 13th century, which resulted in more intensive use of the fields, forests and peat lands surrounding the villages, the *marken* were organised in a more formal way. To prevent the exhaustion of these resources, the farmers in the village started co-operating and establishing the *marke*-organisations. Besides Drenthe, *marken* also originated in the provinces of Groningen, Friesland, Overijssel, Gelderland, Utrecht and Brabant, each with their own regional features. The *marken* in Drenthe, named *boermarken*, managed to evolve with time and they still exist today.

4.3.2 The functioning of the marke

In mutual agreement the farmers determined the borders of the *marke* and the rules concerning the use of the communal uncultivated lands. Everybody that owned more than one hectare of land was allowed to join the *marke*. So without land, one could not become a member: This is a form of exclusion, which is distinctive for a club. Each farmer had a proportional share in the *marke* based on the size of the land they owned: the so-called '*waardeel*' [share]. The amount of '*waardelen*' [shares] determined how many sods of turf one could gather or how much wood one was allowed to cut and how much livestock (cows or sheep) could graze on the *marke*. When a farmer left his farm, he also lost his '*waardeel*'.

The governing body (the board) of each *marke* was formed by '*volmachten*' [representatives], who were chosen by the farmers. The board drew up '*willekeuren*' [arbitrary codes]: ordinances about the management of communal lands. The establishment of these rules is a coordination mechanism in the analogy of the 'Handbook' in Figure 4.1. In the *marke* one can also find elements of the 'Handshake' mechanism. The farmers helped each other, for instance, to secure the harvest or build a new farmhouse. If the farmers did not conform to the norms and values of the community, they were excluded from the community by the *representatives* through social pressure.

4.3.3 The decline of the marke

After the Republic of the Seven United Low Countries was annexed by France in 1795, the liberal ideology of the French acted as a deterrent for the continued existence of the *marke*. At that time the priority lay in increasing agricultural productivity. This left no place for the *marke*, which was portrayed as a 'backward remainder of the past'. Nevertheless, local attempts to 'dissolve the *marke*' failed, due to the conservative collective's desire for self-preservation (Van der Woud, 1987). Consequently the national government had no choice but to force the 'dissolution of the *marke*' with the *Markewet* (*Marke*-law) of 1886. Only relics of former *marke*-organisations are left, like the present *boermarken* in Drenthe.

4.3.4 The survival of the boermarke

In spite of the 'dissolution of the *marke*' many *boermarken* in Drenthe have survived to manage the remaining communal lands and country roads. There are still about 120 active *boermarken* in Drenthe. Their responsibilities are narrowed to communal agricultural machines, shared 'spoelplaatsen' [washing areas] and the issuing of hunting rights. The existence of the *boermarke* was only *really* jeopardised by the '*ruilverkaveling*' [agricultural property re-allocation] in the 60's and 70's of the last century. Farmers left their small village properties to start new farms outside of the village, with the increase in scale that came with it. The number of active farmers within the village communities dropped drastically and with that the membership of the *boermarken*.

At the moment, the coordinating association of *boermarken* is contemplating new functions for the *boermarke* and co-operation with other local organisations. They consider modernising the centuries-old organisational structure because of the many changes that they are confronted with in the rural area. Could this mean the rise of a contemporary *boermarke*?

4.4 Three present-day self-organisation initiatives with a historical correlation

In this paragraph three self-organisation initiatives with a historical correlation are portrayed and analysed. The theoretical framework is used for the analysis of their quest for local collectivity for the organisation of the rural landscape.

4.4.1 *'Nieuwe Marke in de Rijssenervallei'* [New Common Land in the Rijssen Valley]

The *Nieuwe Marke in de Rijssenervallei,* as part of a larger 'landinrichtingsproject' [rural planning project] in the Rijssen valley aims to improve the quality of the landscape by expanding the role of the River Regge and it's water system. Because of the fragmented pattern of ownership and the small number of active farmers, a programme of land acquisition would be too costly. The search for alternative strategies to improve the quality of the landscape focused, therefore, on fostering conditions for cooperation between private landowners and the remaining active farmers. The concept of *New Common Land* is an alternative solution to the planning and management of the landscape. This initiative resulted from the joint effort of an artist, jurist and landscape architect and is based on the principle of exchange of rights and responsibilities, analogous to the historical *marke.* They expected that this principle could stimulate the many different landowners to co-operate (Van Westen, 2004).

The *Nieuwe Marke* is a private organisation whose members commit to realising landscape objectives in exchange for the right to develop. However, the current 'bestemmingsplan' [land designation plan] does not make any allowance for the exchange of rights and responsibilities.

New Common Land wants to establish these landscape objectives in a landscape plan in cooperation with the Rijssen and Wierden municipalities. This would give conditional rights to landowners to build in exchange for landscape maintenance. Essential to this initiative is the joint objective established in a landscape development plan. Private ownership of land is combined with agreement on how to use and maintain the land. Communal ownership isn't an option because there are too many practical problems. Shared ownership is apparently a difficult concept for governments and (financial) institutions. Banks are reluctant to issue loans to individuals involved in shared ownership because they are uncertain about the value of the collateral should there be a problem with the repayments.

Spatial and economic problems are the driving forces behind this initiative. Ideals concerning preservation of small-scale rural landscape are secondary. The realisation of an agreement between the members in cooperation with the municipalities is a first step in the building of trust. Ideally, the 'Handbook' mechanism becomes unnecessary once mutual trust is established.

4.4.2 *'Meent in Hof van Twello'* [Common grazing at Twello Farm]

Hof van Twello is a farm searching for alternatives to compensate for the decreasing returns from the land. Their ideal is to preserve the existing small-scale rural landscape as best they can. The urban context, the existing joint ventures with regional businesses and the positive experiences with the organisation of activities for the public, offer perspectives for cooperation with the community. *Twello Farm* seeks cooperation with the local inhabitants for the development and management of an attractive and profitable landscape. It is believed that in order to organise the (future) participation of members of the community in a sustainable way, a return on investment is needed. The concept of the *meent* (see Box 4.1) is used as a tool to organise this. In return for their investment through labour or financial contributions, the participants in the *meent* get a say in the decision-making process or a share in the agricultural produce (internal benefits). The rights that they earn and the tangibility of the contribution and the returns, form the basis for the commitment of the (future) club members.

Box 4.1. *Marke and meent.*

The *marke* is an organisation of *eigenerfden* (farmers with a farm in full ownership), who collectively owns, uses and manages uncultivated lands. This community has the collective right to use the pastures and the heath-, wood-, peat and marshland that belong to the *marke*. On the other hand, they have responsibilities such as the maintenance of the lands, fences and enclosures (De Monté Verloren and Spruit, 1982). The *meent* is an association of landowners that manages the undivided communal pasture.

They are still looking for the best way to organise this preferred collaboration to achieve a profitable and attractive landscape. Collective use and management of the land and the manner in which rights and responsibilities are exchanged form aspects that have to be considered in more detail. Research is needed into the degree these aspects can be applied within the current institutional context. For instance structural agreements on the contribution of labour by the citizens (e.g. landscape-maintenance) in exchange for goods (e.g. wood, fruit) seem to encounter fiscal problems.

Self-interest *and* social involvement are driving forces for the initiator and the (future) participants. The initiator expects a good profit from the co-operation, because the business operates optimally in the urban context. The participants are offered a chance to influence the local rural environment and they profit by being able to harvest their own crop. At the same time the agricultural cultural landscape as desired by the parties involved, is preserved. In view of the variety of motives, the mechanism of the 'Handshake' and the 'Handbook' could coordinate the collective activities. The initiator's negative experiences with self-harvesting emphasise the need for a set of written rules.

4.4.3 *'Naobers van Zudert'* [The Neighbours of Zudert]

As outlined in the introduction, a number of neighbours in Zudert in the southern part of the village Dwarsgracht, formed a cooperative to preserve their own local rural environment (Floor and Salverda, 2006). This unique landscape was the main reason for many new inhabitants to move there and the residents realised that to protect their real estate investment, they had to take action to reverse the decline in the landscape. By collectively organising the maintenance of the landscape, the inhabitants intend to restore and preserve it. Co-operation is a necessity because the inhabitants can't manage the landscape individually. In this case, the existing nature and landscape preservation organisations didn't regard this as their responsibility because private property was involved. The initiators set up a foundation so that they were eligible for subsidies. The naming of the initiative is not related to a local historic *naoberschap* (see Box 4.2), but derives from the fact that they are neighbours who need each other to realise their vision of the landscape and that the historical concept carries a positive connotation nowadays.

Box 4.2. Naoberschap.
In the past, the *naoberschap* formed an entity with rights and responsibilities agreed upon by a group of neighbours regarding important events and periods in human life. People helped each other, for instance, with births, weddings and funerals. In case of illness or at harvest time one could also call on the *naobers*. The number of families in a *naoberschap* varied between 6 and 14 (Ginkel *et al.*, 1998).

Members pay a fixed, yearly, financial contribution. For the maintenance they also pay a sum that is dependant on their income and the size of their land. Two members are paid to do the large-scale landscape maintenance. Besides this, all members voluntarily participate in the actual maintenance a couple of days a year. The co-operation and the annual hay-making festival have brought about strong social cohesion. There is a strong awareness that mutual understanding and trust is crucial to the continuity of the co-operation.

The basis for cooperation is the shared ideal of re-creating an attractive landscape for a better lifestyle and improved social cohesion. However, the protection of the property value should not be overlooked as a major motivation to participate. The cooperation is based fully on enthusiasm and mutual trust, so one can speak of a largely 'Handshake' mechanism.

This initiative has discovered that it is quite difficult for non-farmers to get subsidies for the maintenance of the landscape, because the government sees these subsidies as primarily meant for farmers. They have also found that applying for subsidies is complicated and time consuming. Not only that, a subsidy is a one-off payment so doesn't contribute to maintenance continuity. An interesting line of action, however, is exploring the possibility of capitalising the surplus value of the real estate – to which an attractive village landscape contributes – to fund a landscape trust. This trust would make the foundation independent of subsidies for the continuation of its landscape management.

4.5 Lessons learned: the cases

The theoretic framework and the historical analysis of the *boermarke* are used to analyse the three self-organisation initiatives. The most important findings will be discussed in this paragraph.

4.5.1 Expressions of present-day desires

None of the three initiatives[4] want to revive the historical *marke, meent* or *naoberschap.* They make use of the historic context to express their desire to take collective responsibility to organise the development, use and maintenance of the landscape at a local level. That is, self-organisation instead of institutional organisation (mainly by the government). Lowenthal (1985, 1989) refers to the human characteristic of dealing with the past in a nostalgic way. This often causes a distortion of the image of the past. That is why statements about the past often tell us more about present-day needs than about the past itself. This theory supports the conclusion that the historical references emphasise the modern-day (desire for) collectivity on a local scale.

[4] These initiatives are different from the still existing historical *boermarken* in Drenthe that consider modernising the centuries-old organisational structure because of the many changes that they are confronted with in the rural area.

4.5.2 Timeless aspects of organisation to achieve collectivity on a local scale

Despite the fact that historical terms are used, the initiatives mainly emphasise aspects of organisation and coordination that are, in fact, timeless. It is logical to associate some aspects of these self-organisations with the past; like the small-scale of an organisation that enjoys a lot of local commitment, communal use and management of land and local exchange of rights and responsibilities, as these are not common characteristics found in the organisation of the landscape by the (national) government today.

In this way, the preferred small-scale organisation of landscape development and management with a lot of local involvement can be seen as a timeless aspect placed in a historical perspective. The initiatives are, just like the *marke, meent* and *naoberschap*, small clubs of locals, whereby collective goals, shared ideals and motivation are very important (next to individual interests of course). A second timeless aspect that is used in a historical perspective, is the collective use and management of land, like the examples of the *zelfpluktuinen* ('self-harvest gardens') on *Twello Farm* and the collective maintenance days run by the *Neighbours of Zudert* that are followed by a hay-making festival. Thirdly, the local exchange of rights and responsibilities is used in a historical perspective to bring about the desired modern-day feeling of local collectivity. In principle individuals incur expenses for, or invest in the maintenance of, the local landscape in exchange for individual and collective benefits. For example, in exchange for the payment of a contribution or the supply of physical labour for the maintenance of the landscape one gains an attractive living environment and an increase in property value. Other examples are the acquisition of a building permit in exchange for management measures which increase the quality of the landscape. It is worth noting that communal landownership, as in the historical *marke* and *meent,* is not one of the objectives in the present-day initiatives. This doesn't appear to be necessary to attain the desired objectives.

4.5.3 Self-organisation by clubs

The self-organisation initiatives in this study can be defined as clubs. This is because they are small-scale groups of individuals that voluntarily and collectively (want to) provide a quasi public good (in this case landscape) and derive collective benefits from it. Non-members are excluded from the internal benefits; for instance they do not have decision-making authority on the development and management of the landscape in question. They *do* have partial access to the external benefits; they can enjoy the landscape for recreation. In other words, the product or service of the club also fulfils social needs, because non-members can profit too. The Club-theory argues that in many cases a club can produce a quasi public good more efficiently than a government can. An important reason for this is the high level of involvement of the club members, as a result of the exclusive nature of the club and the short relation between 'paying – deciding – enjoying'.

4.5.4 Self-organisation as a result of changes in the social and institutional environment

According to the New Institutional Economics (NIE) the origination of and changes in types of organisations can be explained by changes and shortcomings in the social and institutional environment. Self-organisation can be understood better by watching present-day social transitions. A number of explanatory changes in the social environment and the institutional environment are outlined in Table 4.3.

The rural area is changing because of increasing non-agricultural claims for the realisation or preservation of an attractive and characteristic landscape as an environment for housing, living and recreation. As a result of this shift in preferences the rights associated with landownership by farmers are changing too. One of the consequences is that farmers will have to invest more in the landscape to retain their 'licence to produce'. Another result of this shift in preferences is the establishment of local co-operations of residents and (active) farmers. Because of this the residents gain more decision-making authority over the landscape. Residents are willing to pay or to participate in the maintenance in order to create a landscape they appreciate. They want to have a say (more rights) and in exchange

Table 4.3. Changes in the social and institutional environment that induce self-organisation.

Institutional levels	Period of change (years)	Social change as a cause for self-organisation
Social embeddedness	50-100	Urbanisation of the countryside (decreasing role of farmers, increasing role of urban dwellers) Changing role government (withdrawing and facilitating) Changing preferences of society regarding rural area More 'urban' norms, values and mentality in rural area: more assertiveness and initiative by the urban dwellers Dissatisfaction with the existing instruments (development plan)
Institutional environment	10-50	More land owned by urban dwellers in the rural area New policy and rules due to a changing role of the government: 'take care *that*, instead of take care of' Shift in allocation rights of ownership in rural area
Institutions of governance	1-10	Origination of local co-operations by citizens and/or clubs for the organisation of the landscape

for this they are prepared to invest (more responsibilities). This could be called the 'right to consume', which is 'earned' by exerting themselves for the landscape. Residents, so to speak, want to get involved in this quasi public good called landscape, so they can protect and enjoy the local rural environment more.

4.5.5 Motives and coordination mechanisms

Taking care of a quasi public good by a club is based on a mix of two types of motivations for co-operation: one based on shared norms, values and emotions and the other based on the rational self-interest of individuals. In practice the coordination of an initiative of self-organisation is a mix of two coordination mechanisms: the Handshake and the Handbook (see Figure 4.1 in paragraph 4.2). One explanation for the club members' desire to take care of the landscape themselves, instead of leaving it to the government, is the close relationship between the 'paying – deciding – enjoying' of the landscape, which is considered to be attractive. The relationship between the resident and the government – described here as the 'costs' (taxes) and 'benefits' (the result of the landscape policy of the government), is felt to be too distant, too abstract and too indirect. Another explanation for self-organisation is that the independent, highly educated 'urban' residents increasingly determine the course of life in the rural area and consequently the desired organisational relationships. In this way one can see that societal changes, enhanced by the revolution in information technology and the increase in social fluidity, help to explain why more and more individuals want to be involved in organising their own living environment. Local co-operatives are a logical step because an individual doesn't have the means, land, knowledge, experience and time to develop and manage the landscape on the desired scale.

4.5.6 Bottlenecks due to the existing institutional environment

The self-organisation initiatives want to realise collective objectives through the local exchange of rights and responsibilities but this isn't always attainable in the institutional environment. The local land designation plan, for instance, records different land uses but can't attach development-rights to them, let alone link such rights to management obligations. Furthermore, the exchange of goods for labour encounters problems with the fiscal system. Likewise, if initiatives were to desire communal (land) ownership, they would encounter difficulties with the existing institutional environment. Banks for instance hesitate to finance common property.

4.6 Lessons learned: the theory

The theoretical framework, developed in this explorative research, is used to analyse a historical type of organisation and present-day self-organisation initiatives.

The theory of the four levels of institutions (the NIE) has turned out to be a useful means to gain a better understanding of the origination and motivation of present-day self-organisation initiatives. The rise, evolution and decline of the historical *marke* and the potential modernisation of the *boermarke* can also be explained by means of the NIE model. This touches the subject of the evolution of organisations in time. The use of evolutionary theories, theories on cyclic processes and other organisation-theories, that give a broader time-perspective, could be interesting for further research.

The club-theory and the theory on coordination mechanisms were used to gain a better understanding of the functioning of self-organisation. It would be interesting to extend the research with theories, for instance, on the exchange of rights and responsibilities, as this mechanism has turned out to play an important role in the self-organisation initiatives in this research.

The theoretical framework has a mainly institutional perspective. To fully understand why people collectively devote themselves to a quasi public good and its value to and impact on their lives, we need more social or anthropological theories. Furthermore, the theoretical framework doesn't concentrate on motives as image and nostalgia as an explanation for the historical references. Therefore further theoretical exploration in the area of marketing or political science is an option for future research.

4.7 Subsequent question

In a follow-up study, it would also be interesting to zoom in on the manner in which self-organisation initiatives want to organise collectivity through the exchange of rights and responsibilities. It would also be interesting to research what paradigm change is needed to enable this. This paradigm change could be the transition from the transaction-model with land designation and contracts, to a transaction-model based on collectivity and the local exchange of rights and responsibilities. With that sort of transition-assignment history will always seem alluring.

References

Bargeman, B., 1996. Associaties in de vrije tijd. In: Vrijetijd studies 2. Hardloopverslaving, verenigingsleven, stedentoerisme. Boom tijdschriften: Meppel, the Netherlands.

Borgen, S.O. and Hegrenes, A., 2005. How can transaction costs economics add to the understanding of new contractual formats in the Norwegian Agri-food System. Working Paper 2005-7, Centre for Food Policy, Norwegian Economics Agricultural Economic Research Institute, Norway.

Cornes, R. and Sandler, T., 1996. The theory of externalities, public goods and club goods. Second edition. Cambridge University Press: Cambridge, UK.

De Monté Verloren, J.Ph. and Spruit, J.E., 1982. Hoofdlijnen uit de ontwikkeling der rechterlijke organisatie in de Noordelijke Nederlanden tot Bataafse omwenteling. Kluwer: Deventer, the Netherlands.

Douma, S.W. and Schreuder, H., 2002. Economic approaches to organization. Third edition, Prentice Hall International: London, UK.

Floor, L. and Salverda, I.E., 2006. Zelforganisatie in het landelijke gebied. Signalen uit de praktijk. Wageningen Universiteit en Researchcentrum, Wageningen, the Netherlands.

Komter, A.E., Burgers, J. and Engbersen, G., 2000. Het cement van de samenleving. Een verkennende studie over solidariteit en cohesie. Amsterdam University Press: Amsterdam, the Netherlands.

Lowenthal, D., 1985. The past is a foreign country. Cambridge University Press: Cambridge, UK.

Lowenthal, D., 1989. Nostalgia tells it like it wasn't. In: Shaw, C. and Chase, M. The imagined past. History and nostalgia. Manchester University Press: New York, USA.

North, D., 1993. Institutions and credible commitment. Journal of Institutional Economics 149: 11-23.

North, D.C., 1994. Economic performance through time. The American Economic Review 84 (3): 359-368.

Van der Woud, A., 1987. Het lege land. De ruimtelijke orde van Nederland 1789-1848. Bert Bakker: Amsterdam, the Netherlands.

Van Ginkel, R., Jansen, H., Maenhout, J., Müller, Th. and Roukens, K., 1998. Het verloren paradijs. Sociografie van stad en land. SISWO, Instituut voor Maatschappijwetenschappen, Publicatie 400. SISWO: Amsterdam, the Netherlands.

Van Huylenbroeck, G. en Slangen, L., 2003. Nieuwe institutionele arrangementen in het landelijke gebied. Tijdschrift voor Sociaalwetenschappelijk Onderzoek van de Landbouw 18 (2): 107-121.

Van Westen, J. (ed.), 2004. De nieuwe marke in de Rijssener Vallei. Een eigentijds organisatiemodel voor eigendom, gebruik en beheer in het landelijk gebied. Pilot landinrichting Rijssen, the Netherlands.

Williamson, O.E., 2000. The new institutional economics: taking stock, looking ahead. Journal of Economic Literature 38: 595-613.

Chapter 5

Transition starts with people: self-organising communities ADM and Golf Residence Dronten

Rosalie van Dam, Jasper Eshuis and Noelle Aarts

Abstract

In this chapter we explore the transition of societal organisation from heavy reliance on the state towards self-organisation by citizens in communities. We explore how this transition manifests itself by analysing two cases of self-organising communities in the Netherlands. The case studies of the ADM squatter community in Amsterdam and Golf Residence in Dronten show how these communities of self-organising citizens created their own residential arrangements and took the initiative in developing a unique spatial environment. By looking closely at these two forms of self-organising communities, we can enhance our knowledge about transitions and public management.

Keywords: governance, self-organisation, community, transitions

5.1 Introduction

5.1.1 Self-organising communities in the Netherlands

The 'Amsterdamse Doe-het-zelf Maatschappij' [Amsterdam Do-it-yourself Company], is located in the western dock area of Amsterdam Harbour. The ADM property, which used to be a large dry dock, was occupied by squatters in 1997. Since then, the ADM has grown to be a vibrant community of people who live an alternative lifestyle and organise their own residential arrangements. The resident population is about 100 – with about 60 of them living in trailers and houseboats moored at the site. The rest live and work in the old offices. As a self-organising community, the ADM has created a unique residential environment and built up a reputation for art, craftsmanship and cultural activities. It is an example of a community of people organising their own social and spatial environment, relying less on governmental activities.

At a more rural site in the Netherlands, another self-organising community can be found: the privately managed residential development called 'Golf Residence Dronten'. This community is located at the southern edge of the municipality of Dronten. The Golf Residence consists of 360 detached villas, 90 apartments and a golf course on 86 acres. Together, the 450 households own and take care of the area, including things such as infrastructure, common grounds and even the streetlights (which are normally maintained by the local council). Through self-organisation, the community has created it's own residential environment, characterised by a high quality golf facility in a well groomed park with a lot of greenery. This development concept allows for an unusual and distinctive residential design, and a higher standard of facilities than in normal neighbourhoods.

These two very different examples of self-organising communities reflect a transition in the relationship between citizens and government. Although the notion of the state has changed from the state being 'the interventionist' to the state being 'one of the interventionists', transition and transition management are often related to state interventions. The examples of the self-organising communities mentioned above, show us that changes might also originate from the general public. In this chapter, the transition, related to the changing role and position of government and citizens is discussed by focusing on the phenomenon of self-organising communities. Preserving a certain amount of autonomy in social and spatial organisation, self-organising communities can be seen as an alternative practice in relation to housing and residential environments. Members of the public are taking responsibility for the direction of their own lives and residential arrangements, by organising certain (public) matters for themselves. The examples of ADM and Golf Residence show us how the transition in the relationship between citizens and government manifests itself in new residential arrangements and special spatial environments that are created by communities of private citizens.

Because this chapter addresses a transition in the way people organize themselves socially and spatially – preserving a certain amount of autonomy – we will introduce some theoretical notions concerning changes in the roles played and positions held by government and private citizens. After discussing the changes from government to governance, we add another dimension to the idea of private citizens taking charge of their own destiny, by exploring Becks' sub-politics and Giddens' life politics (paragraph 5.2). Private citizens mostly do not take the initiative alone but organise themselves into groups or communities. This is the reason why theoretical notions about self-organising communities are discussed. Then we take a closer look at the above mentioned examples of self-organisation, the squatter community ADM (paragraph 5.3) and residential community Golf Residence Dronten (paragraph 5.4). We analyse both cases by comparing the organisation and forms of self-organisation (paragraph 5.5). In the following section (paragraph 5.6), the cases are related to the notion of governmental and social transition. In the last section (paragraph 5.7), the contribution of a self-organisation perspective to the knowledge about transitions and public management will be discussed.

5.2 Theoretical notes on governance, self-organisation and transitions

5.2.1 Changes in relations between government and citizens

The term 'government' mostly refers to the formal institutions of the state and their monopoly of legitimate coercive power. The notion of government points to the state's ability to make decisions and to enforce them. Using the term 'governance' signifies 'a change in the meaning of government, referring to a new process of governing; or a changed condition of ordered rule; or the new method by which society is governed' (Rhodes, 1996: 652-653). Although often interpreted differently, there is general agreement that governance refers to the development of governing styles in which boundaries between and within public and private sectors have become blurred. The changes in the position of governmental organisations can be seen as a transition, often referred to as the development from government to governance (Kooiman and Van Vliet, 1993; Rhodes, 1996). This transition is characterised by the diminishing importance of the national government as the centre of society. The development takes place on roughly two dimensions. On the vertical dimension, one can see changes in power and policy leading both to regional and local government and towards trans-national governments such as the European Union. As for the horizontal dimension, the shift is one from public to private actors, varying from private organisations to individual citizens.

The transition from government to governance manifests itself in a search for ways to involve citizens in policy processes and a search to determine which responsibilities should be public and which responsibilities should be private. This changing relation between government and citizens puts the latter in a position in which they are expected to organise certain (public) matters for themselves. In other words, private citizens are increasingly expected to take responsibility for the direction of their own lives.

In this study, we look at the idea that governance involves societal management which is not only executed by the central government, but also from a variety of other centres. We see self-organising communities as centres that are initiating and carrying out spatial and social developments and transitions.

5.2.2 Citizens taking charge

In relation to societal dynamics, Beck *et al.* (1994) mentions de term 'sub-politics' to denote societal activities and developments outside traditional political institutions which are politically relevant because of their societal influence. According to him, politics takes place where we normally do not look for politics: for example at our work where the relationships between companies and their employees are negotiated; in our private lives where the relationships between the sexes are preserved or changed; or at the supermarket where consumers influence food chains and the environment. The essence of sub-politics is that political developments are realised outside traditional political 'centres of power' such as the national parliament. Elaborating on Beck's ideas, Holzer and Sørensen (2003) emphasise the relevance of social initiatives which influence society, while at the same time being largely independent and distinct from the political system. Sub-political initiatives can be deliberate and active strategies meant to influence 'formal politics'. But it can also include initiatives that are not aiming to influence formal politics, but do so unintentionally as a consequence of their societal influence (passive or unintended sub-political strategies). Recognising this societal influence, formal politics 'have to' deal with or do something in relation to these initiatives. According to Holzer and Sørensen (2003) it is exactly the non-political character that gives 'sub-political' phenomena their significance.

Giddens (1991) also notices an increasing need among citizens for self-actualisation and for control of their own lives. No longer controlled by traditional institutions of church and state, individuals can, and indeed must, plan their lives. There is a shift in the locus of authority from external structures to individual choice. Giddens calls it 'life politics'. In life politics, political and societal goals are combined with lifestyles oriented towards self-actualisation. Political involvement of people is increasingly connected to moral issues and social relationships which focus on self-actualisation in emancipated social circumstances and to single issues which have relevance to everyday life. Life politics is not a substitute for classic policy, but complementary to it. Whereas the latter emphasises universal social rights and the realisation of social equality, life politics' concentrates on plurality in the light of social rights and focuses on the particular within the universal.

Politics is increasingly found in networks and within groups of actors that have various forms of autonomy in relation to the state. People organise themselves around subjects and events in daily life. These practices can be understood as sub-politics when they influence interests and power relations in society.

5.2.3 Self-organising communities

Today's society seems to invite people to organise themselves in order to get things done. Self-organisation often goes together with community building; citizens organise themselves into communities to assert their rights and to pursue their shared objectives. A community is characterised by the fact that people within the community believe they have something in common (Mercer, 1956; Sennett, 1971; Willmott, 1986). This may be a shared place (i.e. a 'place community'), or a religious belief, sexual orientation or occupation (i.e. an ' interest community') or a sense of attachment to a place, group or a set of ideas (e.g. 'communion') (Crow and Allan, 1995; Lee and Newby, 1983; Willmott, 1986). Within the fields of sociology and anthropology, communities are traditionally perceived as systems of social relations (Tönnies, 1887 [1955]), usually small groups with dependencies and a similar way of life (Delanty, 2003). Communities were perceived as small units characterised by a feeling of 'belonging together' based on mutual experiences, relationships and sharing the same space.

Later on, the idea of people within a community having a shared identity became more important (Castells, 2004; Sennett, 1971). The effects of sharing symbols and rituals were also acknowledged (Anderson, 1983; Cohen, 1985) and it was stressed that a shared identity was reflected in and reproduced through shared symbols; it created insiders and outsiders and enhanced feelings of 'us' and 'them'.

The communities discussed in this chapter are spatially based communities: communities in which people are not only bonded by relational factors or mutual interest, but also by the place where they live, work or spend a substantial amount of time (Van Dam *et al.*, 2005). Self-organisation is understood to be the way (groups of) people organise and shape their own environment; the areas where they live and/or work.

5.3 Squatter community ADM in Amsterdam

The ADM property comprises about 45 acres of land located remote from residential areas in the western harbour area of Amsterdam. The terrain is fenced and has a locked gate. It is currently owned by the family of a property developer called Lüske who bought the ground in 1997. The first ADM squatters entered the property in the same year and occupied the land and its main building. Over the years, the inhabitants of the ADM have formed a community that clearly distinguishes itself from the rest of society. The community is a system of social relations which can be quite strong. People experience a shared identity as being part of the ADM community and they have a definite connection to the place (Van Dam *et al.*, 2008).

ADM can be described as a free cultural haven for alternative lifestyles. The inhabitants are mainly artists and artisans (e.g. mechanics, dancers, photographers, actors). Most of them work and live at the ADM. There are several garages and workshops, where all kinds of objects are created or restored, varying from works of art, to classical busses, boots, decors and props for theatre productions or baking bread. Most ADM people share a strong criticism of our capitalist society. They view themselves as 'dissidents' and 'freebooters'. For many ADM inhabitants the 'establishment' (read: governmental institutions) is something to be avoided as much as possible. Another characteristic of the ADM community is its own micro-economy. As a result of their anti-capitalist views, ADM inhabitants try to live and work in an alternative economy, in which labour and material goods are valued differently than in the mainstream economy 'outside the fence'. People charge different prices for material goods, there are several exchange mechanisms and people have different ideas concerning owning material things. Although everybody has to pay a contribution; the collection of this money is very flexible: people pay when it suits them.

ADM's relationship with the authorities has been difficult, particularly in recent years. The city council is increasingly interested in developing the property according to its own goals and from time to time this leads to tense relations. But in general, state interference is minimal, even counting the recent past. Some of the ADM people experience the relationship with the local government as problematic, others emphasise the challenge and approach the relationship as a game or dance.[5]

5.4 Residential community Golf Residence Dronten

The Golf Residence Dronten is a privately managed residential development. The integrated golf course and the private ownership and management of the terrain is characteristic for this neighbourhood. The residents have a certain amount of private property around each villa, and the rest is communally owned private land. The total area is surrounded partly by a watercourse and partly by a fence. Golf Residence is a 'residential area with access restricted in such a way that normally public spaces have been privatised' (Blakely and Snyder, 1997). Because a relatively large number of residents have had tertiary education (university), one can classify the neighbourhood as upper middle class (Van Dam *et al.*, 2008).

Although relationships between members are not always very close, Golf Residence can certainly be classed as a community. There are several binding elements that create a community feeling. The residents feel common attachment to the place, and many of them relate to the place in the same way and refer to it as 'our park'. Despite the fact that a large number of residents doesn't play golf, golfing is still considered to be the main symbol for the Golf Residence. An intrinsic feeling of community is stimulated because the inhabitants look after each other to some degree, and they share responsibility for the property. Many inhabitants see themselves as living amongst 'their kind of people' and value this like-mindedness.

A private developer initiated the plan in the early 1990's. As more residents moved in they gradually took over responsibility for the residence. Nowadays the maintenance of the common property within the Residence, including the infrastructure, clubhouse and greenery, is organised through a homeowners association. Membership of the homeowners association is obligatory for all residents. When buying a house in the Residence, one signs a contract with the association and thus agrees to abide by the statutory rules of the association and the private bylaws that apply to the property.

As it is a private residential development where the residents themselves are in charge, the residents have a high degree of freedom to shape both their spatial and organisational environment. But this also involves tasks and obligations, including financial obligations. Not

[5] External threats, such as the risk of eviction by the local authorities, brought about a sense of unity for the inhabitants (Elias & Scotson, 1965 [1994]) and resulted in what Castells calls resistance identities (Castells, 2004). Resistance identities are produced by groups of people who feel discriminated against or excluded by oppressors. Resistance identity leads to the formation of groups or communities whose objective is to oppose circumstances they find untenable. Such an identity helps people to survive based on other (sometimes opposite) principles than principles inherent to the dominant institutions in society. With the threat of eviction, the government is seen a mutual enemy and this strengthens the resistance identity of the ADM community.

only do they have to pay local taxes but they also have to pay a contribution to the homeowners association. Golf Residence has private bylaws instead of public ones as is the case in ordinary neighbourhoods. This brings with it extra responsibilities. There are extra community tasks to be done and, although taking part in activities at Golf Residence is voluntary, there always seem to be enough people who are willing to take part in the organising committees. The residents realise that they have to work together and also solve problems together. They therefore recognise that it is important that the mutual relationships are positive and healthy.

5.5 Self-organisation in the ADM and at Golf Residence

In each case, self-organisation plays a different role. The differences in both these forms of self-organisation have to do with (1) how they were established, (2) the role self-organisation has in the inhabitants' motivation, (3) the physical appearance of both communities, (4) the demands the inhabitants place on their residential arrangements and (5) the organisational structure of the communities.

Regarding the establishment of the self-organisation, the squatters themselves took initiative to start a self-organising community at the ADM property. They appropriated the property and they themselves are the driving force behind the ADM in its current form and organisation. In policy terms, one would call it a bottom-up initiative. At Golf Residence, the property developer was the driving force in realising the community. It was the property developer and his interpretation of the future residents' desires concerning the residential arrangements that led to the concept of Golf Residence. It was not a group of people who initiated the Golf Residence. The self-organisation started *after* the planning phase when the inhabitants became involved. Nowadays, the inhabitants organise their residential environment themselves. The guidelines for the design and appearance of the property originally came from the architect but, as time went by, they have been taken over or modified by the residents.

For the squatters at the ADM, self-organisation is a deliberate choice and this choice is related to their identity and way of life. Self-organisation is connected to other important motivations that drive ADM's inhabitants, such as freedom and autonomy. ADM's physical environment and the way ADM inhabitants organise themselves, provides more freedom and autonomy than elsewhere in society. Inhabitants have, literally and figuratively, the space to create their own residential environment. The prevailing critical attitude towards society is an important reason why people want to live at the ADM. Many ADM inhabitants cannot identify with the values present in the rest of society and feel less at home there. The financial side is also important. Life at the ADM is much cheaper than outside. The inhabitants of Golf Residence have other reasons for choosing the Golf Residence, such as quality of life, golfing and its central location in the Netherlands. But social cohesion and safety are also part of the reason for living there. Self-organisation is seen as part of the deal, a necessity if one wants a certain type of living environment. For the greater part, self-organisation is seen more of a means of achieving a certain quality of life than as an end in itself. One could say that self-organisation is not a deliberate choice of the inhabitants. At one point, the residents even tried to convince the local town council to take over the maintenance of their park. But, in the main, it must be said that most people are very pleased with how the neighbourhood looks and is organised. For most, the advantages far outweigh the disadvantages of living at Golf Residence.

The way each community organises itself is vastly different. At the ADM they have far less formal rules than elsewhere in society, whereas at Golf Residence they have more formal rules. At Golf Residence, most of the rules concern the management and appearance of the park. There are many committees and the residents value clear guidelines particularly bearing in mind the forming of possible precedents. The rules and norms manifest themselves in constitutional articles made by the homeowners association, relating to the infrastructure, domestic maintenance issues and about the golf course. These articles are signed by all inhabitants. Each member is allowed to present proposals to alter the constitution at the Annual General Meeting. The organisational structure of the ADM is typified by a minimum of formal rules, certainly in comparison to other communities. Informal rules evolve in informal exchange and can change or differ per person and situation. Individual freedom and the tailor-made rules that take into account the differences between inhabitants and situations is more important than the notion of equality of treatment. The most important rule is to not hinder others. Another important factor in the organisation of the ADM is voluntariness: ADM community members fulfil tasks and roles that are related to their own special talents on a voluntary basis. Collectivity is also emphasised: you need to want something as a group. Moreover, neither formal hierarchy nor central leadership exists at the ADM. The organisation is based on direct interaction. Some people have more influence than others, often based on how long one has lived on the ADM. The community as a whole determines to some extent who is allowed to live at the ADM and who is not, without a 100% guarantee of control. Unwritten rules, manifested as social norms, are essential for the organisation at the ADM.

In relation to the organisation of self organising communities, the size of the community is also an important factor. Evolutionary psychologists Aiello and Dunbar (1993) argue that the maximum size of a well functioning group of people is about 150 members. According to these authors, this has to do with the brain capacity of people. In a group of 150 people, all people can know each other and it is clear who has what position in the group. When it exceeds 150 members, the organisation needs formalisation to keep things going. The organisational forms represented by the ADM and Golf Residence seem to corroborate this theory: The ADM community never exceeds 150 inhabitants and the organisation is quite informal, whereas Golf Residence has about 1500 inhabitants and has a very formal structure.

The physical appearance of both self-organising communities varies a lot. Looking at the premises itself, ADM inhabitants let nature take its own course, and as a result the former dock area has evolved into a unique ecological environment. The Golf Residence inhabitants on the other hand, have clear directions and committees for maintaining the shared grounds.

As a result the communal grounds look very well groomed. When looking at the residential arrangements, the ADM inhabitants' material lifestyle ambitions seem to be considerably less demanding than those of the Golf Residence residents. ADM inhabitants live in trailers, caravans, old boats or in the old offices at the main building. ADM looks typically messy and nature has been left to take its course in the grounds. Most of the people living at Golf Residence inhabit luxurious villas and the properties look well groomed. Besides differences in residential preferences, the difference in status is an important factor. In the case of ADM, the housing and surroundings look temporary partly because the future is anything but certain. There is no guarantee that the ADM will continue to be tolerated by the authorities. The uncertainty manifests itself in ADM's appearance, but also in a sometimes problematic and ambiguous relationship between different governmental authorities and the ADM. In the next section, we will further address the relations between government and self-organising communities.

5.6 Self-organisation as governmental and social transition

In this chapter self-organisation is seen as a transition in societal organisation, particularly in the relationship between private citizen and government. We have explored how this transition manifests itself in two communities that have created their own residential arrangements and have taken the initiative to develop their own spatial environment according to their wishes. By creating these unique residential arrangements, new relations with the state have been formed.

Taking a closer look at this government-citizen relationship in self-organisation, it is interesting to analyse how both parties perceive each other. In both case studies it is obvious that neither community is brimming over with enthusiasm about the government. But the reasons seem to differ. In case of Golf Residence, the inhabitants perceive little added value from their local government. The Golf Residence inhabitants pay, for example, a contribution to the homeowners association and taxes to the government, but they can only directly see the effects of their contribution to the homeowners association in the maintenance of their park. The ADM community feels a large degree of mistrust towards their local government. This is because their organisation is not 'recognised' but only tolerated by the authorities, and there have been several attempts in the past to have the community dismantled.

Looking at the governmental organisations' perception of the self-organising communities, one can determine that local authorities didn't stand in the way of the Golf Residence development, in fact, the self-organisation might even have been inadvertently stimulated by them. The authorities withdrew from active intervention and in this manner left the door open for the self-organisation. The Amsterdam authorities' relationship with the ADM has been ambiguous. On the one hand, they value the existence of this sub-culture with its creative arts. On the other hand, they dislike the squatting aspect and do not want to allow criminal activities on the ADM property. In this case, they have pursued policy of 'toleration' which offers no security for the long term. 'Toleration' is a policy that can change at a (political) whim. This promotes a wary attitude towards politicians and civil servants and a more open attitude towards others, manifesting itself in an internet manifesto and media attention in times of supposed threat.

As said previously, self organising communities can be seen as an alternative practice in relation to housing and residential environments. The self-organisation initiatives of ADM and Golf Residence are expressions of how groups of people take the lead in determining and maintaining their own residential environment. This can be labeled as a bottom-up transition. Although the ADM and Golf residence are two very different manifestations of modern residential arrangements they have a certain degree of autonomy in common. Both residential arrangements are the result of experimenting with new types of relations between the citizens involved and between these citizens and governmental authorities. The differences in the residential arrangements, both socially and spatially, addresses the differences in citizens' needs concerning residential arrangements in society in general. There is a need for the creation of more diverse forms of housing. Because citizens have developed different demands with regard to architecture, spatial design, facilities, etc., a differentiated demand for housing and housing concepts has arisen. Both ADM and the Golf residence reflect this need for variety in housing concepts. These self-organising communities fit the transition model with regard to residential and housing concepts and the housing market in general.

5.7 Conclusion and discussion

Beck's theory of 'sub-politics' (Beck *et al.*, 1994) and Giddens' notion of 'life politics' (Giddens, 1991) can be related to the self-organising communities of ADM and Golf Residence. Furthermore, these theories indicate a new direction for breaking the institutional and societal deadlocks surrounding transitions, one of the biggest issues of our time. Beck propounds a normative perspective: sub-politics are not considered a problem, but instead are seen to be part of the solution. Beck pleas for a 'non-institutional renaissance of the political' (Beck *et al.*, 1994). According to him, traditional politics does react positively to sub-political developments but still attempts to steer or manage. It would be better to value the intangible non-institutional politics and to create openings for these kinds of initiatives.

Looking at our cases of self-organisation in relation to public management and transitions, two closing remarks can be made. The first concerns the value of and need for heterogeneity. Today's society is multicultural and pluriform. Furthermore, citizens are taking the initiative in various forms of self-organising communities. Taking into account the number and variety of these private citizens' initiatives, a change of paradigm is needed. The Dutch policy analyst Van Gunsteren (2006), reveals a vision about democracy and self-government in which diversity, confrontation and indirect management prevails over top-down variations. Based on the principle of 'the wisdom of crowds', meaning that 'under the right circumstances, groups are remarkably intelligent and are often smarter than the smartest people in them' (Surowiecki, 2005), Van Gunsteren defends democracy. He mentions several principles of self-organisation and applies them to democracy. Following in the footsteps of Van Gunsteren, Frissen (2007) advocates the value of heterogeneity as well. He sketches a world which is inevitably characterised by differences and he is a fervent proponent of taking this reality seriously in government and politics. According to Frissen, the Dutch preference for equality has had a strong influence on the welfare state which aimed at decreasing or compensating for differences. The Council for Social Development (Raad voor Maatschappelijke Ontwikkeling, 2006) also agrees that the time has come for a paradigm change: private responsibility has gained in importance and as a result the government has to modify it's own attitude into one which allows and, indeed, embraces differences and heterogeneity. In the previous section, remarks have been made about heterogeneity in relation to residential and housings concepts. In this framework, transition should be seen as a *pluriform change*. Transitions often aim for a common goal, a uniform change. This chapter shows that in practice, situations are varied and as a consequence so should (the contexts of) transition.

A second and final remark concerns the socialisation of public management and transition management. Citizens are increasingly taking charge of their own lives and as a result, public management and transition management are no longer a purely governmental matter. Policy making increasingly involves a multitude of different coalitions and arrangements as shown in governance literature (Hajer and Wagenaar, 2003; Pierre and Peters, 2000; Van Tatenhove *et al.*, 2000). The notion of public management and transition management would be enhanced by analysing private initiatives and self-organisation as they take place in society. It draws attention as to how and why actors, other than governmental actors, act in relation to governance and as such gives insight in how governance takes form in reality.

References

ADM, undated. Available at: www.admleeft.nl

Aiello, L.C. and Dunbar, R.I.M., 1993. Neocortex size, group size, and the evolution of language. Current Anthropology 34 (2): 184-193.

Anderson, B., 1983. Imagined communities: reflections on the origin and spread of nationalism. Verso: London, UK.

Beck, U., Giddens, A. and Lash, S., 1994. Reflexive modernization: politics, tradition and aesthetics in the modern social order. Polity Press: Cambridge, UK.

Blakely, E.J. and Snyder, M.G., 1997. Fortress America: gated communities in the United States. Brookings Institution Press: Washington, D.C., USA.

Castells, M., 2004. The information age: economy, society and culture: The power of identity. Blackwell: Malden, USA.

Cohen, A.P., 1985. The symbolic construction of community. Ellis Horwood: Chichester, UK.

Crow, G.P. and Allan, G., 1995. Community types, community typologies and community time. Time Society 4(2): 147-166.

Delanty, G., 2003. Community. Routledge: London, UK.

Elias, N. and Scotson, J.L., 1965 [1994]. The established and the outsiders: a sociological enquiry into community problems. Sage, London, UK.

Frissen, P.H.A., 2007. De staat van verschil. Een kritiek van de gelijkheid. Van Gennep: Amsterdam, the Netherlands.

Giddens, A., 1991. Modernity and self-identity: self and society in the late modern age. Polity Press: Cambridge, UK.

Golfresidentie Dronten, undated. Available at: www.golfresidentiedronten.nl

Hajer, M. and Wagenaar, H., 2003. Deliberative policy analysis: understanding governance in the network society. Cambridge University Press: Cambridge, UK.

Holzer, B. and Sørensen, M.P., 2003. Rethinking subpolitics. Theory, Culture & Society 20(2): 79-102.

Kooiman, J. and Van Vliet, M., 1993. Governance and Public Management. In: K. Eliassen and M. Van Vliet (Eds.) Managing public organizations. Sage: London, UK.

Lee, D., and Newby, H., 1983. The problem of sociology: an introduction to the discipline. Hutchinson, London, UK.

Mercer, B.E., 1956. The American community. Random House: New York, USA.

Pierre, J. and Peters, B.G., 2000. Governance, politics, and the state. Macmillan: Basingstoke, UK.

Raad voor Maatschappelijke Ontwikkeling, 2006. Verschil maken. Eigen verantwoordelijkheid na de verzorgingsstaat. Uitgeverij SWP: Amsterdam, the Netherlands.

Rhodes, R., 1996. The new governance: governing without government. Political Studies 44: 652-667.

Sennett, R., 1971. The uses of disorder: Personal identity and city life. Vintage Books: New York, USA.

Surowiecki, J., 2005. The wisdom of crowds: why the many are smarter than the few. Abacus: London, UK.

Tönnies, F., 1887 [1955]. Gemeinschaft und Gesellschaft. Fues's Verlag: Leipzig, Germany.

Van Dam, R.I., Eshuis, J., and Aarts, M.N.C., 2008. Zelforganisatie. Een studie naar gemeenschapsvorming in de Amsterdamse Doe-het-Zelf Maatschappij en de Golfresidentie Dronten. Wageningen UR: Wageningen, the Netherlands.

Van Dam, R.I., Eshuis, J., Aarts, M.N.C. and During, R., 2005. Closed communities: een verkennend onderzoek naar gesloten gemeenschappen in Nederland. Wageningen UR: Wageningen, the Netherlands.

Van Gunsteren, H., 2006. Vertrouwen in democratie. Van Gennep: Amsterdam, the Netherlands.

Van Tatenhove, J., Arts, B. and Leroy, P., 2000. Political modernisation and the environment: the renewal of environmental policy arrangements. Kluwer Academic Publishers: Dordrecht, the Netherlands.

Willmott, P., 1986. The debate about community. Policy Studies: London, UK.

Chapter 6

Learning in networks in Dutch agriculture: stimulating sustainable development through innovation and change

José Vogelezang, Arjen Wals, Barbara van Mierlo and Frank Wijnands

Abstract

In a world that is increasingly dynamic and global, it is difficult – if not impossible – to unilaterally find solutions to the emerging challenge of sustainability. This challenge is intrinsically complex and unpredictable. It is clear that to survive, Dutch agriculture can no longer lean on obsolete forms of problem solving. In this chapter, the authors introduce the phrase of *learning in networks.* This phrase encompasses multiple forms of learning which may help to unleash creativity to design new systems that can help create a more sustainable world. This chapter sets out the key characteristics of *learning in networks,* illustrated with some case studies where new forms of learning networks in new institutional settings have been put into practice.

Keywords: innovation, learning, networks, sustainability

6.1 Introduction

A completely new vision for the agricultural sector was presented in July 2000 by the then Minister of Agriculture, Mr Laurens-Jan Brinkhorst. Together with his under-secretary, Tjibbe Joustra, he presented radically new policy proposals to the public[6]. The proposals represented a 'definitive turnaround in thinking about the agro-food sector'. With innovation high on the agenda in society as a whole, the Ministry of LNV affirmed that this need for innovation called for an innovative response from the government, the sector itself and the scientific community. In Voedsel en Groen [Food and Green] (LNV, 2000a), the Ministry of LNV revealed the direction in which it was responding to the new challenges facing agriculture. It proposed that food production should no longer focus on farming alone, but on the whole agro-food chain from primary producer to consumer. It also re-defined 'green' as being more than our natural heritage, encompassing quality of life, living conditions, recreation, open space, undisturbed areas and water resources as well. In the light of these ideas, it proposed that agricultural businesses should have to apply for a license to produce whereby not only the product itself but the method of production had to conform to the new demand for an

[6] Visie op de toekomst van LNV. Extra editie *Met Name* 20 juli 2000.

ecologically sustainable and socially accountable agro-food complex. To achieve this goal, the Ministry formulated a new mission statement. *Impuls voor vernieuwing* [Impulse for Renewal] (LNV, 2000b) embodied this new direction for LNV. In this statement, LNV proposed the promotion of quality food production of an international standard – implying continued competitiveness, nationally and internationally based on top level scientific innovation. Not only that, the Ministry highlighted its intention to stimulate the agricultural and horticultural sector to increase its contribution to the quality of the rural landscape.

This strong message had clear implications for the Dutch agricultural research system. Giving an impulse for socially responsible production systems in the agro-food complex requires a whole new scientific approach. This is easier said than done. A preliminary workshop in 2001 involving stakeholders from government, agri-business and research institutions raised other thorny issues such as who should take the lead in system innovations: government, the private sector or research institutions? It became apparent that the vested interests and their effect on the individual role of the participant would have to be defined in order to clarify the issues. Finally, it was not yet clear which methods of approach would contribute to the creation of the necessary system innovations.

The LNV policy statement indicated that the Dutch agricultural food complex needed more comprehensive, integral and coherent strategies and novel ways of looking at issues and challenges. However, this is not simple bearing in mind the large numbers of stakeholders involved in these issues and given their inevitable interdependency. The stakeholders themselves are beginning to realise that in order to remain viable, they need to balance multiple interests and can no longer automatically give economic ones priority over social and ecological ones. Some refer to this balancing act as 'triple P-juggling' – trying to keep the P's of prosperity, people and planet all in the air without dropping one (the triple bottom line). This balancing act requires new ways of thinking, performing and behaving for all parties involved. Existing routines and values no longer suffice to deal with the challenges facing modern agriculture.

A year later after it's launch, Food and Green formed an intrinsic part of the fourth Dutch National Plan for the Environment (VROM, 2001). Since its presentation, a wide range of stakeholders have invested time, money and energy into creating fertile ground for initiating routine-breaking changes and innovations. This is because agricultural entrepreneurs and organisations have been forced by the rapid changes in an increasingly globalised world to search for sustainable new directions in which to develop. As a result, the number of learning-based experiments and innovation strategies is increasing. This chapter describes some key characteristics of learning-based change and innovation as they emerge in the Dutch agricultural sector's attempt to re-invent itself.

6.2 Key issue: entrepreneurs and innovation

Entrepreneurs in primary agriculture face major decisions about the future direction of their businesses. The Ministry of Agriculture, Nature and Food Quality (LNV) typecasts these choices in a paper 'Kiezen voor Landbouw' [Choosing for Agriculture] (LNV, 2006). Summarised they read as follows: bigger, better or different. In other words, farmers have the option to increase their level of production by increasing scale and/or forming co-operatives; improving their efficiency; or by changing the nature of their business by, for instance, specialising (i.e. niche markets) or diversifying (i.e. multiple land-use). Closing down, or emigration are also options. Not choosing is not an option these days, because the changes are occurring too fast. This means that entrepreneurs have to decide what they want, what they are capable of, and how they can turn these choices into concrete actions to shape their future. Studies have shown that what sets innovators apart from the majority, is their openness and their focus on the environment, their ability to involve others in their business and their capacity to learn (see Box 6.1). Not only that, new contacts stimulate the learning process which can lead to new visions for the future and unveil possible routes towards realising them.

Innovation is a central but ill-defined and ambiguous concept in our emerging knowledge-based society. It has increasingly become a pillar of policy in the Netherlands and has become a key word in the European Union. Irrespective of its precise meaning, the likelihood of innovation appears to increase when a number of conditions are met. These include the creation of heterogeneous groups of stakeholders and unlikely coalitions (to provide spontaneous, mostly novel, perspectives on a challenge or problem); the development of mutual trust and social cohesion (openness, honesty, transparency); a communal vision on the future (ownership); and good process management (facilitation utilising a range of creative work methods and inspiring environments for joint learning) (Wals, 2007a; Loeber, 2003; Rotmans, 2003; Guijt, 2008). Another factor is that there is a growing possibility within the limits of legislation and public and/or private funding, for small niche experimentation and stakeholders are given sufficient time to develop something radically new.

6.3 Theory: the role of knowledge and learning in innovation

6.3.1 Knowledge

In the post-privatisation period we can indeed speak of a knowledge-intensive economy, in which all the parties of the former education, extension services and research triptych search for new networks and configurations to realise innovations (see Chapter 3 in this volume). The need for new smart combinations between education, research and business seems to be evident for all parties involved in the Dutch agriculture (Leeuwis *et al.*, 2005). A new range of coalitions has emerged in the Netherlands in order to bridge the gap between knowledge production and its application. These collaborations stimulate innovation. Education and

Box 6.1. Learning from pioneers and their search for viability.
Research has been done on how 20 innovative entrepreneurs from various areas in the open field and horticultural sectors have given shape to their search for viable and sustainable forms of running their businesses (Potters *et al.*, 2007). All the entrepreneurs have achieved considerable success in the area of economic, ecological and/or social sustainability in widely differing and unusual ways. Via in-depth interviews and a workshop with these entrepreneurs, insight has been gained into their experiences and these have been analysed on the basis of *belief systems* (Buurma *et al.*, 2006). Each of these entrepreneurs has his own individual perspective on diverse aspects of sustainability. Their stories reveal five different ways of looking at sustainability: as idealist, grower, designer, pragmatic idealist en businessman (see Table 6.1). The entrepreneurs in this study stood out because of their capability to look at opportunities and possibilities with (roughly) two types of search behaviour; 'from the outside to the inside' and 'from the inside looking outward'. Important catalysts for the first kind of search method are: curiosity, a large network, a broad interest for development in the region and a ability to turn external observations into personal goals. A clear insight into one's own abilities, desires and motivations is an important quality in the search method 'from the inside looking outward'. Inspiration gained from outside the agrarian sector appears to magnify this power to innovate. 'Idealists' and 'growers' tend towards the search 'from the inside looking outward', while 'pragmatic idealists', 'businessmen' and 'designers' search more 'from the outside to the inside'. Another important aspect in the realisation of successful innovations is: the ability to transform opportunities and possibilities into a strategy applicable to one's own situation. Good management skills and the ability to pursue the strategy are necessary to achieve this. For the latter aspect, one needs a high degree of motivation, the ability to cope with set-backs, and dogged persistence.

Table 6.1. Five different approaches to sustainability.

	Sustainability viewed as ...	People viewed as ...	Planet viewed as ...	Profit viewed as ...	Objective
Idealist	Goal	Mankind	Earth	Means by which to live	Balance
Grower	Basis	Society	Environmental production factors	Business survival	Healthy business
Designer	Technical challenge	Horticulturalist and entrepreneur	Efficient input from outside	Living	Innovation
Pragmatic idealist	Pre-condition	Client and entrepreneur	Production criterion	Income, business success	Added value
Businessman	Selling point	Client and entrepreneur	Healthy product	Continuity, income business success	Distinction

training, research and development institutes and often businesses co-operate to this end (Lans *et al.*, 2006). These arrangements are often referred to as knowledge arrangements.

A knowledge arrangement is the smart combination of hardware, software and orgware, viewed from a societal or economical point of view, that leads to effective contributions to the knowledge-based economy' (Nijkamp, 2002; Leeuwis *et al.*, 2006). Knowledge arrangements appear in many forms, like Communities of Practice, learning networks, innovation networks, etc. There is a large body of literature discussing the different ways of conceptualising knowledge acquisition. Based on their research in the Dutch agricultural complex Lans *et al.*, (2006) distinguish three kinds of knowledge processes: knowledge transfer, knowledge sharing or circulation and knowledge co-creation.

Knowledge transfer refers to the transfer of explicit, codified knowledge (information) from 'sender to receiver', a mainly one-way process, which requires detailed insight into the intended receiver and the way in which he interprets the information offered (Nonaka and Takeuchi, 1995). Knowledge transfer appears most fruitful in unambiguous problem situations in which upgrading, optimising or dissemination plays a key role. This knowledge has stood the test of time and is uncontested. Traditional extension 'delivery' methods and formal education and training methods, often supported by ICT, play an important role in the process of knowledge transfer. The singular use of linear knowledge transfer and dissemination lies under scrutiny not only in Dutch agriculture but in other sectors and countries as well. It's no longer a question of matching supply and demand followed by dissemination of knowledge in a linear model – whereby the primary role of developing knowledge lies with scientists (the so-called mode-0 approach).

Knowledge circulation refers to the process of sharing, applying and developing knowledge further in an interactive process usually within heterogeneous groups. An important aspect within knowledge circulation is the interchange of scientific and tacit knowledge found within the different parties involved (mode-1 approach). In Dutch agriculture we are witnessing a range of new networks that seek to link the knowledge and experience bases of a range of players operating within the same sector or chain.

Knowledge co-creation refers to the collective search for direction and answers to questions that involve a re-design of existing routines, practices, values, principles and precepts. Co-creation is in essence a creative search process in a dynamic environment which is functional in ill-defined problem situations characterised by uncertainty and complexity. The result of co-creation is joint awareness of a certain problem but also the generation new, often highly contextual, knowledge.

As early as the 1990's frequently cited scholars, such as Gibbons (1994), were suggesting that the creation of new knowledge in what he calls a 'mode-2 setting' is not a mono-disciplinary linear process anymore, but is instead an interdisciplinary or even trans-disciplinary process.

Rather than the classical research and development route, where new knowledge is acquired and subsequently 'pushed' into society (technology push), knowledge is being actively created and disseminated by all parties in the professional networks. Hence, in the knowledge-based economy, innovation depends on the capacity of organisations and networks to create and exploit knowledge (Boreham and Lammont, 2000). Several networks in Dutch agriculture are taking up the challenge of the co-creation of knowledge.

6.3.2 Learning

The title of this chapter speaks of 'learning in networks'. We use this as a 'catch all' phrase for a range of emergent types of learning that all seem relevant in anticipating change by dealing with complexity and coping with uncertainty. We will briefly describe these types of learning and their distinguishing qualities.

Experiential learning is characterised by an incremental and interactive cycle of action borne out of concrete experiences, reflective observation, abstraction and experimentation (Kolb and Fry, 1975; Lewin 1946). This type of learning occurs frequently. The type of reflection involved is restricted to an individual reflection on the relationship between individual actions and their consequences.

In focussing on organisational learning, Argyris and Schön (1996) have made a well known distinction between *first* and *second order learning*. In first order learning players modify their perspective on solutions and strategies if they notice that the results of their actions don't match their expectations or desires. Second order learning goes, as it were, 'deeper'. Not only is their perspective on solutions and strategies modified, but also their goals, vested interests, standards and values. The changes that go along with this reach further than the incremental changes that are the result of first order learning.

The mind set of the players in first as well as second order learning is action oriented. While first order learning doesn't require more than a self-evaluation of the results of actions, second order learning requires a confrontation with new perspectives. In that confrontation there is a chance that theories that underlie action become interlinked and players develop new forms of action and interaction.

Social learning tends to refer to learning that takes place when people with divergent interests, norms, values and views of reality meet in an environment that is conducive to meaningful interaction. It can take place at several levels, individual, group or organisational, or at the level of stakeholder networks (Keen *et al.*, 2005). The social character of the learning process is but one side of the many ways in which the concept of social learning is used. Another one involves the possible results of interaction and collective reflection. As an alternative to the often discussed consensus or collective action, Grin and Van de Graaf (1996) and Grin *et al.* (1997) propose congruency. If congruency is to be achieved, diverse players in a professional network agree over a new line of joint action as a useful and valuable solution to their problem.

Van Mierlo speaks of *convergent learning* in cases where heterogeneous groups of players cooperating in pilot projects develop congruent visions on the possibilities that become apparent (Van Mierlo, 2002). In complex projects convergent learning may occur if there is an open and creative negotiation process and management geared towards network and negotiation processes. Convergent learning can be distinguished from divergent learning; a process in which participants in pilot projects learn a lot but in diverging directions.

In all forms of learning the conflicts and their underlying sources, need to be faced rather than concealed. By explicating and deconstructing the often diverging norms, values, interests and points of view that people bring to a sustainability challenge, it not only becomes possible to analyse and understand their roots and their persistence, but also to highlight the things they have in common, thereby enabling a collaborative change process whereby possible joint action plans emerge (Wals, 2007b).

With *system learning* a group learns to view the underlying institutional structures within which they operate in a new light. Structures that are normally seen as a given are then open to discussion (Loeber *et al.*, 2007). If the parties involved regard these parameters as a given, they will assume that no change is possible and the structures themselves aren't challenged.

Learning in networks is at the heart of innovation in the agricultural sector. It can be seen as a combination of second order learning by individuals and organisations and convergent and system learning in networks of heterogeneous groups of stakeholders. These types of learning can be stimulated in collective endeavours by providing process criteria within the network. In the following paragraph we will describe the setting in which these learning networks operate, bearing in mind the changes in the Dutch agricultural knowledge system surrounding them, while illustrating these changes with examples of concrete networks.

6.4 Key characteristics of learning-based innovation in practice

The art of working on the transition to a more sustainable agriculture is to stimulate and facilitate new developments; to find major breakthroughs based on desired visions of the future, foster the pioneering work of innovators in practice and to link these things to long term goals. (These two complementary pathways are presented in Chapter 11 by Wijnands and Vogelezang, 2009, this volume.) Learning is not only a by-product of these developments, but is essential to this process; theory and practice working hand in hand.

6.4.1 From blanket solutions to tailor made (system) innovations

In Western production-oriented agriculture, knowledge was traditionally developed in research institutions and subsequently translated into practice via intermediaries (extension services). This linear approach generally worked well in the past when all parties shared the same perspective on the goal and when increasing agricultural productivity through technological advancement was sufficient to achieve this goal. This point of view has in practice gradually come to an end because we have reached limits of growth in our current agricultural systems as is manifested in the enormous adverse impact of 'modern' agriculture on the environment both locally and globally. There is still a great gap between long term sustainability goals and the present pollution (Oenema *et al.*, 2006).

These days there is no longer one simple option for achieving our goals for the future (blanket solutions), but there are several options to achieve a more sustainable future (Horton and Freire, 1990). These various paths require innovations that can differ from each other in nature and character and which are more specific to their own context. (Practical cases are described in Chapter 11, this volume). Knowledge is vital for all these innovations; knowledge that has to be tailored and developed in an interactive process where (explicit) scientific knowledge and (implicit) tacit knowledge are brought together. Not only that, the problems facing stakeholders and entrepreneurs are no longer linked exclusively to technological solutions. The challenges involve finding new forms of cooperation in coalitions and new criteria for sustainable development. It also involves another type of government policy which frames goals (giving a licence to produce) instead of giving detailed guidelines for running the business. These new paths need to be discovered and developed within an entire production chain or a region, because the needed changes transcend the limitations of individual businesses.

6.4.2 From linear knowledge supply to co-creation in networks

Working in innovation networks that primarily engage in knowledge sharing and knowledge co-creation requires somewhat unconventional roles for and new competencies of those participating in these networks. There is a clear and specific new role for intermediaries, scientists and educational institutions. In addition to acting as brokers or facilitators of networks, intermediaries and scientists also need help to create optimal conditions for this communal search for possible paths to solutions to complex problems (Regeer and Bunders, 2007). Moreover, all participants in such systems and networks need to develop a range of new competencies. Understanding the external dynamics, empathy for others, the ability and desire to reflect is crucial for learning in networks (see below: From evaluation to continuous learning).

Achieving coherent change in both the mind set and the actions of network participants is one of the precepts of learning in networks. This idea is in the process of being implemented in the present education system and is already being used in professional networks. (see Box 6.2: 'Leren met toekomst' [*Learning for the future*] and Box 6.4: 'Telen met toekomst'[*Farming with a future*]. These cases demonstrate a new working model based on the ideas of co-creative learning.

Box 6.2. Learning for the future

A team of students from all different levels (stages of their study or even from completely different studies) are working on solutions to a development issue given them by an entrepreneurs network. The team gets support from a researcher, a teacher and, if necessary, from extension services. The researcher and advisers perform as content and process supervisors. They ensure that the project is set up properly, facilitate access to knowledge and other researchers and they safeguard the quality of the project. The teacher coaches the learning process and acts as safety net (safe learning environment). The entrepreneurs from the network not only commission the project, they are the people with the practical experience and contacts with the outside world. In this way the researcher, the teacher, the advisor and the entrepreneur all work together to train the scientists and entrepreneurs of the future. This concept has already been developed as a pilot project for three schools (see Figure 6.1; Potters *et al.*, 2006). At this point in time the process of scaling up the project has begun.

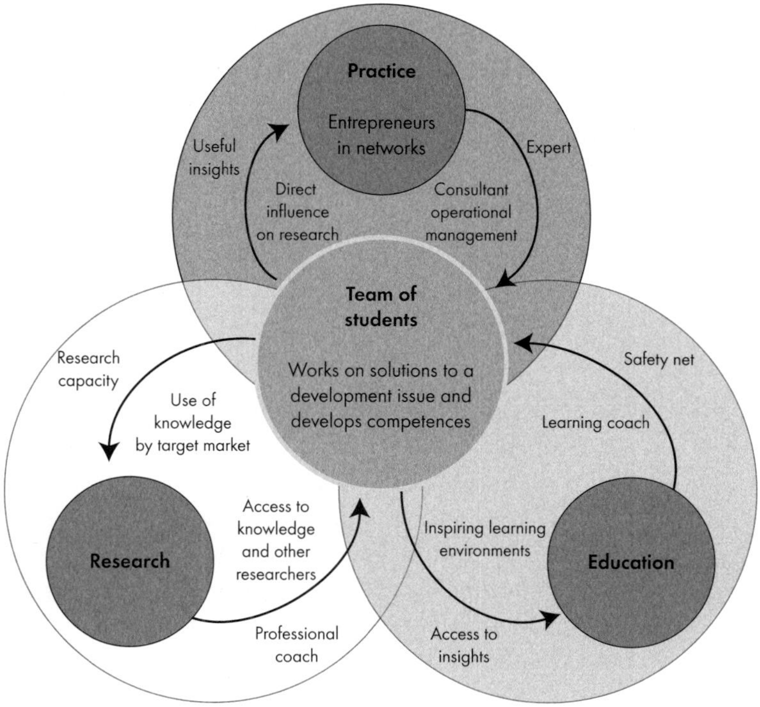

Figure 6.1. Concept of 'Learning for the future'.

6.4.3 From vertical regulation to horizontal partnership

Change and enhancing system innovation is dependent on government agencies assuming a different role. The responsibility for initiating transition processes directed at sketching new perspectives for the future was first given by the Ministry of Agriculture (LNV) to scientific organisations like Wageningen UR's applied research institutes. In 2004 the LNV Ministry changed it's policy: from vertical (regulating) to horizontal (facilitating). Entrepreneurs and entrepreneurial groups now have the initiative and the LNV Ministry is starting to re-think its own role and responsibilities. The LNV Ministry is completely committed to supporting innovative networks initiated by entrepreneurs (see Wielinga and Geerling-Eiff, 2009, this volume: Chapter 7), and it stimulates them to set up their own innovation agendas (see Box 6.3: Bioconnect). This is the opposite of a vertical structure where decisions from higher up are implemented for others and not with them.

6.4.4 From 'poldering' consensus to changing coalitions and respectful dissent

In the '90s, clashes of interest between stakeholders in the Dutch community were mainly overcome using the so-called 'polder' model. In the 'polder' model for conflict resolution a consensus develops over what course should be followed and what measures need to be taken. The advantage of this method is that the course of action is decided by all players involved. In an approach to problems based on transition, however, this consensus model is too limited in that it often leads to half-baked compromises and watered-down solutions, while failing to unleash the power of innovation within the various players. In this situation it is better to look for small areas of shared interests. Through learning networks, these tentative and initially indeterminate 'pockets of innovation' can grow into forms of coordinated action. As Grin and Van der Graaf (1996) explains: when it comes to transition, it is more a question of generating creative congruency – finding win/win constructions – than forcing 'polder' consensus. This process can be speeded up by assessing the driving interests of the various stakeholders right at the beginning and then working on a mutual vision of the future (see Box 6.4: 'Telen met toekomst'[*Farming with a future*]).

The idea of creative congruency conforms with the current insight that to achieve and stimulate change, a constant interaction is needed between creating new perspectives and allowing the creation of an environment where these perspectives can be realised. Etzkowitz and Leydesdorff (2000) have introduced what they call a triple-helix approach to innovation in the knowledge-based society. When the government, the private sector and scientific institutions engage jointly in innovation, flexible networks or alliances emerge creating their own dynamics. They see this as another prerequisite to achieving innovation in a knowledge-based economy. In doing so, they highlight the importance of proper management and step-by-step supervision of these flexible alliances.

Box 6.3. Bioconnect: empowerment of the Dutch organic agriculture sector.

In 1998, a new way to stimulate knowledge development was initiated in the Netherlands. The Practical Network BIOM was set up within networks spanning the whole organic farming sector (Wijnands, 2000). Entrepreneurs, researchers and public relations bureaus set out together with other interested parties to seek new ways to ensure continued growth in this sector. The central goal was knowledge circulation and knowledge transfer to other organic sector entrepreneurs. The financing and management of this network originated with the LNV Ministry with an important executive role for Wageningen UR and extension services.

The choice for a new, more facilitating role for the Ministry of LNV has lead it to decide to put the responsibility for knowledge development into the hands of the organic sector itself. Since 2005 the organic sector itself decides what sort of knowledge it requires and how it should be disseminated into businesses. This new form of demand-driven knowledge development has given rise to a new organisation, *Bioconnect* – the knowledge network for organic farming and food (see Figure 6.2). All the different sectors are represented via product work groups where businesses from all parts of the production chain and social organisations formulate knowledge projects and are then responsible for their implementation. Researchers have an advisory role. The Ministry of LNV is formally the initiator, but assumes that the business sector will act out its role in a responsible way. Aside from work groups from the sector itself, there are also a few thematic working groups dealing with specific issues.

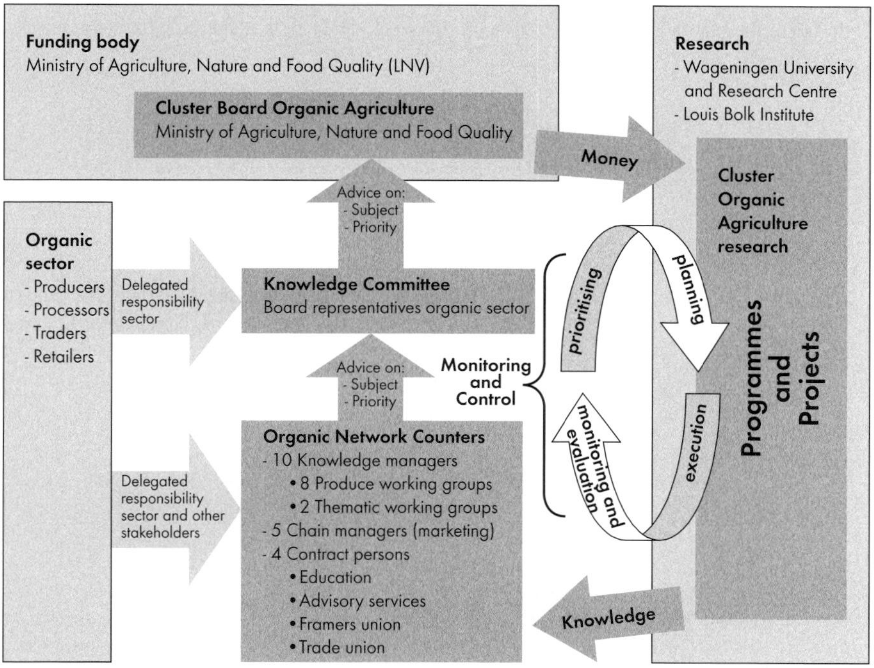

Figure 6.2. Schematic impression of the demand driven organisation of publicly funded organic research in the Netherlands (Bioconnect is in the centre of the figure).

Box 6.4. 'Farming with a future' and stakeholder management.

'Farming with a future' is a national project in the Netherlands (2003-2011). The objective is to implement more sustainable pest and crop nutrition management in arable and horticulture sectors. 'Farming with a future' is a covenant about crop protection and nutrition that has been co-signed by the Ministry of LNV, the Ministry for the Environment, the crop protection industry, the water boards, the surface water authorities and producer boards like the LTO (Land en Tuinbouw Organisatie) The main purpose of the covenant is to realise a reduction of 90% in chemical crop protection and nutrition compared to the levels used in 1989, by 2010.

'Farming with a future' promotes two courses of action. The first involves 35 study groups working together with about 400 entrepreneurs on testing and improving new crop management strategies. The second is ensuring that the strategies that prove effective and feasible are put into practice by the stakeholders.

Many of these new methods and techniques have little positive impact on the bottom line for the growers. That means that their introduction is heavily dependent on the personal motivation of the people involved. 'Farming with a future' seeks out active dialogue and cooperation with stakeholders who can influence the entrepreneurs, like the LTO, the crop protection dealers and the water boards. Via good stakeholder management it is possible to find out who the *blockers*, the *floaters* and the *movers* are (see Figure 6.3). By understanding the underlying vested interests it is possible to seek out win/win situations, those which promote sustainability as well as the interests of the stakeholder. 'Farming with a future' has bound various parties in common actions. Each party is responsible for a specific part of the whole. There are already 230 stakeholders from all arable and horticulture sectors actively involved in this network. These stakeholders form a crucial link in the realisation of a more sustainable agrarian sector.

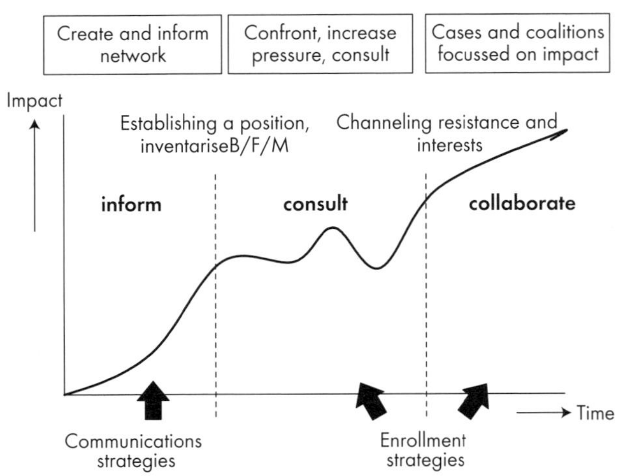

Figure 6.3. The initials B/F/M stand for blockers, floaters and movers. It is important to know which of the stakeholders support the goals and are willing to do something (movers), which ones set conditions for support (floaters) en which ones do not support at all (blockers). Once this is apparent, you can harness the influence these people have on each other. You can mobilise the movers to create the desired conditions for the floaters. You can enter into dialogue with the blockers to isolate the issues which form the obstruction. During this process the role of facilitator shifts. In this case it passes from the project organiser 'Farming with a Future' to influential stakeholders.

6.4.5 From evaluation to continuous learning

The emergent Dutch knowledge system is characterised by the presence of multiple and interdependent stakeholders and multiple levels of learning. It also encompasses different types of learning and different types of knowledge.

Various forms of monitoring and evaluation can be used in the process of learning. On the one hand there are goal- or result-oriented evaluation methodologies and on the other hand constructivist and participatory approaches of monitoring and evaluation. In addition to these two, a third approach is being developed: *reflexive monitoring* (Arkesteijn *et al.*, 2007, Grin and Weterings, 2005; Guijt, 2008). The adjective 'reflexive' has been added to stress that the monitoring should refer back to the learning and innovation processes by challenging current thinking and institutions. A learning system has to be reflexive in order to break away from existing routines, norms, values and interests.

We have experience with using reflexive process monitoring in the project 'Telen met toekomst'[*Farming with a future*] (Mierlo *et al.*, 2007). The monitoring actions are tied to specific planned and actual operations pertaining to the project. This form of monitoring, designed to stimulate learning and self-evaluation not only analyses the degree of success of the project, but contributes to that success by regularly checking back on the quality of the process criteria and the resulting changes in perspectives and practices of the players concerned. The quality and character of learning and its process conditions are monitored by using specific indicators that are grounded in theories on learning and (system) innovation. As such it is a collective theory-driven methodology.

6.5 Lessons learned: theory and practice

We have now been working for several years with diverse participative working methods in the transition process towards a more sustainable agriculture. This new approach is illustrated in this chapter by a number of practical applications in the field. There are more examples in various places throughout this book. The consequences that this approach has for the roles of the various stakeholders and the way in which they cooperate are defined in paragraph 6.4 as key characteristics of an environment that stimulates innovations. Learning is an essential part of the innovation process. In this paragraph we explore in more depth the pre-conditions or critical factors for success in realising innovations, as they are described in other scientific literature (see paragraph 6.2): heterogeneous groups of stakeholders, mutual trust and social cohesion, communal visions of the future and good process management. These factors are placed in the context of the theories regarding the role of knowledge and learning in innovation, as discussed in paragraph 6.3. Critical success factors that have been observed in practice are mentioned explicitly.

6.5.1 Creation of heterogeneous groups of stakeholders

In the post-war period, the drive for change led to the organisation of a rigidly defined and disciplined research/knowledge infrastructure. This can be seen in the organisation of education, extension services and research triptych as well as the organisation of the private business sector within the Dutch animal husbandry, agriculture and horticulture organisation (LTO) where narrow interest groups represent specific sectors. A knowledge infrastructure built along these lines is effective in dealing with mono-disciplinary optimalisation issues, but doesn't suffice for innovation impulses which are needed for co-creation in a mode-2 setting. In the meantime we have experience with a number of innovation networks which straddle diverse sectors: *Waardewerken* [Value works] and *Plattelandsimplus* [Rural impulse], networks in multifunctional agriculture, various innovation networks in greenhouse horticulture and the programme *Netwerken in de veehouderij* [Networks for animal husbandry] (Vogelezang and Wijnands, 2007). While creating groups, a critical success factor was determining the optimal cognitive distance: the participants needed to be different enough to be able to learn from one another, but not so different that they were unable or unwilling to understand each other (Nooteboom, 2000; Heymann and Wals, 2002; Wals and Heymann, 2004). For that reason in the network *Rural impulse,* a project directed at strengthening strategic business competencies within 35 different groups of entrepreneurs, it was decided to choose a mix of entrepreneurs to prevent the discussions from merely descending into a focus on their daily management issues. Our experience with innovation networks has been positive: entrepreneurs indicate that colleagues from other (sub)sectors in fact inspire them and challenge their own preconceptions. It is precisely because the needed innovation processes require a completely new approach, that it is important to stimulate learning to think and act differently through heterogeneous groups. It is thus important for the network advisors to bear in mind that the comfort zone for individuals varies and that they learn to 'read' people. 'Ideally facilitators become skillful in reading people's comfort zones and, when needed, expanding them little by little. An important role of facilitators is to create space for alternative views that lead to the various levels of dissonance needed to trigger learning both at the individual and at the collective level' (Wals, 2007c).

6.5.2 Development of mutual trust and social cohesion

The former education, extension services and research triptych relied predominantly on linear knowledge development and dissemination (see Chapter 3, this volume). Knowledge transfer was achieved through announcements and informing the target group through articles, demonstrations and lectures. But the linear knowledge model doesn't suffice when the innovation challenge requires the building of relationships and coalitions in the chain or in a local or regional context. Knowledge circulation and knowledge co-creation are the dominant forms of knowledge development in this new scenario. This insight has lead to an approach based on facilitating the search process based on a shared communal vested interest by stakeholders in the networks *Farming with a future* (this chapter) and *Networks in animal husbandry* (Chapter 7). In *Farming with a future* by placing the emphasis on stakeholder management, and in *Networks in animal husbandry* by using specific learning instruments to allow networks to realise their goals more quickly and effectively. It is this approach that ensures that the motives and interests of stakeholders are voiced and an atmosphere of openness, honesty and transparency is created. A potential problem lies in the fact that this approach costs a lot of time and energy and that there are few visible results in the short term. In other words, how can you make the 'soft' aspects (improved relationships between players, increase in mutual trust, improved conflict resolution and utilisation, increase in openness and flexibility) of the innovation quest harder or more visible? It is precisely this issue that has made monitoring and evaluation such an important part of these projects, not only from the perspective of learning, but also from the perspective of accountability towards the commissioner of the project. A second critical factor for success is to avoid *disjointed learning*. In practice, it has proved difficult to transfer the learning experiences gained in the networks to people outside of them. It is crucial to prevent a sort of innovation elite forming which has lost contact completely with its own roots. One can speak of disjointed learning when a small, enthusiastic group gets involved in an innovation project but forgets that to realise the solutions they have found, they need other people who were perhaps not as intensively involved, or not involved at all in the innovation process. These people who are perhaps unable or unwilling to take part, don't learn or don't learn at the same tempo as the participants in a learning innovation network. A gulf can develop which ultimately stands in the way of the realisation of innovations. This doesn't mean that everyone has to take active part, this is not only impossible but also undesirable from the point of view of process management, but it does mean that there must be continuous communication with the participants' support base and the commissioners of the project. In fact, the commissioners of the project (in this case the government) should also take part in the network, so that possible changes in policy can also be included in the lessons learned. Innovation networks supported by government demand a government that is also willing to change.

6.5.3 Communal perspective for the future

In a linear knowledge model, the confrontation between theory and practice occurs in a very late stage of knowledge development. Ownership by the target group for the developmental solution pathways is weak in this situation and the application of the knowledge can stagnate. Knowledge circulation and knowledge co-creation allows for the interchange of scientific and tacit knowledge thus promoting a shared ownership of the necessary innovation task and the resulting solutions. The search for a communal vested interest can be stimulated in different ways. In *Value works*, a network of pioneer entrepreneurs in multifunctional agriculture, the drawing up of a vision for the future and the resulting innovation agenda contributed to a large degree to the creation of a mutual feeling of responsibility for the further professionalisation of the sector which surpassed the vested interest of the individual member. In *Bioconnect* the responsibility for setting the knowledge agenda has been placed by the government in the hands of the organic sector itself, which can be seen as government intervention to transfer ownership to the private sector. The open application procedure for networks in the programme *Networks in animal husbandry* can be seen as an instrument allowing people in the field to set the research goals. An area of tension is holding interest in the development in the long term. Entrepreneurs have as their priority the continuity of their business and sector, whereby daily problems tend to claim their attention. Retaining enough attention for long term goals has to translate into a balanced innovation agenda where short term benefits make up an integral part of the pathway to the future.

6.5.4 Good process management

Networks conform to a trend that can be seen throughout society, and that is the forming of fluid and changing alliances. No formal engagement to one another but a temporary relationship between people who want to realise a specific goal. Learning is an important aspect within these networks. In our projects and networks a great deal of attention is paid to creating a safe and inspiring learning environment. A number of these aspects have been highlighted in this discussion already. Specially tailored *Innovatieve leeromgevingen (ILO)* [Innovative learning environments] have been set up for specific learning objectives which can't be realised within an authentic environment (Gielen *et al.*, 2006). Collaboration in the context of new roles has major consequences for the network participants. Do they have the right competencies and skills as researcher, advisor or entrepreneur to make contacts in the context of your new role? This means that in the selection process for managers, Wageningen UR has placed emphasis on their skills in process supervision. It has also given rise to a new training programme for project and network supervisors, which is designed to develop the skills necessary to supervise successful innovation projects and networks.

6.6 Conclusion

Moving towards sustainable agriculture requires changes: a radical yet coherent renewal of content, procedures, work methods and conditions. Learning is a key factor in fostering the broad range of innovations required. Learning in networks has been introduced in the Netherlands as a strategy to deal with the complexity and unpredictability of the challenges surrounding the issue of sustainability. It encompasses a whole range of learning forms and activities. We have discussed a number of features and principles of the various types of learning and described some of the practical applications of the theory now taking place in the field. Learning in networks embraces the fact that solutions lie in facing and dealing with the heterogeneous nature of the problems, the players and the possible paths forward for everyone involved. Learning environments that are conducive to learning have been reviewed, as well as appropriate forms of monitoring and evaluation.

Acknowledgements

Research was financed by the Dutch Ministry of Agriculture, Nature and Food Quality (LNV).

References

Argyris, C. and Schön, D.A., 1996. Organisational learning II: theory, method and practice. Reading, MA: Addison-Wesley.

Arkesteijn, M., Van Mierlo, B. and Potters, J., 2007. Methoden voor monitoring en evaluatie van innovatieprojecten. Brochure Praktijkonderzoek Plant & omgeving (PPO). PPO: Lelystad, the Netherlands.

Boreham, N. and Lammont, N., 2000. Work process knowledge in technological and organisational development. Final Report, Framework Programme IV, Targeted. Socio-Economic Research Programme, European Union Directorate-General XII (EU DG XII). EU DG XII: Brussel, Belgium.

Buurma, J.S., De Buck, A.J., Klein Swormink, B.W., Stokkers, R. and Munneke, F.J., 2006. Innovatieprocessen in de praktijk: vorming van socio-technische netwerken. Report 6.06.08 Landbouw Economisch Instituut (LEI). LEI: Den Haag, the Netherlands.

Etzkowitz, H. and Leydesdorff, L., 2000. The dynamics of innovation: from nations systems and 'mode 2' to a triple helix of university – industry – government relations. Research Policy 29: 109-123.

Gielen, P., Biemans, H. and Mulder, M., 2006. Inspirerende leeromgevingen voor ondernemers: aanwijzingen voor ontwerpers en begeleiders. Brochure department of Social Sciences Chair group Education and Competence Studies. Wageningen University: Wageningen, the Netherlands.

Gibbons, M., Limoges, C., Nowotny, H., Schwartzman, S., Scott, P. and Trow, M., 1994. The new production of knowledge: the dynamics of science and research in contemporary societies. Sage Publications: London, UK.

Grin, J. and Weterings, R., 2005. Reflexive monitoring of system innovative projects: strategic nature and relevant competences. Paper for the session 'Reflexive governance for sustainable development' at the 6th Open Meeting of the Human Dimensions of Global Environmental Change Research Communicate, 9-13 October 2005, University of Bonn, Germany.

Grin, J. and Van de Graaf, H., 1996. Handelingstheorieën en beïnvloeding in netwerken: Ongelijksoortige rationaliteiten en congruente betekenissen. Beleidswetenschap 4: 349-366.

Grin, J., Van de Graaf, H. and Hoppe, R., 1997. Interactive technology assessment: een gids voor wie het wagen wil. Werkdocument 57 Rathenau Instituut. Rathenau Instituut: The Hague, the Netherlands.

Guijt, I., 2008. Seeking surprise: rethinking monitoring for collective learning in rural resource management. PhD Study. Wageningen University: Wageningen, the Netherlands.

Heymann, F. and Wals, A.E.J., 2002. Cultivating conflict and pluralism through dialogical deconstruction. In: Leeuwis, C. and R. Pyburn (Eds.) Wheelbarrows full of frogs: social learning in resource management. Van Gorcum: Assen, the Netherlands.

Horton, M and Freire, P., 1990. We make the road by walking: conversations on education and social change. Temple University Press: Philadelphia, USA.

Keen, M., Brown, V.A. and Dyball, R. (Eds.), 2005. Social learning in environmental management: towards a sustainable future. Earthscan: London, UK.

Kolb, D.A. and Fry, R., 1975. Toward an applied theory of experiental learning. In: Cooper, C. (Ed.) Theories of group process. John Wiley & Sons: London, UK, pp. 33-57.

Lans, T., Kupper, H., Wals, A.E.J., De Beuze, M. and Geerling-Eijff, F., 2006. Alles is kennis? Report DLO-programma 420 Kennisdoorstroming en –circulatie in groen onderzoek en onderwijs. Wageningen UR: Wageningen, the Netherlands.

Leeuwis, C., Smits, R., Grin, J., Klerkx, L.W.A, Van Mierlo, B.C. and Kuipers, A., 2006. Equivocations on the post privatization dynamics in agricultural innovation systems. In: Smits, R.E.H.M. (Ed.), The design of an innovation-enhancing. Transforum Agro & Groen: Zoetermeer, the Netherlands, pp. 3-5.

Lewin, K., 1946. Action research and minority oroblems. Journal of Social Issues 2(1): 34-46.

LNV, 2000a. Voedsel en groen. Het Nederlandse agro-food complex in perspectief. Publication Ministerie van Landbouw, Natuur en Voedselkwaliteit (LNV). LNV: The Hague, the Netherlands.

LNV, 2000b. Impuls voor vernieuwing. Organisatieontwikkeling bij LNV. Publication Ministerie van Landbouw, Natuur en Voedselkwaliteit (LNV). LNV: The Hague, the Netherlands.

LNV, 2006. Kiezen voor landbouw. Een visie op de toekomst van de Nederlandse agrarische sector. Publication Ministerie van Landbouw, Natuur en Voedselkwaliteit (LNV). LNV: The Hague, the Netherlands.

Loeber, A., 2003. Inbreken in het gangbare. Transitiemanagement in de praktijk: de NIDO-benadering. NIDO: Leeuwarden, the Netherlands.

Loeber, A., Van Mierlo, B., Grin, J. and Leeuwis, C. (2007). The practical value of theory: conceptualising learning in the pursuit of a sustainable development. In: Wals, A.E.J. (Ed.) Social learning towards a sustainable world. Wageningen Academic Publishers: Wageningen, the Netherlands.

Nijkamp, P., 2002. Groei, kennis en overheid, een meta-analytische verkenning naar kennisinfrastructuur. Publication Koninklijke Nederlandse Akademie van Wetenschappen (KNAW). KNAW: Amsterdam, the Netherlands.

Nonaka, I., and Takeuchi, H., 1995. The knowledge creating company: how Japanese companies create the dynamics of innovation. Oxford University Press: New York, USA.

Nooteboom, B., 2000. Learning by interaction: absorptive capacity, cognitive distance and governance. Journal of Management and Governance 4: 69-92.

Oenema, O., Van der Kolk, J.W.H. and Groot, A.M.E., 2006. Land en milieu in transitie. Publication 2 Wettelijke Onderzoekstaken Natuur & Milieu. Wageningen UR: Wageningen, the Netherlands.

Potters, J., Buurma, J. and De Buck, A., 2007. Inspiratie voor transitie – leren van pioniers en hun zoektochten. Brochure 364 Praktijkonderzoek Plant & Omgeving (PPO). PPO: Lelystad, the Netherlands.

Potters, J., Van der Hoeven, N. and Gielen, P., 2006. Leren met toekomst: versterken van de samenwerking tussen onderzoek en groen onderwijs. Brochure Praktijkonderzoek Plant & Omgeving 353 (PPO). PPO: Lelystad, the Netherlands.

Regeer, B.J and Bunders, J.F.G., 2007. Kennis co-creatie: samenspel tussen wetenschap en praktijk. Publication V.10 Voorstudies en achtergronden Raad voor Ruimtelijk, Milieu- en Natuuronderzoek (RMNO). RMNO: The Hague, the Netherlands.

Rotmans, J., 2003. Transitiemanagement: sleutel voor een duurzame samenleving, Koninklijke Van Gorcum: Assen, the Netherlands.

Van Mierlo, B., 2002. Kiem van maatschappelijke verandering. Verspreiding van zonnecelsystemen in de woningbouw met behulp van pilotprojecten. PhD study. Aksant: Amsterdam, the Netherlands.

Van Mierlo, B.C., Arkesteijn, M.and Leeuwis, C., 2007. Reflexieve procesmonitoring en telen met toekomst, de eerste fase. Report Praktijkonderzoek Plant & Omgeving (PPO). PPO: Lelystad, the Netherlands.

Vogelezang, J. and Wijnands, F.G., 2007. Leren in netwerken. Syscope 13 Kwartaalblad Systeeminnovatie-programma's. Wageningen UR: Wageningen, the Netherlands.

VROM, 2001. Een wereld en een wil: werken aan duurzaamheid, vierde Nationaal Milieubeleidsplan (NMP4). nota 01.0433 14548/176. Ministerie van Volkshuisvesting, Ruimtelijke Ordening en Milieubeheer (VROM), Den Haag, the Netherlands.

Wals, A.E.J. (Ed.), 2007a. Social learning towards a sustainable world. Wageningen Academic Publishers: Wageningen, the Netherlands.

Wals, A.E.J., 2007b. Learning in a changing world and changing in a learning world: reflexively fumbling towards sustainability. Southern African Journal of Environmental Education 24 (1): 35-45.

Wals, A.E.J., 2007c. Epilogue: creating networks of conversations. In: Wals, A.E.J. (Ed.) Social learning towards a sustainable world. Wageningen Academic Publishers: Wageningen, the Netherlands, pp. 497-507.

Wals, A.E.J. and Heymann, F.V., 2004. Learning on the edge: exploring the change potential of conflict in social learning for sustainable living. In: Wenden, A. (Ed.) Educating for a culture of social and ecological peace. SUNY Press: New York, USA, pp. 123-145.

Wielinga, E. and Geerling-Eiff, F., 2009. Networks with free actors: an ecological approach on innovation and transition. In: Poppe, K.J, Termeer, C. and Slingerland, M. (Eds.) Transitions towards sustainable agriculture and food chains in peri-urban areas. Wageningen Academic Publishers: Wageningen, the Netherlands, pp. Chapter 7 in this Volume.

Wijnands, F.G., 2000. BIOM, a pilot farm network of organic arable and outdoor horticultural farms in the Netherlands. In: Proceedings 13th IFOAM Scientific Congress. Zurich, Switzerland.

Wijnands F.G. and Vogelezang, J.V.M., 2009. Two complementary pathways to support innovation towards sustainable developments in Dutch agriculture. In: Poppe, K.J, Termeer, C. and Slingerland, M. (Eds.) Transitions towards sustainable agriculture and food chains in peri-urban areas. Wageningen Academic Publishers: Wageningen, the Netherlands, pp. Chapter 11 in this Volume.

Chapter 7

Networks with free actors: an organic approach to innovation and transition

Eelke Wielinga and Florentien Geerling-Eiff

Abstract

Innovation is an autonomous process: it cannot be forced. It occurs if conditions are favourable. The same is true for transitions. Major shifts in the orientation of actors within networks can't be realised in a relatively short period either. However, it is possible to foster the necessary conditions for innovation by using new insights and methodologies that have been designed to create these conditions. In the search for ways of speeding up the process of change, we have to determine the factors that influence these conditions in both a positive or a negative way. In practical terms this translates into determining the obstacles that prevent individuals taking the initiative. This chapter focuses on the role of 'free actors' in networks for innovation and transition. Free actors have the position and the capacity to do what is necessary to create a network conducive to innovation. As in all organic systems, evolutionary development towards task division, specialisation and more complex forms of organisation will occur as long as all parts of the system remain interconnected. In times of transition, dysfunctional or strangulating connections have to be replaced by new ones. It takes actors with freedom to act to work on making new connections and do whatever is necessary to ensure that these connections flourish.

Keywords: animal husbandry, ecological view, free actors, networks

7.1 Introduction: the experiment 'networks in animal husbandry' (2004-2007)

The 'networks with free actors' approach emerged from a large scale experiment with networks of entrepreneurs in animal husbandry in the Netherlands. In the period 2004-2007, each year some 50 networks of entrepreneurs were assisted by 25-35 knowledge workers; researchers and advisors – the so-called 'free actors' (Wielinga *et al.*, 2008). The programme 'networks in animal husbandry' was set up to foster innovations for sustainable production systems, by assisting networks of entrepreneurs with expertise. The pre-requisite for assistance to the network was that the farmers themselves had to take the initiative. In total, the programme has assisted 120 networks. These networks have generated an impressive stream of ideas, publications, demonstrations and new developments.

In this chapter we begin with an introduction to the experimental programme, 'networks in animal husbandry'. We review the historical background to the development of the knowledge market – looking specifically at the famous 'education-extension-research triptych' in the period 1945-1990. Network approaches can be seen as a response to the shortcomings of former systems. The ecological view of knowledge as described in the third paragraph of this chapter, provides a useful theoretical framework for the third generation system for development. We believe that the current trend towards network approaches to solve developmental problems might lead to a third generation of organisational principles behind the Dutch agricultural knowledge system. Finally, we return to the experiment and the network approach that emerged from it: the tools that were developed and the tangible results. We conclude with a reflection on possible implications for influencing future transitions.

7.1.1 Around the kitchen table

Six poultry-farmers sit down around the kitchen table, together with an expert from Wageningen University and Research (WUR). They review the serious problems affecting their flocks. These include feather pecking and cannibalism. Although they don't like it, trimming the bird's beaks is the only way to reduce the damage. The government intends to prohibit this practice from 2012 onwards. Together, the farmers analyse the situation. What causes the problem? Frustration because of the housing systems? Components in the feed? Genetics? Why is one flock more affected than another? Do Dutch farmers have more problems than colleagues abroad and if so, why? What can they do about it?

The farmer who took the initiative for this meeting had noticed the invitation to submit proposals for sustainable innovations in animal husbandry within the framework of the programme (funded by the Ministry of Agriculture, Nature and Food Quality) 'networks in animal husbandry'. He asked fellow farmers from his study club to join him, because one of the requirements was that the initiative should be represented by a network. Although he had hoped for more, five farmers joined him. They discussed the issues with an expert, who was sent by the programme. This expert didn't have ready-made answers and the answers weren't tucked away on a shelf somewhere at the research institute either. To find solutions they had to take action themselves and motivate others to get involved as well. As one of the first steps in the programme, the WUR expert helped the farmers make a network analysis. Did this network have a broad enough basis, or did they need to involve others to arrive at viable solutions? Because they suspected that sensitivity to the problems developed early in a chick's lifetime, they decided to invite representatives of the hatcheries that bred the chicks to join them, as well as a veterinarian who had many large scale poultry-farmers amongst his clients. The network grew to 11 participants who engaged in a 'search and learn' process for one year.

7.1.2 The programme

This is just one example of a network that took part in the research programme 'networks in animal husbandry'. The programme was funded by the Ministry of Agriculture, Nature and Food Quality and the money was used to hire in expertise including a facilitator, communication support, action research and programme coordination, adding up to roughly €2 million annually. A number of networks were assisted for longer than one year and a total of 120 networks have enjoyed government support through this programme.

The programme did not start with a theory or a method but with an ambition: to bridge the gap between research and farmers in a new manner, with the initiative resting by the farmers themselves. Networks of farmers were considered to be a better vehicle for innovations than individuals. These networks would be supported by experts in the relevant field from Wageningen University and Research in order to link the demand for knowledge to current scientific expertise.

Through announcements in professional magazines and websites, farmers were invited to submit proposals to develop sustainable innovations in their sector and to present themselves as a network of at least three entrepreneurs. Selection took place after representatives from WUR had visited the networks. Important criteria were: (1) innovativeness and relevance of the initiative for sustainable agriculture and society, (2) enthusiasm and the time that the network participants were prepared to make available for the initiative, and (3) their need for assistance. After admission to the programme, a researcher was assigned to each network. In the second and third phase, farm advisors also joined the team of facilitators. A communication group supported the networks in making publications, organising events, maintaining a website and sending weekly 'eye-catching emails' to interested subscribers with newsflashes about network results and links to relevant articles and contact persons. The number of subscribers rose quickly to more than 6,000. An action research team assisted the facilitators with language and tools for working with networks, and organised regular small group meetings of five to seven facilitators to reflect on experience gained and plan further action. This team[7] also analysed what this new network approach could contribute to policy and sector goals as well as the possible effect it could have on the current scientific discourse on the innovative capacity of knowledge systems.

7.1.3 Facilitating is more than knowledge transfer

It soon became clear that the work was more complicated than just adding a technician to a network who would ask what farmers wanted and then find the right shelf with scientific knowledge waiting there to be used. The linear view of the knowledge chain, with researchers as the source of new knowledge and farmers as target audience for knowledge dissemination

[7] The authors of this chapter were part of this action research team.

proved insufficient in practice. In many cases the knowledge that farmers were looking for was simply not available, or at least had to be adapted to their specific circumstances. Appropriate knowledge had to be created together. Moreover, innovation was not only a matter of applying or creating new knowledge, but also of motivating other stakeholders, like partners in the production chain, policy makers, conservationists, animal welfare activists, etc. This also implied that the market view of knowledge fell short: the farmers networks were not clients for knowledge products, but the central force in the process of moving towards innovative change. The facilitators were active in stimulating this process within the network. Although networks were selected for their initial enthusiasm, each network invariably hit a low period. At this point, facilitators played a major role in restoring the momentum. In short, three essential functions for facilitators were identified:

- *The linkage function:* connecting relevant actors (experts, opinion leaders, policy makers) to contribute to the development from mere initiative to application in practice.
- *The process function:* maintaining connection (the lines of communication) between the participants in order to optimise collaboration.
- *The strategic function* stimulating momentum within the network environment to foster the necessary innovations for change.

7.2 Where former knowledge systems failed

7.2.1 The gap in the Dutch agricultural knowledge system

In the last few years the seemingly successful Dutch knowledge system in the agro-food complex has become inadequate to deal with the present demands of society. Although it has undergone rapid changes in the past two decades, these changes have not proved sufficient to address the sector's current needs. It is useful to analyse the knowledge transfer systems of the recent past to present a rounded picture of the issue that this research programme is designed to address. This programme was one of a variety of efforts to bridge the gap that has grown between farmers and research. In this paragraph we look at the nature of this gap and the lessons that need to be learned to fill it. In other words: what kind of transition we are looking for?

7.2.2 The education, extension and research (EER) triptych (1945-1990)

The Dutch agricultural sector used to have a world wide reputation of being highly efficient and innovative. It was generally believed that the agricultural knowledge system, with its short lines between farmers, researchers, policy makers and industry, was a major factor for its success (see Chapter 3). This knowledge system has gone through several major transitions since 1945. In the years of reconstruction after the Second World War, Dutch agrarian policy focussed on rapidly increasing food production. To this end, the government created an optimal environment for farmers to earn a decent income from producing food (Vermeulen, 1989). One of the measures was to invest in knowledge that was applicable to the large majority of small family farms (Mulder, 2004). Education, extension and research were government facilities, all under the responsibility of the Ministry of Agriculture. All parties interacted well (Van der Ban, 1987). The three entities together formed the EER-triptych (Leeuwis *et al.*, 2005). Another important measure was to stimulate the formation of farmers organisations and to share responsibility for the development of the sector with them. Farmers cooperatives played an important role in reducing self-defeating competition between individual farmers. When, at the end of the 50ties, the Dutch market became saturated, the joint effort shifted its focus to the world market. This implied a transition towards rationalisation and increase of scale. The joint effort was geared on the one hand towards creating optimal conditions for the viable family farms and on the other to create favourable conditions for those with a limited chance of survival to close down (Zuurbier, 1984).

In this climate of shared responsibilities and ambitions, it became normal to share knowledge in an 'open knowledge system'. The government run agricultural extension service was not only active in advising farmers, but also in supporting farmers study clubs and sector organisations, in negotiations with policy makers at different levels, in assisting in agricultural education and in conveying farmers experiences back to the researchers. The positive effect was that the sector became remarkably innovative and efficient in conquering an impressive share of the world market in agricultural products. There were less positive effects as well. The governing network of leaders in the farming world, captains of agricultural industry and policy makers – the 'Green Front' (Frouws 1993; Wiskerke and Van der Ploeg 2004) – became so powerful that it could ignore the growing concerns in the community about serious environmental problems until it was too late. The system was not able to deal with the problems of overproduction and pollution in the 80's and that eventually led to its collapse (Wielinga, 1999, 2001).

The tryptich was linear in the sense that research was given a high status. The agricultural community believed in one common goal and that was to produce ever-increasing quantities of milk, chicken meat or flower bulbs, for example, supported by scientific research. This did not mean, however, that new knowledge came from research and trickled down to the users: the direction for new developments emerged from interaction between farmers, research and farming leaders and, together, they had a heavy influence on politics (Vijverberg, 1996). The problem was that the system did not sufficiently interact with its surrounding world.

Within the instrumental paradigm of knowledge, people were able to ignore this problem. In this paradigm, the world is seen a huge machine. Science analyses how it works and determines the best way to manipulate it. For example, the response to the manure problem was to invest heavily in research to find technical solutions that would obviate the need to take drastic measures such as reducing the number of livestock in the country. But in a climate of opposing interests, great uncertainty and complex problems it becomes hard for scientists to produce truth that is uncontested. Policy makers had to deal with the uncertainty of the future, balance the interests of different pressure groups and address the urgency of the problems to be solved. The knowledge system worked well in times of production increase but failed to produce satisfactory answers when the negative consequences of that increase had to be curtailed and new actors appeared in the arena. Rabbinge and Slingerland (2009, this volume: Chapter 3) describes in detail the downfall of the EER-triptych.

7.2.3 The knowledge market (1990–present)

In the 90's the orientation of the knowledge system shifted towards market principles. Knowledge became a product, with researchers producing it and farmers and policy makers as clients. The time of shared responsibility was over and the market had to do its work instead. The extension service was privatised and its only chance to survive was to become demand driven. The government became a client in the market, buying research and extension programmes for issues of public interest, instead of just paying the salaries of knowledge workers.

This paradigm solves a number of problems. Knowledge can be of value, even if it is not uncontested. The client determines its value for his own use in production, trade, or as a weapon in acquiring position in a conflict of interest. Instead of a machine, the world is seen as a marketplace where individuals compete, or sometimes collaborate in win-win situations, but ultimately look to their own survival. We could also use the metaphor of the arena where 'the best man wins'. The inhibiting need for everyone to agree on the best solution to a problem is replaced by competition, allowing much more diversity. This was deemed necessary not only because of the growing conflicts of interests but also because of the ongoing process of specialisation and task division in the agricultural sector, which could no longer be adequately served by the centralised government facilities for education, extension and research.

This dominant paradigm in the 90's did not solve all the problems, however, and new problems emerged. Once the open public knowledge system came to an end – since there was no public extension service anymore linking all actors together – competition between farmers and also between knowledge workers, made people more reluctant to share knowledge. Commercial advisors were inclined to go for quick and safe wins rather than risky innovative projects and farmers complained that they could not find the right counterpart anymore to deal with issues that went beyond the farm gate. Collective issues are hard to address in a competitive market. The government had to impose rules and regulations for reducing overproduction and pollution that were unpopular to farmers. These measures have limited impact if there is no broad support base, or there is no common understanding of their urgency. Furthermore, sustainable systems have a price. Who is going to invest first when individual actors are all fighting to survive?

One method adopted by the government was to invest in research-extension programmes for mitigating the negative environmental effects of agricultural production systems, e.g. the surplus of manure. Research focussed on improving farm practices, to be tested at experimental farms. The knowledge gained would then be transferred to knowledge brokers such as advisors, and then to the front runners in the farming community, after which these practices were supposed to trickle down to the majority of farmers. Although there was interaction between the different layers in this knowledge pyramid, it was a linear and top-down system. The awareness of different technological options increased because of the programme, but there were other factors that prevented the advice from being followed (Geerling-Eiff *et al.*, 2004; Hubeek *et al.*, 2006). This experience does not stand alone. In fact, the introduction of the knowledge market to the Dutch agricultural knowledge system caused most research institutes to try to get public funding, rather than 'selling' their skills and products to the market that their research should serve.

7.2.4 The network solution

The usual management approach starts with a mission statement that is formalised using SMART formulated targets and a log frame. Competence profiles are made for the people to do the job, and they are monitored by performance indicators. It seems logical to expect that such an approach should apply to networks as well. Evaluating fourteen case studies on networks for creating sustainable food supply chains throughout Europe, Roep and Wiskerke (2006) recommend that 'such initiatives need a clear and coherent strategy', and 'have good communication and coordination based on strong alignment'. No one doubts that they are essential but it cannot be assumed that this coherence and alignment is already present in the early stages of a network initiative. Their case studies show how much effort it costs before the networks actually conformed to these criteria. How to get there is precisely what network facilitation is all about.

A shared vision of the future cannot be taken for granted when a network has just been set up. Someone with a good idea might inspire others to join but they each have their vested interests and these may not be exactly compatible with one another. The actors have to compromise, allowing for flexible targets. Innovation processes are also unpredictable which may force the network to change its focus along the way. The same is true for pursuing plans that are made. In networks there is no hierarchy to force the issue. If participants lose interest, they drop out. It is crucial that the creative energy in the network is maintained. Facilitators in the networks for innovation require a new set of tools that enable them to help networkers find their way in this hitherto unexplored territory.

7.2.5 Challenges for network approaches

Leeuwis *et al.* (2005) mention that: 'one of the positive tendencies of the functioning of the current Dutch agricultural knowledge infrastructure is the involvement of new categories of stakeholders. New knowledge arrangements have indeed resulted in some space for non-agricultural stakeholders to become involved in the knowledge infrastructure which is progressively less agricultural. Representatives of, for example, environmentalists, animal welfare organisations and nature conservationists are now operating in committees that decide about the allocation of funds, and also have become participants in a range of projects and programs. Thus, progress was made regarding the policy objective of opening up the knowledge infrastructure to new parties.' However: 'it is inherently unclear who should be paying whom in a multi-party innovation process. Instead it is better to speak of a process of "innofusion" in which innovation and diffusion coincide and actors perform various roles depending on the stage the innovation process is in.' In addition, Leeuwis *et al.* see that: 'the kinds of services being demanded and supplied (with some exceptions) still resemble those that existed in the EER-tryptich era in that they still focus very much on "knowledge", respectively that services that would be more in line with an innovation perspective are not articulated sufficiently. Therefore there is reason to worry about the capacity of the current knowledge infrastructure, including its relations with users and other actors involved in innovation processes, to effectively support innovation towards sustainable agriculture and multi-functional land-use.'

The present focus on networks as a vehicle for stimulating innovations might turn out to be the transition that takes us to a third generation of organising principles behind the knowledge system. In a network, people interact because they are aware of their dependence on each other to realise their ambitions. Knowledge plays a crucial role in finding common ground for concerted action.

One major challenge for the network approach is facilitating interaction between stakeholders in issues that affect public interest, such as the competitive power of a sector, sustainable production systems or farmers acquiring a new 'licence to produce'. New solutions aren't imposed by, but emerge from, such interactions. This challenge implies a different concept of what knowledge is. If all stakeholders are to contribute, descriptions of knowledge as 'the scientifically validated truth' or 'a product in the market' are insufficient. Experiences, convictions and theories held by all actors have to be taken seriously. This fits into constructivist theories, stating that all individuals construct their own knowledge (see for example Derrida, 1978). Only information can be transferred; knowledge cannot because it is internalised individual property. Concerted action requires social learning processes where individual constructs of reality are shared and gradually converge into communally accepted knowledge as a basis for action. The description of the learning environment becomes the metaphor of a village with a community of faith. In order to survive, the inhabitants have to acquire *agency* by learning together.

An additional challenge is to develop network methodologies that go beyond mere consensus. Interactive and participatory methods for multi-stakeholder processes are not new (Engel, 1997). They proved effective as long as stakeholders were aware of their interdependency and willing to collaborate. In this method, every new step is legitimised by consensus within the network. In the quest for sustainable agriculture in an environment where multiple actors compete for resources in the rural space, such consensus cannot always be reached by common participatory methods. Is there an alternative when the binding authority of science or the power of the investor and the state are insufficient?

7.3 Theory: an ecological view of knowledge and networks

7.3.1 Human networks as living organisms

The machine metaphor of how knowledge is acquired describes a linear concept of knowledge/technology transfer and dissemination methods. The market or arena metaphor is used for the concept of knowledge products with emphasis on efficiency and effectiveness, and control-oriented management approaches. In the village metaphor, the concept of knowledge acquisition is pictured as knowledge co-creation that depends heavily on voluntary contributions and consensus of all the 'villagers'. Because all these metaphors have their limitations in working with networks, the programme gradually adopted the metaphor of living organisms for knowledge creation (Wielinga, 2001). Human networks can be seen as living systems, just like cells, plants, animals or ecosystems. The Gaia theory (Lovelock, 1979) postulates that the entire world is actually a living system, keeping an ecological balance by an extremely complex mechanism of networks within networks, all interrelated by feedback mechanisms. In this sense we can see social networks as living organisms that generate creative energy when they are healthy. Such networks can also be sick and even die, to make place for others. This metaphor fits well in an ecological paradigm, as Röling and Jiggins (2000) called for when they argued that a new paradigm is needed in which people who are aware of the damage they cause their ecological environment, develop the necessary *agency* to restore the ecological cycles that feed life.

Capra (1996) summarised insights from ecologists, physicists, mathematicians and neurophysiologists and concluded that all living organisms can be seen as networks of interrelated components, each forming in turn a network in itself. Some of the most important conditions for life are the following:

- the components are interrelated by *feedback mechanisms*;
- the network has an *identity* distinguishing it from the outside world;
- the components can exist and *reproduce themselves* by means of the network;
- the *network is reproduced by the interaction patterns* of its component parts.

Under these conditions organic structures can develop an ever-increasing task division and complexity over time. At times old structures have to die in order to make room for new ones, and entire structures may have to pass points of instability before reaching a higher order of complexity. If we look at a human network this way, its healthiness depends on the quality of the connections between the individuals constituting the network. In other words: if they don't take each other seriously, the network is ill. If they do and find that interaction is worthwhile, they become more willing to modify their individual behaviour to achieve the necessary collaboration within the network. This generates energy and makes the network a more attractive proposition.

7.3.2 The biological function of knowledge

The metaphor of organic networks is inspired by the Santiago theory on the biological function of knowledge (Maturana and Varela, 1987). All living organisms are able to perceive signals, to make sense out of them, and to generate a response. Through this 'cognitive cycle' organisms are structurally coupled to their environment. Higher organisms have developed the ability to distinguish more signals and they have a wider range of possible responses at their disposal. Through evolution, humans have developed the ability to communicate using symbols and to form complex and abstract images of reality. This makes it possible to reach higher degrees of task division and adapt to extreme conditions. The biological function of knowledge is social coordination (Maturana and Varela, 1987; Capra, 1996; Varela, 1999). It includes the ability of an organism to respond adequately to its environment. For this purpose people use much more than explicit knowledge, they also use implicit constructs of reality, in which experience, behavioural patterns, hope, anxiety, intuition, etc. play a role. Maturana and Varela (1987) define knowledge as 'effective action in the domain of existence'.

In this ecological view, the focus is on the responsive capacity of a system. This capacity to respond adequately to changes in the environment depends on the coherence of the system: the way task division, specialisation and structural couplings allow for adequate mobilisation of physical and social resources for concerted action. The quality of the connections between all relevant elements of the system is crucial. Connectedness is not the same as consensus. For example, Walker *et al.* (2002) state that: 'decision makers in social-ecological systems must make decisions based on imperfect knowledge, with limited resources. Furthermore, decisions do not solely concern the consumption of goods and services. Agents often do not make income-maximising decisions and the utility functions used to represent agent behaviour must be sufficiently rich to include this. Utility depends upon social context. In economics terminology, agents are boundedly rational.' But disagreement and a certain degree of conflict is, in fact, useful to create awareness of various actors' stakes (Van Dongen, 1996; Van Gunsteren, 1998; Termeer, 2006). Competition is healthy within the limits of accepted rules of the game that ensure that competing actors stay connected. It goes wrong when some monopolise the scene and others are excluded. At this point the structural coupling is lost.

Pór (2001) claims that: 'the vibrancy of a corporation's knowledge ecosystem is a more reliable indicator of its future performance than its financial might. Money indicates only its past performance on the market. The vitality of its knowledge ecosystem tells about its potential to meet rapidly moving, strategic challenges and opportunities. To meet them, organisations must rely on the shared knowledge and intelligence of their members. A company can have billions of dollars in the bank, but if that company is not in tune with the requirements of the knowledge-based economy, it's very likely that it will not survive.'

7.3.3 Implications for change

The ecological view has important implications for the way measures to promote change are legitimised. In the instrumental paradigm, taking the world as a machine, expertise legitimises change. Objective scientific knowledge feeds rational decisions about what to do. In the strategic paradigm, taking the world as a market or an arena, whether or not the change serves the interest of clients legitimises its implementation. For issues of public interest, government is the client, setting the rules, being legitimised by its democratic mandate. In the communicative paradigm, taking the world as a global village, actions are legitimised by consensus amongst the stakeholders. In the ecological paradigm, actions for change are legitimate in as far as they serve the responsive capacity of the system.

The responsive capacity of a system depends on the quality of its fundamental connections. According to Malhotra (2002), 'within the knowledge ecology, focus on people does not only imply understanding of knowledge exchanges and relationships based on such exchanges, it also implies understanding of how such knowledge influences action or potential for action based on such exchanges. Knowledge ecology is made up of knowledge nodes and knowledge exchanges and knowledge flows. In knowledge ecology, the basis for cooperation and survival is differentiation and similarity between the knowledge nodes. Highly differentiated nodes can collaborate to accomplish specific actions and may dissolve thereafter. However, collaboration between such nodes would require that they are able to relate to one another under an overarching mission or theme. Knowledge ecology treats knowledge creation as a dynamic evolutionary process in which knowledge gets created and recreated in various contexts and at various points of time.'

In a network, there are three levels of connection: (1) connections between stakeholders, (2) the connection between the network and the outside world in which it has a function and (3) the connection stakeholders have to their own authentic identity, giving them their sense of meaning. The notion of connectedness is not without complications. Weick observes that connections within human systems can become too tight, thus hampering their capacity to respond to changes (Weick, 1997). He distinguishes tightly coupled systems from loosely coupled systems and uncoupled systems, and suggests that for an innovative climate, loosely coupled systems work best. So, somewhere between too tight and too loose there is an optimum level where people have freedom to act while still being conscious of their connectedness. This comes close to the role of the free actor as we defined it in the experiment. One complication is that a free actor might feel legitimised to intervene, even without consensus, to improve the quality of connections while the other actors involved are opposed. Who determines what is quality, and who determines the difference between a healthy and a sick network?

7.3.4 Vital space

All human interaction has at least two dimensions. In the relations dimension, the optimal level of freedom can be understood as a certain position in a space. At one extreme all interaction is dominated by the network requirements, leaving no space for authentic individual contributions. At the other extreme every individual does what he wants without considering the network. Somewhere between 'we' and 'me', the value of the network can be felt, where individuals feel free to act and where they can benefit from the synergy of the network. The second dimension concerns the content of the relationship; the subjects that matter, the understanding people have and the goals they strive for. This dimension has two extremes as well: they range from 'similarities'; when people have a common understanding and common interests, to 'differences', when they do not share a common understanding and have no common interest. Here too, there is an optimal space where people can be both open to new ideas and ambitious for themselves. Where the optimal level falls in both dimensions, interaction is worthwhile and rewarding: it generates energy. This is called the 'Vital space' (Wielinga, 2000, 2001). Outside this vital space, interaction is not rewarding and it drains energy.

The circle of coherence visualises vital space as a circle placed in two dimensions with four quadrants, each of them partly within and partly outside of the circle. Figure 7.1 presents the circle in its simplest form. For use as a tool for networkers, the extended version of the circle shows interaction patterns: four stimulating ones within the circle and four regressive varieties outside of it. Discussing the model to its full extent, goes beyond the scope of this chapter.

The important issue here is that this model shows the organic nature of human interaction in networks. The steering mechanisms are inherent. Too many differences can lead to confusion. The natural response to this is that people are inhibited in their perception as to what they can

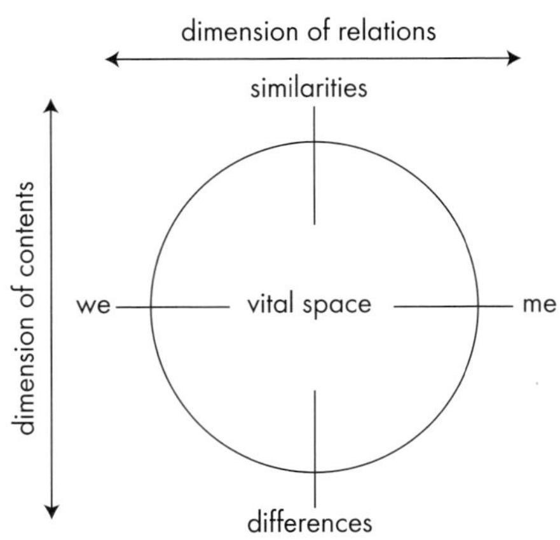

Figure 7.1. The circle of coherence.

cope with. When there are too many similarities, people respond by seeking out the inevitable differences to make it more interesting. Too little space for individual action promotes aggressiveness, while too little attuning leads to loss of collective protection and added value. This causes anxiety. Aggression might stimulate the enlargement of individual space but anxiety leads to more openness to the strengths of the group. The limits need to be probed all the time to find the optimal balance. This is basically what children learn while playing. In a healthy situation, children are inquisitive and want to play. The same goes for adults.

If we see the evolutionary development of the world as occurring within one huge ecosystem then this concept is only logical. If there were no inherent steering mechanisms that lead the evolutionary process to ever more complex forms of ordering, life would never have passed the stage of the primordial soup. There is no reason why the same principles should not apply to human systems.

The consequence of this view, however, is far reaching. It means that we can assume that development takes a healthy direction towards more specialisation, task division, synergy and harmony as long as the mechanisms that keep all elements connected grow along with the complexity. Although we think this is an inspiring idea, the question whether mankind will be in time to restore a healthy connection with the carrying capacity of ecology is cause for deep concern. This is basically the quest for sustainability.

The basic element in vital space is trust. An actor can concentrate on his own task as long as he can trust that others do their share. He can engage in a learning process as long as he feels confident that the risks are limited. Trust cannot be constructed or created, although it needs to be tested continuously. Elster calls it a by-product (Elster, 1983): 'the more someone tries tot achieve it, the less likely he will succeed. This is also true for spontaneity, creativity, innovativeness, courage, passion, love: in short, all things in life that really matter.' Following the logic of self-organising organic systems, we can assume that trust will grow autonomously if the conditions are right. This gives us a perspective for possible action. If we recognise that the lack of *connection* is the limiting factor for trust to grow and we succeed in making that connection, the chances for vital space to develop will increase.

As every interaction has two dimensions, the relationship (meta-communication) factor as well as content, interventions for improving interactions in a network can include both communication and repositioning. Communication (e.g. arguing, negotiating, mediating, etc.) has an impact on the way actors understand the situation. Repositioning makes use of power which changes positions of influence. In everyday life we always make use of both measures to get things done. The use of power is risky, however, because it can easily lead to an unhealthy struggle. It makes a lot of difference, however, if power is being used to win or restore connection.

7.3.5 Responsive capacity and free actors

According to Rotmans (2003) 'social renewal demands new forms of governance and interaction. Transitions and system innovations can only exist and develop through interaction and collaboration. None of the actors or parties have absolute power nor control. Of essential importance is the formation of a collaborative network for innovation, a transition arena in which all parties involved create the culture and conditions for radical change and in which every party and actor has its own role and responsibility.'

If interaction in a network is healthy, there will always be actors who sense imbalance and act to restore vital space. This is how the inbuilt steering mechanisms work. The thresholds for individuals to take up such leadership roles vary. They depend on personality, culture, task division, formal and informal authority, etc. The qualities they need vary for differing situations; for protecting and promoting individual interests, for improving interaction within a network, or facilitating connection between the network and the outside world.

Systems can evolve into more complex systems and every level of ordering requires its own mechanisms to keep the elements connected. Castells argues that society has entered a new stage, which he calls '*the network society*'. Mechanisms, rules and institutions that were developed in the past, lose their meaning and their effectiveness in this new era (Castells, 1996). However, he offers no clear idea how the a new generation of mechanisms work. If one accepts the ecological view, then individual initiative and leadership will become more important than ever before. Leaders will have to learn to reflect on their own behaviour, that of the group and the process itself. Hobbs and Njoya (2005) state that 'as an antidote to the limits of traditional regulatory interventions, reflexive governance methods seek to achieve their ends indirectly by adopting a procedural orientation which is intended to induce actions by social actors and encourage autonomous processes of adjustment, in particular by supporting mechanisms of group representation and participation. The objective is to structure decision processes without trying to control the substantive outcome of any decision. The role of reflexive governance is thus to stimulate and facilitate a process of self regulation by providing a frame or steering mechanism.'

The network society needs free actors to maintain its responsive capacity. In this respect, Latour (2005) stresses the importance of spokespersons who express what the group wants. Groups are never uniform, there are always different voices and opposing views. There should be at least one person who keeps building its identity. Likewise we see the function of free actors in keeping vital space in a network.

Pór (2001) believes that to promote a healthy knowledge ecology, free actors must do the following things to extract the best possible contribution from the members of the network:

- Be master learners and model better practices for individual and collaborative learning in situations where people interact in environments optimised to support their collaboration.

- Elicit the leadership qualities in all who work with them.
- Become future-responsive; anticipating and articulating future possibilities.
- Model all of the above for others so they can learn to do the same.

For the research programme 'Networks in animal husbandry', the ecological view on knowledge has led to focus on the following aspects:

- *Co-creation of knowledge.* Knowledge emerges from interaction. Scientific input can inspire but not prescribe. Specific situations require specific solutions. Every actor has to go through his own learning process.
- *Energy.* For building up trust and coherence it is crucial to work on the willingness of participants to invest in the network and to attune to others. Enthusiasm is not a prerequisite nor stable factor.
- *Connection.* Trust requires constant feeding and grows in connection. Keeping connection lines open is a major task for networkers.
- *Free actors.* Networks require free actors: people with the overview and the capacity to do whatever is necessary to keep the network healthy.

7.4 The experiment revisited: results of the network programme

7.4.1 Tools for stimulating energy and connection

The experiment did not start with a conceptual framework or methodology but with an ambition: to bridge the gap between farmers and research. During the process itself the tools for making it work were sought out or developed by the researchers and shared with the group. Reflection was found to be one of the crucial tools for success.

It was remarkable to see how often the facilitators intuitively did the right thing. They only realised what they had done in retrospect when analysing their experiences with the networking tools in the programme. The tools were not collected, developed and used to replace their intuition but to sharpen it. It can be compared to the master in martial arts who does not think about his actions in a fight but responds instinctively in a split second. This response has been sharpened by years of experience, reflection and training so that the right one is available when required. The networking tools provide a language for reflection on experience, insights in network dynamics and offer a range of possible interventions for specific situations. The following tools were in use by the end of the programme:

- *The network analysis:* distinguishing *positions of involvement*: partners, links, suppliers and users. This tool helps to assess the viability of the network and gives a direction to follow in order to create or improve connections (Poorthuis, 2006).
- *The spiral of innovation:* distinguishing *stages of an innovation process*: initial idea, inspiration, planning, development, implementation, dissemination and embedding. In each stage the role of knowledge differs. For example, in the development stage, finding knowledge is important. In the implementation phase knowledge is used as a weapon

to force the environment to accept change. Each stage also requires connections with different actors. For example, supporters are needed at the inspiration stage and experts at the development stage (Zaalmink *et al.*, 2006; Wielinga *et al.*, 2007).

- *The triangle for change:* distinguishing *change agents, gatekeepers* and *survivors.* Only if change agents have generated sufficient energy can they effectively negotiate with gatekeepers. It is a waste of energy to start with convincing survivors: they will only accept change if the risk to their position is reduced to a minimum (Wielinga, 2001, 2004).
- *The circle of coherence:* distinguishing *patterns of interaction.* In healthy interaction there is 'vital space' where people build up trust, openness to new ideas and commitment. Each pattern can turn into an unhealthy pattern, leading to a loss of motivation and coherence. For each pattern, the tool indicates 'warm' and 'cold' leadership interventions in order to restore connection and vital space. Warm interventions, like inspiration, negotiation or humour, work through communication. Cold interventions, like regulation or strategies, work through positioning (Wielinga 2001, 2004).
- *The time line method:* a simple but effective way to discuss progress in a network with the participants. It focuses on critical moments and personal discoveries (Zaalmink *et al.*, 2006).
- *The learning history:* adding analysis to the results of the time line method, making use of the models mentioned above. This tool is useful for reflection as well as reporting. The concept of Learning Histories, developed at MIT by Kleiner and Roth (1997) provides context specific information that makes results and lessons learned more transferable to other contexts.

Instead of being used as targets and instruments for controlling the process, these tools allow people with initiative to assume a leading role. They help monitor the quality of connections between key stakeholders and indicate what can be done to improve them if necessary in different stages of network development and in different situations that might occur. Ultimately such interventions contribute to the self steering capacity of the network and its responsive capacity towards its environment.

7.5 Results

7.5.1 Examples of results in the field

Just to give an impression of different types of results the networks achieved, some examples are listed below:

- Developing techniques and new applications:
  - a breeding programme for self-moulting sheep, which makes them low maintenance;
  - a 'claw check' for sows in group housing in pig husbandry;
  - a tool for measuring parasites in poultry housing;
  - a mobile milking robot for cows grazing outside.

- Opening up new markets and products:
  - a brand for chicken meat of high quality that has been produced in an animal friendly way;
  - a marketing chain for horse milk as food supplement;
  - energy supply by biogas;
  - biofuel from grass.
- Developing tools for exchange of experiences:
  - a handbook for entrepreneurs in pig husbandry;
  - an internet tool for monitoring the use of labour in dairy farms;
  - an internet tool for recognising vector diseases in pigs;
  - a database for toppers and bloopers in animal husbandry;
  - a tool for collegial consultation by multi-point video conferencing.
- Scaling up efforts:
  - a network, discovering there were no easy answers to a parasite problem in poultry, convinced the sector organisation and the Ministry of Agriculture of the magnitude of the problem, resulting in a much larger fund for research;
  - inspired by the tool for recognising vector diseases in pigs, the sector organisation for poultry production is now investing in the development of a similar tool for chickens;
  - a network of 11 dairy farmers who developed a 'cow coach' tool (including social venture aspects) for collegial consultation on high quality milk production for a special brand of ice-cream has been contracted by a cooperative aiming to apply this approach within its own organisation consisting of 550 members. The facilitator left his research job and joined the cooperative to become manager of this project.
- Developing new forms of cooperation:
  - entrepreneurs in biogas production have formed an association to defend their interests – with considerable impact on policy makers;
  - a network of dairy farmers is working on merging their farms into one big enterprise, allowing for task division, specialisation and advantages of scale; two of them have effectively decided to merge already;
  - two other networks seek private investors who are willing to share responsibility for maintaining the quality of the landscape;
  - farmers acting as 'school coaches', making the farm an authentic learning environment for students.
- Stimulating dialogue:
  - a number of networks opened up dialogue with organisations managing nature reserve areas, promoting the combination of farming with conservation;
  - networks for animal friendly housing systems for dairy cows and also for rabbits, involved animal welfare organisations in their search for improvements.
- Influencing rules and regulations:
  - the rules for preventing specific disease in sheep did not allow for a proper breeding programme for self-moulting sheep. The network managed to convince policy makers to change the interpretation of the rules, allowing for improvement programmes.

- in another case, the newly formed association of biogas producers listed specific problems and possible solutions for the introduction of bio-digesters. This document had impact on government officials at both local and national level.

Not all the networks in the programme have been successful. In each cycle a few networks stopped prematurely because of a lack of motivation amongst the participants to continue (no energy). This had various reasons: (1) some farmers expected a more directive approach and could not make the switch to self-organisation; (2) the initiator got distracted by other concerns before other participants had really become partners; (3) conflicts of interests within the group could not be bridged; (4) disappointment in the perspectives for their individual problem. In 9 cases assistance had to be withdrawn early. We believe that this is not a bad score out of a total of 120 networks.

The Ministry has regarded the approach successful enough to start a subsidy programme for 'Networks of entrepreneurs in animal husbandry' open to initiators amongst farmers and advisors. Furthermore, Wageningen Business School now offers a 3 day course for networkers – for both initiators and facilitators – to familiarise them with the networking tools from the programme. The first course in 2008 was fully booked (24 participants). The booklets outlining the approach and the tools have already been reprinted several times. Obviously, this approach is making its mark already.

7.5.2 Strategic space

The list of results achieved in so many networks could easily be much longer. With a technical issue as a point of entry, the facilitators assisted in processes of knowledge co-creation, involving not only the initiating farmers but also others like experts, policy makers, representatives, etc. The result for individual farmers was that they increased their 'strategic space' (Huber, 1991). This can be defined as the variety of options an actor has at his disposal to deal with challenges; options being the alternatives for behaviour on either a professional, operational or strategic level. An increase in variety provides a range of behavioural alternatives an entrepreneur can choose from, although it might also close down some alternative routes that appear to be unproductive. The larger the strategic space, the greater the innovative capacity of a network to solve specific business problems (Weick, 1997). Entrepreneurs can enhance their strategic space by gathering information and by learning and interacting with colleagues and stakeholders. External influences – social, economical, political and technical factors – can influence the strategic space as well, i.e. can increase or decrease options for change (Geerling-Eiff *et al.*, 2005; Van Baalen *et al.*, 2005; Hubeek *et al.*, 2006). It can happen that because of competitiveness between the farmers themselves they are unwilling to exchange knowledge that they have. Like any transition, this is a long process and it requires trust, both between the actors involved [social trust] and in the value of the knowledge being created [cognitive trust] (Cohen and Levinthal, 1990; Kollock, 1994).

7.5.3 Facilitators as free actors

The assistance provided by facilitators has been especially valuable for networks on the precarious developmental track which turns a group of initiators into a fully fledged network that knows what to do, how to do it and whom to mobilise for getting it done. One year of assistance was too short to realise major innovations. The contribution of the programme was nevertheless important. Even though there are many worthwhile initiatives that die prematurely because of a lack of leadership, the desire to nurture initiatives cannot be imposed. It takes people who can look beyond their own interests and do whatever is needed for the network. They have to be able to see what is needed and be in the position to respond appropriately. Instead of managers who are assigned to reach predetermined targets via instruments of control, it requires free actors who know how to find their way in unfamiliar territory to keep interaction healthy. The programme provided these free actors on a temporary basis. Their role was important in the process of change within the network since innovation requires changes in external relationships, interaction patterns and role performance of the participants. Generalising the experience, it is has been shown that any network needs at least one free actor to keep it functioning well. Continuity of the networks after the assistance had ended could be predicted by the availability of a free actor amongst the network participants. The approach that emerged from the experiment is therefore called 'the FAN approach': Free Actors in Networks.

7.6 Implications for processes of innovation and transition

7.6.1 A review of history from an ecological perspective

By observing 'thresholds for individual, horizontal and vertical leadership' in six different periods after 1945, it was possible to explain why the innovative capacity of the agricultural sector varied over time, and also why the position of the sector in society became problematic (Wielinga, 2001). In the period of the 'education-extension-research triptych' the sector was assisted, free of charge, by a large number of knowledge workers with a lot of freedom to do what was thought to be necessary for the farmers. This 'army of free actors' stimulated the development of a huge network 'avant la lettre'. In this period the threshold for horizontal leadership was low.

In the same period, the threshold for vertical leadership was high. For a long time nobody felt the need to ensure that developments in agriculture were socially accountable. Agriculture was performing well in feeding the country, and later, by improving the international trade balance. It had a reputation of being modern and innovative. This changed in the 70's when groups in society started to criticise the sector for its negative effects on the environment. In contrast to the finely-mazed communication lines between stakeholders within the sector, the same connections had not been built up with the outside world: the threshold for vertical leadership was high and became even higher when criticism became louder.

After the government extension service had been privatised, the transition from the period of the triptych and their shared responsibility for agricultural policies changed rather drastically within a few years. The communication lines between the policy makers at the Ministry of Agriculture and the sector, that had been maintained by the extension service, suddenly were cut off. This lowered the threshold for vertical leadership. On the other hand, under market conditions, thresholds for horizontal leadership became high. In a competitive market, where most actors have to struggle to survive, free actors become a rare species. The pyramid, through which government tried to stimulate sustainable production systems, failed to connect the bottom with the top.

7.6.2 The free actors approach

The challenge was to develop a new generation of steering mechanisms for the network era. Much depended on the quality of connections, that took a different shape than in previous periods: they are less structured and often temporary. The responsive capacity of society would depend on the responsive capacity of its constituent parts. Since networks cannot be managed as organisations, they have to be led in another way.

The results so far suggest that the approach with temporary free actors, focussing on energy and connection, offers a good perspective, at least at the micro level. The crucial element is trust. Building up trust requires a good balance between providing space and using intervention instruments when that space is not being utilised properly.

Table 7.1 summarises the theoretical part of this chapter. The essential difference between the ecological paradigm and the previous ones is one of control. Instead of deciding what should be done and subsequently choosing instruments for getting people to do it, it places its faith in the self-organising power of human systems. It calls for a shift in focus – determining possible blockages to desired change, and building up a repertoire of interventions to deal with them when they occur.

7.6.3 Agenda for the future

The experiment has been one step on the long transition pathway leading to the third generation rural knowledge system in the Netherlands. There is still a lot of work to do. In this final part of the chapter we reflect on what is still needed.

Firstly, once the usefulness of free actors is recognised, the question arises who should pay for this role. In the experiment, the free actor was offered temporarily as part of a research programme. Although farmers came to appreciate this assistance in most cases, they are more inclined to pay for clear and short term advice than for uncertain, long term processes of searching and learning. While innovation networks have proved to be positively influenced by free actors, their work should probably continue to be funded collectively.

Table 7.1. Schematic overview of visions on knowledge and legitimacy for actions in different paradigms.

Paradigm	Instrumental	Strategic		Communicative	Ecological
Metaphor for the world	machine	market	arena	village	organic system
Knowledge as	uncontested truth	product	weapon	construct	responsive capacity
Legitimacy for action	expertise	client	interest/state	consensus	connection
Strengths	objective, impartial	diversity	order	voluntary collaboration	synergy
Weaknesses	ineffective in conflict or complex issues	collective issues hard to address	limits to force, evokes opposition	prone to obstruction, equity not always present	requires free actors with authority

Closely related to this is the question: who should be paid? When we look at innovation networks as communities for the co-creation of knowledge, the common supplier-beneficiary model does not apply anymore. This changed situation should be reflected in the assignment of costs and benefits amongst the actors involved. In the experiment, funds were made available to pay for knowledge workers (researchers, advisors) as facilitators and experts. In practice, facilitating tasks were often shared between the knowledge worker and an active chairman of the network.

Another issue that needs further attention is monitoring and evaluation (M&E). Early in the experiment, attempts to set up a M&E framework failed because we were unable to set markers for success and indicators for progress that were sufficiently unambiguous for all actors involved. Gradually we shifted towards collecting both negative and positive feedback on what participants and facilitators had found important and ensured that the analysis of their stories was included in the learning histories. M&E has a function in managing the activity as well as in justifying both the process and the results to the investor. Short communication lines with policy makers who took interest in the experiment kept them updated, which resulted in trust and willingness to invest further in network approaches. From the scientific point of view however, it remains a challenge to supplement these subjective, narrative kinds of learning experiences with solid, objective data.

The role of free actors can be very satisfying, but also demanding. It cannot be learned from a book. It has been found that reflecting on experience and exchanging ideas with peers is crucial for professionalising, as well as maintaining a personal energy balance in times that the network goes through difficult periods. However, the value of collective reflection is often only appreciated once they do it and this can be overlooked in the initial stages in the development of a network. Facilitators without this experience are not inclined to devote time for it in their planning and managers and investors are quick to skip this reflection time when it comes to economising. How to create space for collective reflection is a continuing challenge.

It should be noted that the experience gained in this experiment occurred mainly during the stages inspiration, planning and development, according to the Spiral of Innovations (paragraph 7.4.1). A few networks also reached the stages of dissemination and embedding. We expected this when we began the selection process for new initiatives. Going through all the stages from the initial idea to embedding usually takes longer than one year. We noticed that the shift from development to realisation was often difficult. Looking for new knowledge and experimenting with new practices can easily be done when it takes place in a relatively safe (niche) environment, whereas the *realisation* of the new idea forces the network out into a larger arena where many other actors, power play and conflicting interests can determine behaviour. In the network dealing with feather pecking and cannibalism, for example, a conflict of interests between the poultry farmers and the hatcheries came to the surface. Interaction with various scientists pointed out that solutions could be expected along two different pathways. One lay in the composition of the chicken feed, the other lay in breeding practices. However, the breeders kept on insisting that only feed could be held responsible, afraid as they were for possible claims if something would appear to be wrong with their breeding programmes. Although the network learned a lot, more work has to be done to create conditions which would encourage the hatcheries to take an active part in the search for solutions. Further experiments should validate this approach and the tools used by it, under such challenging circumstances. We think that the results so far are promising.

References

Capra, F., 1996. The web of life: a new synthesis on the matter of mind. HarperCollins: London, UK.

Castells, M., 1996. The rise of the network society. Blackwell Publishers: Malden, USA

Cohen, W.M. and Levinthal, D.A., 1990. Absorptive capacity: a new perspective on learning and innovation. Administrative Science Quarterly 35: 128-152.

Derrida, J., 1978. Writing and difference. University of Chicago Press: Chicago, USA.

Elster, J., 1983. Sour grapes: studies on the subversion of rationality. Cambridge Press: Cambridge, UK.

Engel, P.G.H. and Salomon, M.L., 1997. The social organisation of innovation: a focus on stakeholder interaction. KIT: Amsterdam, the Netherlands.

Frouws, J., 1993. Mest en macht [Manure and power]. Dissertation Wageningen University, the Netherlands.

Geerling-Eiff, F.A., Hubeek, F.B. and Van Baalen, P.J., 2004. Kennis en gedrag: een studie binnen het kader van de Eindevaluatie Actieplan Nitraatprojecten. Project 30177, rapport 7-04-03. LEI: The Hague, the Netherlands.

Geerling-Eiff, F.A., Hubeek, F.B. and Van Baalen, P.J., 2005. Naar nieuwe kennisarrangementen: aanbevelingen voor de inrichting van kennisnetwerken. Wageningen UR: The Hague, the Netherlands.

Hobbs, R. and Njoya, W., 2005. Regulating the European labour market: prospects and limitations of a reflexive governance approach. Annual Review Article 2004. British Journal of Industrial Relations 43 (2): 297-319.

Huber, G.P., 1991. Organizational learning: the contributing processes and the literatures. Organization Science 2 (1): 88-115.

Hubeek, F.B., Geerling-Eiff, F.A. and Van Baalen, P.J., 2006. Supply- versus demand-driven knowledge dissimination: a focus on 'strategic space'. In: Proceedings ISPIM congress (11-14 June 2006), Athens, Greece.

Kleiner, A.and Roth, G., 1997. Learning histories: a new tool for turning organizational experience into action. Paper provided by MIT for Coordination Sciences, New 21st Century Working Paper Series, nr. 002.

Kollock, P., 1994. The emergence of exchange structures: an experimental study of uncertainty, commitment and trust. American Journal of Sociology 100 (2): 313-345.

Latour, B., 2005. Reassembling the social, an introduction to actor network theory. Oxford University Press: Oxford, UK.

Leeuwis, C., Smits, R., Grin, J., Klerkx, L.W.A., Van Mierlo, B.C. and Kuipers, A., 2005. Equivocations on the post privatization dynamics in agricultural innovation systems. In: Smits, R.E.H.M. (Ed.) The design of an innovation-enhancing environment. Transforum: Zoetermeer, the Netherlands, pp. 2-58.

Lovelock, J.E., 1979. Gaia: a new look at life on Earth. Oxford University Press: Oxford, UK.

Malhotra, Y., 2002. Information ecology and knowledge management: toward knowledge ecology for hyperturbulent organisational environments. In: Encyclopdia of Life Support Systems (EOLSS), Unesco/ Eolss Publishers: Oxford, UK.

Maturana, H., and Varela, F. (1987, 1992 revised edition). The tree of knowledge. The biological roots of human understanding. Shambhala: Boston, USA.

Mulder, M., 2004. Agricultural education – building competence for innovation in the agri-food complex. Wageningen University: Wageningen, the Netherlands.

Poorthuis, A.M. (Ed.), 2006. De kracht van netwerkbenadering. Van Gorcum: Assen, the Netherlands.

Pór, G., 2001. Management education and knowledge ecology. Updated version of the article originally published in the November-December 2001 issue of: BizEd, Association to Advance Collegiate Schools of Business (AACSB). Available at: http://www.co-i-l.com/coil/knowledge-garden/kd/ke_4_biz.shtml.

Rabbinge, R. and Slingerland, M. 2009. Change in knowledge infrastructure: the third generation university. In: Poppe, K.J, Termeer, C. and Slingerland, M. (Eds.) Transitions towards sustainable agriculture and food chains in peri-urban areas. Wageningen Academic Publishers: Wageningen, the Netherlands, pp. Chapter 3 in this Volume.

Rotmans, J., 2003. Transitiemanagement: sleutel voor een duurzame samenleving. Van Gorcum: Assen, the Netherlands.

Roep, D. and Wiskerke, H., 2006. Nourishing networks: fourteen lessons about creating sustainable food supply chains. Reed Business Information – Agriboek: Doetinchem, the Netherlands.

Röling, N.G. and Jiggins, J., 2000. Agents in collaborative management: The logic of collective action. In: Buck, L.E., Geisler, C.G., Schelhas, J. and Wollenberg, E. (Eds.) Biological diversity: balancing interests through adaptive collaborative management. CRC Press: Boca Raton, USA.

Termeer, C.J.A.M., 2006. Vitale verschillen: Over publieke leiderschap en maatschappelijke innovatie. Inaugural speech, Wageningen Universiteit, the Netherlands.

Van Baalen, P.J., Geerling-Eiff, F.A. and Hubeek F.B., 2004. Kennisdiffusie en strategische ruimte. Tijdschrift voor Sociaal Wetenschappelijk Onderzoek van de Landbouw 19 (4): 195-207.

Van den Ban, A.W., 1987. Communication systems between agricultural research and the farmers – the Netherlands way. Journal of Extension Systems 3: 26-34.

Van Dongen, H.J., De Laat, W.A.M. and Maas, A.J.J.A., 1996. Een kwestie van verschil [A question of difference]: conflicthantering en onderhandeling in een configuratieve integratietheorie. Eburon: Delft, the Netherlands.

Van Gunsteren, H.R., 1998. A theory on citizenship: organizing plurality in contemporary democracies. Westview Press: Boulder, USA.

Varela, F., 1999. Ethical know-how: action, wisdom, and cognition. Three lectures. Stanford University Press: Stanford, USA.

Vermeulen, W.H., 1989. Europees landbouwbeleid in de maak: Mansholts eerste plannen 1945-1953. Historiae Agricultura. NAHI: Groningen, the Netherlands.

Vijverberg, A.J., 1996. Glastuinbouw in ontwikkeling: beschouwingen over de sector en de beïnvloeding daarvan door de wetenschap. Thesis. Wageningen University: Wageningen, the Netherlands.

Walker, B., Carpenter, S., Anderies, J. Abel, N., Cumming, G. Janssen, M. Lebel, L., Norberg, J., Peterson, G.D. and Pritchard, R., 2002. Resilience management in social-ecological systems: a working hypothesis for a participatory approach. Conservation Ecology 6(1). 14.

Weick, K., 1997. The social psychology of organizing. (2nd edition). Random House: New York, USA.

Wielinga, H.E., 2000. Rural extension in vital networks: the changing role of extension in Dutch agriculture. Journal of International Agricultural and Extension Education 7(1): 23-36.

Wielinga, H.E., 2001. Netwerken als levend weefsel. Een studie naar kennis, leiderschap en de rol van de overheid in de Nederlandse landbouw sinds 1945. [Networks as living tissue. A study on knowledge, leadership and the role of government in Dutch agriculture since 1945]. PhD thesis Wageningen University. Uilenreef Publisher: 's Hertogenbosch, the Netherlands.

Wielinga, H.E., 2004. The response-ability of networks: healthy and sick agricultural knowledge networks in the Netherlands. In: Christovão (ed.) European farming and society in search of a new social contract – learning to manage change. Proceedings of the sixth European IFSA symposium, Volume 2. UTAD: Vila Real, Spain, pp. 483-496.

Wielinga, H.E., Zaalmink, B.W., Bergevoet, R.H.M., Geerling-Eiff, F.A., Holster, H., Hoogerwerf, L. and Vrolijk, M., 2007. Netwerken met vrije actoren. Stimuleren van duurzame innovaties met netwerken in de veehouderij. Wageningen University: Wageningen, the Netherlands.

Wielinga, H.E., Zaalmink, B.W., Bergevoet, R.H.M, Geerling-Eiff, F.A., Holster, H., Hoogerwerf, L. and Vrolijk, M., 2008. Networks with free actors: encouraging sustainable innovations in animal husbandry by using the FAN approach. Wageningen University: Wageningen, the Netherlands.

Wiskerke, J.S.C. and Van der Ploeg, J.D., 2004. Seeds of transition: essays on novelty production, niches and regimes in agriculture. Van Gorcum: Assen, the Netherlands.

Zaalmink, B.W., Smit, C.T., Wielinga, H.E., Geerling-Eiff, F.A. and Hoogerwerf, L., 2007. Netwerkgereedschap voor vrije actoren. Methoden en technieken voor het succesvol begeleiden van netwerken. Wageningen University: Wageningen, the Netherlands.

Zuurbier, P.J.P., 1984. De besturing en organisatie van de landbouwvoorlichtingsdienst. PhD thesis, Landbouw Hogeschool Wageningen: Wageningen, the Netherlands.

Chapter 8

Collective analyses of barriers to and opportunities for sustainable development using the Innovation System Framework

Barbara van Mierlo and Marlèn Arkesteijn

Abstract

This chapter focuses on system analysis as a potential tool for use in innovation networks whose purpose is to stimulate sustainable development in a particular sector. By looking at the system as a whole, researchers and project teams can isolate the systemic factors which might be a barrier to progress or provide a potential opportunity for desired change. On the basis of its use in a few case studies, we look at the problems and the potential of system analysis. We begin by outlining the reasons why system analysis might be relevant to innovation networks and the challenges that system analysis should address in this context. We discuss existing approaches for carrying out a system analysis and explain why we chose to use the Innovation System (IS) Framework developed by Klein Woolthuis and coworkers. We look at the differences between expert-led and collective analysis and along the way we identify possible improvements to system analysis based on our experiences in the case studies. We describe how, over time, the character of the analyses changed from mainly expert-led to entirely collective, while at the same time their quality and influence on the ongoing innovation process changed as well. We will look at how these changes are interrelated, and we will also look at the way system analysis itself can actually encourage learning. We conclude by suggesting that although the empirical evidence is still thin, the framework appears to be a good instrument to stimulate system learning, especially if it is used in a collective, cyclical, step-by-step analysis.

Keywords: innovation system, system analysis, learning, innovation, sustainable development

8.1 Introduction

The programme, *Farming with a future,* started in 2001 as an extension project to test and disseminate Best Practices to growers with the aim of contributing to sustainable crop production through integrated pest management and sustainable use of manure. After a while, it was concluded that growers' practices would only change if other actors around them, like consultants, traders in pest management and manure suppliers and manufacturers would change the way they worked as well. The researchers and extensionists of *Farming with a future* had far less influence on growers than these actors; the pesticide suppliers particularly influence growers' practices greatly, not only because they act as consultants when they visit farms, but also because growers usually have only one supplier. From 2004 onwards, in some sectors, networks were built to involve a range of relevant stakeholders that were willing to cooperate to develop more sustainable practices in crop production. A typical feature of these networks was that in the different plant sectors local groups of interested growers met regularly to get advice from researchers and to implement and test so called Best Practices. In many of these networks one or two representatives from other stakeholders were also involved. The strawberry network in West-Brabant for instance consisted of strawberry growers, consultants/retailers, a contract worker, two representatives of the Agricultural Producers Board (LTO) and a representative of the Water Board. One of their activities was to test ways of composting, whereby detrimental micro-organisms would be destroyed.

After a few years of network building, the managers of *Farming with a future* wanted to know whether they were successful in their endeavour to stimulate new dynamics within the networks and also, whether their own activities were effective. The ultimate goal was to become an important player in the Dutch food cultivation sector by showing how to work on sustainable crop production. It was because of this that they looked for a monitoring methodology founded on scientific theory and research that would support the processes of learning and innovation needed to reach their objectives. More specifically, the managers wanted to know whether the network activities helped develop new joint visions and practices in these networks and whether the activities help to overcome systemic barriers that currently hinder sustainable development in the sector.

The authors decided to reflect on the question whether the network activities focused on systemic barriers or opportunities, by conducting a system analysis. It was hoped that such a reflection would also help to develop joint visions on desirable and feasible interventions, thereby enhancing the chances of stimulating an innovation process towards sustainable development.

8.2 The challenge for interactive system analyses

Until recently, the dominant innovation model in Dutch agriculture showed a linear process starting with invention by scientists, followed by adoption by farmers and then distribution within society through education and extension in order to ensure sufficient food production for a reasonable price. This view of innovation and its institutionalised and material structures has facilitated the economic expansion of the sector in the past. However, this economic development was achieved at the cost of many unintended, undesirable side-effects such as a diffuse pollution of the surface water, soil dehydration, animal diseases, and problems with animal welfare. The consensus in Dutch society for the ongoing development in the sector, disappeared in the face of these severe and intractable problems (Leeuwis *et al.*, 2006). In this context of social disapproval, the traditional manner for promoting innovation has become increasingly replaced by network based initiatives, such as *Farming with future,* that embrace a more complex approach to innovation processes, which of their nature are fraught with scientific uncertainty, social ambiguity and unpredictability.

A good analysis of the systemic factors that play a part in the persistence of environmental and other problems could help to design new forms of joint action within a network whose aim is more sustainable development. By analysing which systemic factors are the cause of the problems – for example, the existing physical infrastructure and dominant cultural values – the system analysis may help focus network activities on those factors. Various analytical tools and frameworks are recommended and used to conduct these analyses. Some have been developed specifically for a system analysis involving sustainability issues. Others that originate from innovation science and were used originally to identify general barriers to national or regional innovation, have been translated for use in the field of sustainable development. Most of these system analyses are conducted by scientific experts. Simultaneously, there is a call from scientific and policy domains for participatory or interactive analyses, to enhance the quality of the assessment by incorporating diverging and possibly conflicting perspectives, in that way also enhancing the effectiveness of policy measures which need the cooperation of stakeholders (see for instance Quist and Vergragt, 2006; Van de Kerkhof, 2006). In addition, we presume that simply doing a joint analysis might help develop mutually supportive ideas and desires for future developments. If a system analysis is conducted collectively it may stimulate reflection on the relation between practice and systemic features found in various participants in an innovation network and thus create chances for double-loop learning and coherent structural changes.

We decided that it would, therefore, be interesting to find out to what extent a system analysis can be effectively carried out by the actors in the innovation networks and/or the project teams themselves. The question whether it would help them to critically examine structures and underlying values instead of taking them for granted, was of particular interest. In their evaluation of Innovation Network programmes dealing with sustainability issues, Grin and Van Staveren (2007) suggest using combinations of scientific-technical system analyses for which, amongst other things, ecological expertise is required, for an analysis in which actors' experiences and perceptions are reconstructed. The latter, which requires social scientific capacities, does not necessarily require a collective approach, though it may appear to do so. It can be done by a collective (the project team or broader network) or by an individual social scientist. That means that in the field of system analysis based on a reconstruction and comparison of the perceptions of actors involved, a distinction can be made between expert-led and collective analyses. In this chapter we will explore in more depth the degree of interaction between actors in an innovation network and researchers in social scientific system analyses[8].

[8] With the term 'social scientific' we want to emphasise that social scientific ways of data collection and analysis (interpretation) are involved. Not that they can only be conducted by social scientists.

We experimented with the use of the IS framework in different ways. In the first period the researchers analysed the system based on interviews with the actors involved and their perception of barriers and opportunities without giving them feedback on these analyses. In this period the methodology suggested in an article by Klein Woolthuis *et al.* (2005a) was used: a researcher interviewed project leaders and participants, then interpreted the interview results in terms of the framework. However, questions arose about the quality of the system analysis in relation to the perceived barriers which were mentioned in the interviews, because of the select nature of the group of interviewees.

The second and third period provided an opportunity to test our expectation that a reflection on the relation between practices and systemic features among various participants in an innovation network could create chances for double-loop learning and coherent structural changes. In the second period the analysis also began with interviews and analysis by the researcher, but ended with a collective reflection in the project teams. In the third period the project teams and the researcher conducted the analyses jointly.

At the end, the experiences with the use of the Innovation System framework were linked and scrutinised in light of the following questions about the consequences of the differences in interactivity:

1. What are the qualitative differences between collective and expert system analyses? More specifically, are the collective system analyses less in-depth than the expert analyses if the same analytical tool (the Innovation System framework) has been used?
2. What are advantages and disadvantages of collective and expert system analyses for further learning and innovation?

8.3 Analysing systemic barriers and opportunities

In this paragraph some perspectives on analyses that shed light on the barriers to and opportunities for sustainable development in a certain sector are discussed in view of their contribution to innovation networks. All approaches face similar challenges:

1. How to reduce complexity in order to make analysis possible and the outcome comprehensible.
2. How to deal with scientific uncertainty.
3. How to deal with social ambiguity.

Without suggesting that our list is complete, we briefly review the following approaches: integrated assessment; participatory backcasting; functions of innovation systems and the innovation system framework.

Integrated Sustainability Assessments (ISA or IA) are a scientific way of modelling cause and effect relationships stemming from complex system thinking in response to mono-disciplinary ways of explaining the causes of sustainability problems. Well-known models like IMAGE and TARGETS were developed at RIVM (National Institute for Public Health and Environment) and ICIS (International Centre for Integrated assessment and Sustainable development, Maastricht University). They are:

> '*...an interdisciplinary and participatory process of combining, interpreting and communicating knowledge from diverse scientific disciplines to allow a better understanding of complex phenomena. IA has two main characteristics: (1) it should provide added value compared to insights derived from research within a single discipline, and (2) it should offer decision makers useful information*' (Rotmans and De Vries, 1997: 12).

At the core of an integrated assessment is the notion of 'sustainability' in the economic, ecological and socio-cultural sense. The main concepts used in the models are stocks and flows. The stocks are the reservoirs that society draws from: economic stocks such as natural resources, machinery and knowledge, socio-cultural stocks like demography, health and quality of living and ecological ones like biodiversity. The amount, quality and function are the characteristics that define the status of a stock. A flow is the rate at which a stock is depleted. Economic stocks are for instance the resources for the income and production flows. The models are, therefore, intended to provide a coherent framework for assessing trade-offs between social, economic, institutional and ecological determinants and impacts. In Figure 8.3 an integrated analysis of the overall production and consumption system is shown as an example. Although it is not really representative and does not illustrate the possible value of a good integrated assessment, this one has been chosen because it was made for one of the cases we will discuss. We will explain the reason for this later.

An ISA helps to define the interrelationships between the status and values of the stocks for a specific domain or system in their historical context and is meant to explore future developments as well. It is assumed that the status of the stocks may be uncertain and ideas on their quality may diverge. According to the scholars involved, the problems surrounding sustainability issues are so complex that insights from a multitude of disciplines and stakeholders are needed. Moreover, successful environmental policy is dependent on the involvement and actions of heterogeneous actors. Since the often controversial character of sustainability issues is recognised, a pluralistic approach is suggested implying that an IA model should comprise of a *set* of perspectives with estimations on selected salient uncertainties instead of a single perspective (Rotmans and Van Asselt, 2001). Several interactive forms of IA emerged in which it was suggested to complement the expert assessments on complex environmental issues with participatory sessions involving citizens or stakeholders on the matter (Van der Sluijs, 2002; Rotmans, 2006). In one of the first participatory assessments (named ULYSSES), procedures were developed allowing interfaces between expert models of environmental change on one hand, and lay participants in focus group discussions on the other (Jaeger and Kasemir, 1999). Information from the expert analyses was brought into the focus groups. Using focus groups in the IA increased the opportunities for citizens to articulate their views on environmental issues (Van der Sluis, 2002). Recently, however, one of the founding fathers of the IA stated that '...the evolutionary transition of decision-support tools to user-friendly, exploratory, participatory, heuristic tools has not succeeded' (Rotmans, 2006: 44).

Stakeholders' involvement in participatory integrated assessment is mainly to provide a broader picture for the decision makers, policy makers, by enhancing the quality of the IA:

> *'Participation may help to bridge the gap between the way the problem is defined by members of the scientific community and the daily experiences and practices of the actors who have to contribute to the solution'* (Van de Kerkhof, 2001: 4).

In this way a participatory integrated assessment is not seen as part of an innovation process within and around a network itself, but the assessment is taken as a starting point which should provide further information for decision makers. It is a valuable tool to unlock ecological and economic expertise for stakeholders, as well as simultaneously revealing the stakeholders' perspectives that are relevant for public decision making. It is, however, too complicated for our purposes when conducting collective system analyses, since an Integrated Assessment requires a considerable degree of expert knowledge on the specific domain and modelling in general.

Instead of it being a mere supplement to an expert analysis, interaction with stakeholders is the core of the *participatory backcasting* methodology (Van de Kerkhof, 2006; Quist and Vergragt, 2006). Backcasting has been developed to break trends as a response to approaches that are designed to predict the future or explore possible future scenarios, because those approaches seem to simply prolong existing trends. In this methodology, desirable future scenarios are formulated in interaction with stakeholders, such as future images of Dutch society with 80% less greenhouse gas emissions (compared to 1990), or views on sustainable households. By focusing on a long term future, participants are expected to be able to transcend their usual implicit assumptions. The participation of stakeholders is seen as important not only because of their context specific knowledge, but also their role in achieving approval for results and realising the proposed action agenda and follow up.

The aims central to participatory backcasting are more or less the same as those of (participatory) integrated assessment, which is to explore future pathways that might be desirable as well as feasible, and to support public decision-making. The analyses are a small part of a wide-ranging methodology in participatory backcasting which begins with clarifying the desires for the future and diverging values of actors involved and then explores what (governmental) interventions are needed to realise the future vision, working backwards to the present situation[9]. These analyses of cultural, institutional and other changes that are needed to realise the desired future scenario give insight into the obstacles and opportunities along the way. Quist and Vergragt (2006) suggest conducting three scenario assessments: economic, ecological and consumer acceptance. These should be done by experts, who interview the stakeholders, amongst others.

Although the overall process is participatory, the analytical parts, which are our main interest, are less so. In addition the focus is not necessarily placed on systemic barriers and their interrelations. This is in contrast to what we envisage the function of a collective system analysis to be; that is, a support tool for a group of stakeholders, who are taking the initiative to contribute to a sustainable development, by helping them design collective plans of action and to redirect their planned activities if needed in the light of systemic barriers.

Other relevant frameworks for a system analysis can be found in literature on innovation systems. Many innovation scientists argue that knowledge and learning are at the heart of modern economies. Lundvall (1992) for instance speaks about a 'learning economy', that is, an economy in which the pace of the creation and destruction of knowledge has become very fast. The core notion here is that (national) innovation systems may either stimulate or slow down processes of learning and innovation. Such a system is 'constituted by elements and relationships which interact in the production, diffusion and use of new and economically useful knowledge' (Lundvall, 1992: 2). These are, for example, the relationship between producers and consumers; the institutional set-up within firms, between firms and in policy-making. These elements of the innovation system can reinforce each other in promoting processes of learning, or they can block such processes because of system imperfections. This is in sharp contrast to the view that sees institutions as merely hindering innovation and the rate of institutional change as lagging behind other – e.g. technical – changes. Here, we would like to mention two different ways of doing system analyses that build further upon the concept of a system of innovation: the system functions approach and the innovation system framework.

The *'functions of innovation systems'* (in short: system functions) approach has been developed as a response to the fact that classical studies on national innovation systems do not sufficiently explain how and why technological innovations occur because they focus on the current structure with typical indicators like R&D efforts, patents and patent applications. The system functions approach purports to map the dynamics within a system as well. In order to reduce complexity (as in national systems) the chosen unit for analysis is a technological innovation system (TIS). This allows for the opportunity to map the dynamics in the system

[9] According to Quist and Vergragt (2006) important steps are: (1) strategic problem orientation, including setting normative assumptions and goals, and achieving agreement on normative assumptions among stakeholders; (2) construction of sustainable future visions or scenarios; (3) backcasting, exploring the difference between the future visions and existing situation to see what is needed to come to the desired future; (4) elaboration, analysis and defining follow-up and action agenda; (5) embedding of results and generating follow up and implementation.

in addition to looking at its structural elements. System functions are considered the key processes that influence the development, diffusion and implementation of (new) technology. One set of functions that has been determined is the following: (1) entrepreneurial activities; (2) knowledge development; (3) knowledge diffusion through networks; (4) guidance of the search; (5) market information; (6) resources mobilisation; and (7) creation of legitimacy/ counteracting resistance to change (Hekkert *et al.*, 2007; Negro, 2007).

These functions are supposed to interact with and reinforce one another causing virtuous cycles (propelling a technology to a higher level of diffusion) or vicious cycles that hinder higher levels of diffusion. The analysis of these functions in their historical setting is used to determine, in retrospect, the weaknesses and strengths of the emerging innovation system around a new technology and the dynamics in this system. It is performed by the researcher(s) and the results are verified by other experts in the field. See Table 8.1 for the kinds of methods and indicators used for the analysis in, for instance, Negro's PhD study (2007). It is thus entirely expert-led, leaving no room for social ambiguity. It is a model for looking at the key processes that influence the development, diffusion and implementation of (new) technology. The final results are used to determine an optimal strategy for innovation. This preferred methodology has two main disadvantages. The first is that it takes up a lot of time and expertise in the specific technological-scientific domain. The numerous innovation initiatives in Dutch agriculture can't be expected to invest the large sums needed for this. The methodology does not seem to allow the involvement of 'non-experts' either. A second disadvantage may be that this approach focuses on systemic features of the system around a new technology and, because of that, systemic features are overlooked that have evolved around incumbent technologies in the prevailing system(s).

Although adjusting the system functions approach to include a collective element comparable to the one found in integrated assessment may be deemed feasible, for our own work we decided to put our faith in the *innovation system framework*.

We have set out to use the Innovation System Framework (IS Framework) in our case studies, in our attempt to conduct a collective system analysis. As we have chosen not to use the existing collective approaches – participatory integrated assessments and participatory backcasting – our choice will need some further explanation. The IS framework is grounded in a thorough overview of the systemic factors hindering innovation, known in innovation science as 'system failures'. In our view barriers to innovation are strongly related to barriers to sustainable development, although the problem definition in these innovation scientific approaches is different from ours. Many innovation systems studies have an improved economic performance as their goal but, for our purposes, the goal is development in a more sustainable direction. We do however endorse the other concepts underlying the IS approach. First of all innovation does not take place in isolation but is embedded in a context of formal and informal institutions and a market structure, that form the 'rules of the game' which reduce uncertainty for the actors involved in a dominant system (Edquist, 1997). These rules

Table 8.1. The method and indicators for the analysis of systems functions (Hekkert et al., 2007).

Function	Method and indicators
Entrepreneurial activities	This function of turning knowledge into concrete action is analysed by mapping the number of new entrants, the number of diversification activities by incumbent actors (current regime), and the number of experiments with the new technology.
Knowledge development	Three typical indicators to map this function over time are: 1) R&D projects, 2) patents and 3) investments in R&D.
Knowledge diffusion through networks	The exchange of information is analysed by mapping the number of workshops and conferences devoted to a specific technology topic, and by mapping the network size and intensity over time.
Guidance of the search	This function (a body taking the lead and setting the parameters) can be analysed by mapping specific targets set by governments or industries regarding the use of technology and by mapping the number of articles in professional journals that raise expectations about new technological development.
Market information	This function can be analysed by mapping the number of niche markets that have been introduced, specific tax regimes for new technologies, and new environmental standards that increase opportunities to implement new environmental technologies.
Resources mobilisation	It is advisable to map the function of mobilising financial and human resources by means of interviews to see if actors perceive access to sufficient resources as problematic.
Creation of legitimacy	This function can be analysed by mapping the rise and growth of interest groups and their lobbying activities.

are shaped by actors, even though they may be perceived as structures by other actors. In evolutionary processes variety is generated, selections are made across that variety, and there is feedback from the selection process to the creation of variation. The creation of novelties is necessary to maintain the diversity that makes selection possible (Nelson, 1993). This process of novelty creation is the result of constant interaction among heterogeneous actors in a population, whereby cooperation and interactive learning are regarded as important processes (Lundvall, 1992).

In all of the above situations, system imperfections may occur that block learning and innovation by actors while slowing down the innovation system as a whole. Klein Woolthuis *et al.* (2005a) summarise these imperfections or 'failures' into four basic categories:

1. *Infrastructural failures*: the physical infrastructure, such as railways and telecommunication systems.
2. *Institutional failures*: hard norms (laws, regulation) and soft norms (values, implicit rules of the game).
3. *Interaction failures:* a too strong or too weak interaction in networks.
4. *Capability failures*: like entrepreneurship and labour qualifications.

The authors believe that the framework is useful for analysing the causes of lock-ins that hinder innovation and, because of that, it can help to evaluate interventions, since interventions need to address these system failures in one way another. They also think that it may help to focus intervention designs. In our view, another main strength of the framework might be that it not only looks at systemic features but also the actors that cause and reproduce these barriers in their daily practices. In this way the system analysis offers a good perspective for action. By determining which actors are involved in the main perceived barriers or windows of opportunity, it may lead to actions to either involve them in the innovation network or to try to influence them as outsiders. For these reasons the IS framework seems relevant for the monitoring and evaluation of results of innovation projects as well as for designing interventions. We also expect that it could be especially appropriate to integrate a system analysis into an innovation network's activities with the aim of contributing to a collective learning process, because:

- It may motivate people to learn and stimulate them to change if a match is sought between their perception of barriers and of opportunities. In this sense this concept of 'a match' can be seen as a substantive equivalent of the general 'sense of urgency' which is seen as an important condition for people to learn and change (Leeuwis, 2004).
- It can be carried out without natural scientific experts or modelling expertise and thus by all kinds of relevant actors.

Finally, it might provide a good basis for a more collective system analysis, grounding the analysis in interviews with participants. The scholars who developed the framework propose to interview policy makers, project leaders and participants about the major barriers they perceive with regard to their industry and the system in which they operate and whether the project addresses the right issues and all relevant parties. The interview results can be triangulated with all project documentation. The final result is supposed to be a good analysis.

> *'We see that the framework proves to be a valuable instrument for determining where exactly the bottlenecks lie and how they are addressed (which actors and failures)'* (Klein Woolthuis *et al.*, 2005a: 618).

Although the analysis leans heavily on the reconstruction of experiences and perceptions of actors, the factors of social ambiguity and scientific uncertainty are not considered. For that reason we decided to give the IS framework an additional twist, by using it as an instrument to stimulate a collective process of interpreting situations and developments. We have made another modification to the original IS framework by adding 'market failures' to the list of barriers, because the market structure greatly influences the practices of actors and their access to novelties.[10] The infrastructural barriers have been broadened to include not only physical ones but the 'soft' knowledge infrastructure as well, because a dominant knowledge infrastructure such as the former triptych of research, extension and education in Dutch agriculture could form a hindrance to the creation of novelties (Van der Ploeg, 1999). 'Capabilities' were dropped as a category because these are related to single actors and have no systemic characteristics in themselves. The resulting Innovation System framework used in the analyses is shown in Figure 8.1.[11]

The most explicit criticism of the Innovation System Framework comes from Bergek and others (2008: 409) and deals with the static character of the framework:

> *'... all the four types of system failures identified by Klein Woolthuis et al. (2005) in their recent synthesis and re-categorisation of previous system failure literature are related to structural components: infrastructural failures (related to actors and artefacts), institutional failures (related to institutions), interaction failures (related to networks) and capabilities failures (related to actors). However, it is difficult, if not impossible, to evaluate the 'goodness' or 'badness' of a particular structural element or combination of elements without referring to its effects on the innovation process.'*

	Consumers	Producers	Policy makers	Knowledge institutions	Interest organisations
Infrastructure - knowledge					
Infrastructure - physical					
Institutions - hard					
Institutions - soft					
Interaction					
Market structure					

Figure 8.1. Applied IS framework.

[10] The concept of an innovation system was primarily developed to complement theories on market failures which hinder innovation. Leaving it out of the IS framework might, therefore, prove to be a mistake by the authors.

[11] Some other minor changes have been made: the type of actors have all been reformulated in terms of their social function and the system failures in the first column have been reformulated as systemic features that could prove to be a system failure or provide a window of opportunity.

This criticism is however, not entirely justified, because the IS framework does include dynamics in the context of actors (re)producing the systemic barriers or changing them via their practices. Moreover, for small scale initiatives it is quite legitimate to consider features that are perceived by participants in small innovation networks as inflexible institutional arrangements or structural settings.

It is precisely by reflecting on the actors that reproduce systemic barriers that we think innovators may be stimulated to question these 'given' conditions and start seeing them as changeable. By this kind of system learning, a project may help actors to challenge and redefine the very structures that hinder their aspirations for more sustainable practices; that is to regard the relationships between the structures in which they operate and their own practices in a new light (Loeber *et al.*, 2007).

To stimulate dynamics, however, a collective analysis done with the IS framework should encourage participants not just to consider barriers but also windows of opportunity in the institutionalised context of an innovation project. That is why it was decided to reformulate the barriers in the first column of the table in the original work into systemic features that may form either a barrier or an opportunity or may not be important at all in a specific case (see Figure 8.1). Taking all this into account, the authors of this chapter present the abovementioned system functions approach as an alternative that does take account of dynamics. In fact, we were less concerned about the possible static nature of the IS framework than it's fragmenting effect, since all factors and actors are put into different cells of a matrix, while the very idea of system thinking is to look at interrelations within the system and emergent properties.

8.4 The use of the Innovation System framework in different cases

8.4.1 Experience with expert-led system analyses

Our first experiences with system analyses relate to two cases, i.e. two of the programmes initiated by NIDO (the Dutch Initiative for Sustainable Development) which was a public and privately funded collaborative programme aimed at stimulating leaps in system innovation. The core of the empirical research consisted of in-depth interviews with almost all participants within one project as part of the selected programmes (11 interviews). A second round of interviews was conducted with actors who participated indirectly, in this case the sounding board committee of the programmes that functioned at a national level (an additional 9 interviews). The two main topics of the interviews were learning and system imperfections. Stakeholders were also asked in the interviews to identify what they thought to be the key problems in the system that the project dealt with. Afterwards, the researchers typified these problems in terms of system imperfections and projected them into the IS framework. On the basis of relevant documentation, such as programme proposals and evaluation reports they also typified the NIDO interventions and the innovation networks' activities and compared them with the (system) barriers as experienced by the participants (Klein Woolthuis *et al.*, 2005b).

The overall goal of the first programme, *Value of water,* was to give an impulse to the development of sustainable water management in cities. This was considered to be of great importance because of the excessive usage of drinking water for functions other than drinking; diffuse contamination of surface water; and contaminated sewerage sludge. Furthermore, water resources' undervalued potential, such as its recreational and ecological value, were to become more prominent, visible and capitalised upon. Although these barriers in the local urban water management system were expressed separately, they were all apparently linked to each other and to the major cause underlying the environmental problems as described in the programme plan: the rigid division of responsibilities for water management in the Netherlands between different parties.

NIDO's activities in this programme adequately addressed the system imperfections as seen both by the project participants in Zaandam-Oost and by the programme manager. Parties who usually manage separate parts of the water chain were stimulated to cooperate by jointly performing a sustainability study and a pilot project. The stakeholders were confronted with (and became actively engaged in dealing with) many system imperfections, such as the longevity of the unsustainable physical infrastructure, the costs involved in the water chain's management and maintenance, as well as the dominant, prevailing thinking about water which hadn't integrated the ecological aspects.

The second case study was the NIDO-programme *Market opportunities for sustainable products* (2001-2003). Its aim was to transform existing niche markets for sustainable products into mainstream markets. This was to be achieved by increasing producers' understanding of marketing and by educating consumers to distinguish between sustainable and less sustainable products. The programme was conceived by a group of experts who acted on the assumption that a large number of producers and consumers were dissatisfied with existing production modes and consumption patterns. In the most important project[12], the choice was made to bring together different green niche company marketers, assuming that the learning effects of the programme would then be spread in different fields of application. NIDO deliberately avoided networks in which heterogeneous parties were already interdependent in an organic way or might become so, as is the case within a branch, an industry or a market[13].

In order to do an integrated assessment of this programme, several variants had been suggested, ranging from an analysis of the complete production-consumption system, to the system surrounding one of the specific products in the programme; to the marketing system. When the first one was done very roughly, it revealed which small part of the complete production-consumption system the programme was aimed at (see Figure 8.2).

[12] The 'Companies for companies' project.

[13] Marketing itself was considered to be a (cross sectoral) system with its own dynamics and explicit or implicit rules, but only the actors who NIDO considered to hold key functions in this system were brought together.

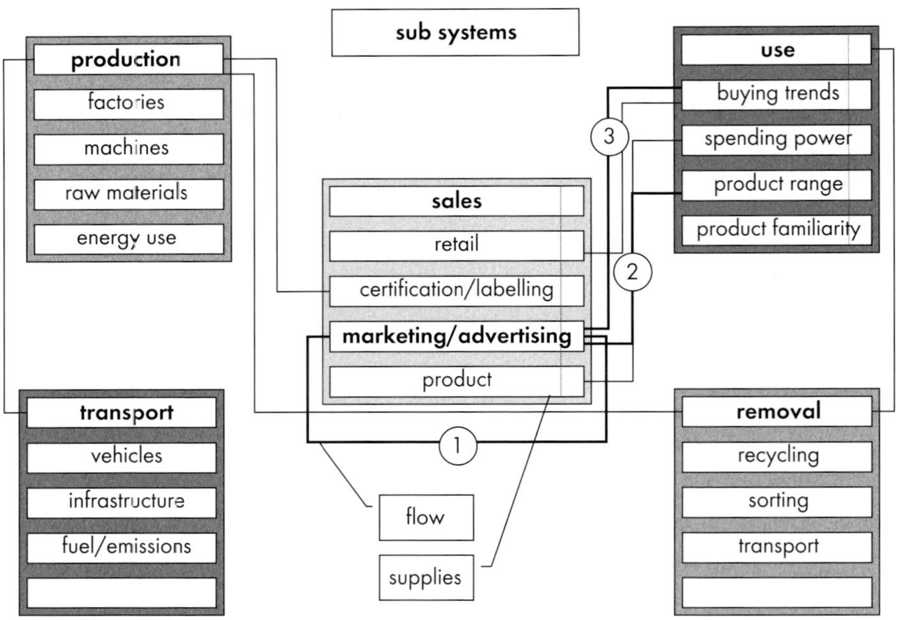

Figure 8.2. Rough integrated assessment of the production-consumption system for consumer goods (adapted from Rotmans et al., 2004). Squares show 'stocks'; thin lines indicate flows in the system; thick lines indicate flows which the programme wishes to influence; circles: the three projects.

In general, the problems that were addressed were not those mentioned by participants themselves. The perceptions regarding barriers and opportunities were very diverse, such as a lack of knowledge about potential customers and (niche) markets in the future; consumers' lack of willingness to pay more for sustainable products; and the unclear definition of 'sustainability'. The programme's activities focused instead on a lack of professionalism in the marketing of green products by niche players and took the form of education on marketing for the green companies, by a marketing expert.

It appeared that second order learning occurred in the case where deliberate interventions focused on system imperfections as identified by a project's participants. This didn't happen in the project in which the activities did not target the system imperfections as perceived by the participants. However, even if there is a match between perceived imperfections and interventions, an innovation programme manager may not wish to 'settle for' just resolving the imperfections mentioned by a project's participants. This approach might set the standard for the envisaged innovations or transformations of the system at an undesirably low level and the current, relatively stable set of social arrangements and structures may remain unchallenged.

It was not possible to investigate whether system learning could be enhanced by reflection on the system analysis resulting from the interviews since the results were fed back to the programme managers only after the projects were over.

8.4.2 Experience with feedback from system analyses to project teams

In the next phase feedback on the results of a system analysis was given to project teams in order to stimulate both the learning and innovation processes. In order to illustrate the effects of the feedback, we will draw on our experience with two cases that were part of the project *Farming with a future*. This project seeks to play an important role in a transition towards sustainable crop production through integrated pest management and sustainable manuring practices in many crop sectors in the Netherlands (see Vogelezang *et al.*, 2009, this volume: Chapter 6). As explained in the introduction, the project management wanted to monitor the networks' activities. Together with *Farming with a future's* management, two networks were selected to be monitored: the fruit network in Zeeland and the strawberry network in West-Brabant. Both these networks were supported by a team of *Farming with a future* network managers, a plant scientist and an extensionist. A system analysis and a collective reflection on the systems in which these networks operated were part of the monitoring design (Van Mierlo *et al.*, 2007).

The core of the fruit network consisted of a study group of fruit growers and a retailer in pest management who discussed and tested best practices. A broader group with representatives from other retailers/consultants on pest management, the interest group for growers in the south of the Netherlands (ZLTO), and a Water Board came together with the study group less frequently. The aims of the fruit network in Zeeland were to apply integrated pest management (IPM) in general and to disseminate best practices in IPM among growers in particular. The network managers did not talk about the nationally agreed goal in the covenant sustainable crop protection (to reduce the environmental impact of chemical pesticides on surface water with 95% in 2010 compared to the year 1998), because they were afraid it would cause resistance from within the study group.

The strawberry network in West-Brabant of growers and several other stakeholders aimed to stimulate sustainable manuring practices in this sector in which pest control is hardly an issue but manure management all the more so. In practice, however, the focus was not only on sustainable manuring practices (e.g. by testing ways to compost) but also on contesting the new Dutch norms for the use of manure. The growers that were dependent on manure were convinced that these norms were too rigid to assure a high strawberry production in the long run. For that reason, a test to examine this assumption was set up together with the other actors in the strawberry network. Although the ministry of agriculture had meanwhile increased the usage norm for nitrogen (N) from 90 kilo per acre to 170 kg per acre as a result of lobbying by some growers from the Agricultural Producers Board, it was hoped that this test would legitimise a further increase.

All members of the two networks were interviewed by the researcher to explore their practices and perceptions of barriers to sustainable pest management/manuring in their sector. Moreover, a number of relative outsiders were interviewed (a representative from the auction, a national representative from the Agricultural Producer's Board technical department [LTO-Groeiservice] and Zeeland's Environmental Federation) to get a broader view on the institutional barriers. The selection of these external actors was based on a network analysis via a snowball technique asking informants which actors influence growers in their pest control and manuring choices and application practices.

Barriers and opportunities that were mentioned by more than three interviewees, as interpreted by the researcher, were entered into the innovation system framework. The framework was presented to the project teams in a workshop to check the analyses and to stimulate the team members to reflect on the systemic features of the sector they worked in. In both cases, since the conclusions of the analyses were more or less acknowledged by the project teams, the second step quickly followed. This involved a discussion on the analysis of the match between perceived barriers and the project activities, which was also prepared and presented by the researcher. Although this analysis was also accepted, the project teams responded defensively to some of the 'gaps' shown, i.e. barriers that were not addressed by project activities. Although it was explained that not all system barriers could or should be addressed and that the model was merely meant as a tool to reflect on the relationship between the focus of the project and systemic barriers, this hardly seemed to reassure the team members.

The quality of these system analyses can be supposed to be better than those in the NIDO projects. One reason for this is that a network analysis was done in the initial stages of this project to get a complete list of actors that might have had an influence on the barriers to and opportunities for sustainable development in the specific sector, while for NIDO only the participants were interviewed.

In other respects the system analyses were very much alike: in both the NIDO and the *Farming with a future* analyses, the interviews were screened by the researchers for comments that were interpreted to fit a certain cell within the matrix. The validity check in both situations was that more than one interviewee should have mentioned an item. Moreover, the analyses were all member checked with the network managers. However, we were not quite satisfied with the system analyses because they had a fragmented pattern and it was impossible to determine the relative importance of the barriers or their interrelationships.

The network managers said that they had obtained valuable new ideas from the feedback sessions on the system analyses but they did not stimulate a reflection on the planned activities, which was the intention. Although the auction (the Greenery) was seen to be an important actor, that should be involved in the fruit network, most of the network managers felt insecure about this new way of working. Indirectly, the system analyses had some influence on the overall project management, because they demonstrated the large gap between the projects' ambitions and the networks' activities, which still focussed on testing and trying to diffuse best practices instead of creating novel dynamics in heterogeneous networks. The results of the analyses in combination with other stimuli gently guided the project managers towards the new role that they envisaged and made them decide to select some networks to be spearheads for applying the new approach.

8.4.3 Experience with collective system analysis

One of these spearhead cases provided the opportunity to conduct a collective system analysis with the project team in the second phase of monitoring in the project *Farming with a Future*. The initiative for the sub-project arose out of concerns about the emissions of chemical pesticides into surface water by greenhouse cultivation (neo-nicotinoïdes such as Imidacloprid). In 2005 a RIZA report (The National Institute for Water Management and Waste Water Management) showed that although the norms for 2010 for the use of these agrochemicals had already been met, as set down in the Covenant Sustainable Crop Protection, the emissions were still too high (RIZA, 2005). Norms for usage per acre proved to be insufficient for achieving the water quality norms. The sense of urgency among important stakeholders became more pressing when the plan for regional expansion of greenhouse cultivation in Southern Holland was quashed because the greenhouse sector was not able to guarantee an improvement of the water quality. In addition to other initiatives, such as research on emission routes by the chemical company, Bayer, the management of *Farming with a future* decided to build a network with the parties involved to design a collective solution. A dilemma for the management was that it had promoted Imidacloprid because it was suitable for integrated crop protection well. The focus was therefore, to reduce the emission of this and other agrochemicals into surface water and postpone the search for non-chemical alternatives to a later date. The envisaged approach was to run a pilot project in a greenhouse area to reduce or eliminate emissions in cooperation with relevant actors like the water board, chemical manufacturers and the Agricultural Producers Board (Van Mierlo *et al.*, 2009).

As this greenhouse cultivation project was still being set up, the system analyses could be integrated nicely into the design of the network interventions. In actual fact, a system analysis was conducted twice, facilitated by the researcher in cooperation with the project team consisting of six members in total: four people who coordinated growers' study groups, a process manager and a market expert. The first one was thoroughly prepared: the members of the project group interviewed all relevant stakeholders on their definition of the problems and ideas about solutions. Then each team member was asked to read one or two of these interviews prior to the meeting in which the analysis was conducted. During the meeting itself the interview reports were analysed for the (systemic) barriers and windows of opportunities. The team was dissatisfied with the final results of this exercise. One reason was the poor quality of these initial interviews, because in this early contact between team member and stakeholder, priority was given to building trust. According to the team members themselves, another reason was that their conceptual understanding of 'system failures' was inadequate to carry out the analysis. This was confirmed by the researcher who judged that the selection of items from the interview reports had a strong technical-breeding bias.

It was decided that it would be worthwhile to do another analysis, but this time it was prepared differently. All team members read some comprehensible literature on the system barriers in the Dutch agricultural system. The final collective analysis facilitated by the researcher consisted of the following steps:

1. All team members were asked to write down what they saw as the main barriers to more sustainable crop protection in greenhouse cultivation, as well as the main windows of opportunity in the dominant, prevailing system; each barrier and opportunity on a separate post-it. They were asked to draw from their knowledge, experience and contacts in the field, including the interviews conducted a few months earlier.
2. One by one the team members presented their perceived barriers and opportunities by sticking their post-its in one of the fields of the IS framework and explaining it to the others.
3. The others asked critical questions about the character of the barriers mentioned, to analyse whether it was a system barrier or merely a symptom of a problem, in this way a selection was made.
4. In addition the facilitator-researcher asked the team members to consider the underlying causes of the barriers mentioned. It was stated, for example, that the growers were unwilling to try growing without emissions. By analysing the underlying causes it was concluded that there were neither incentives nor disincentives, to change growing practices. In this step, a selection of barriers was made, and the definition of the barriers changed.
5. In the next step the empty cells in the framework were considered to see whether the analysis was complete. It transpired that the systemic feature 'hard institutions' was neglected by mistake. In the final analysis this factor was considered to be one of the main system barriers in greenhouse cultivation: the inability to achieve the norms set down in the Waste Water Disposal/Greenhouse Horticulture Act [*het lozingenbesluit glastuinbouw*]. Since emissions into surface water are measured at only a few places, they cannot be traced back to a single farmer.
6. At that point the team decided collectively that the analysis of system barriers and opportunities was quite complete.
7. Finally, the pilot project was measured up against the IS framework as well to see whether there was a match between its approach and the system analysis. Two system barriers had been missed out; one of which the team wanted to address in their future work. In this way the functioning of the market became relevant to the project, along with the opportunity represented by the auction, because of demands for food without pesticide residues. This step also stimulated the participants to think about how their approach could become institutionalised.

Figure 8.3 shows the final result of the system analysis (The match with the approach is left out for reasons of clarity). It demonstrates how the team was able to cluster particular

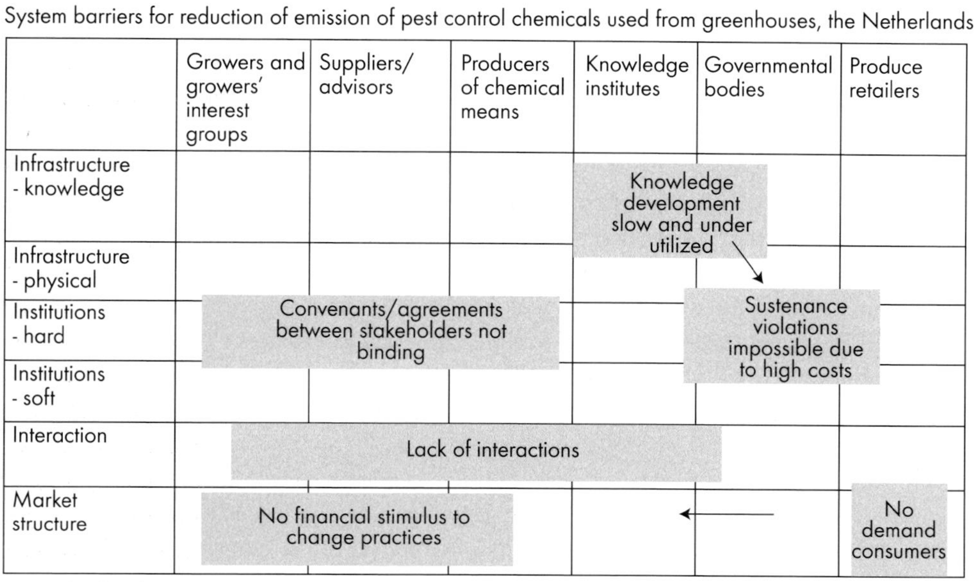

Figure 8.3. System barriers to and opportunities for reduction of emissions in greenhouse cultivation.

barriers and to reflect on relations between the barriers. An important advantage of the collective approach compared with feeding back the results of the analyses in the former phase was that the project team was very much involved with the content of the analysis. The step-by-step methodology helped make a selection out of all the items that came up in the first step, which made the analysis more in-depth. According to all participants, the meeting generated a lot of energy, when the new insight into the problems of maintaining the norms emerged and also when, nearing the end, it became clear that the regional pilot project's envisaged approach would address many of the perceived systemic barriers. It was encouraging, because it supported the ideas that had been developed (seemingly) intuitively thus far. The idea was to start building the network as it was envisaged before the meeting took place, supplemented with representatives from the auction and supermarkets, in line with the outcomes of the analysis of influential actors.

In addition to these positive direct results, the system analysis had some influence on further developments at the time of writing. The opportunity represented by the auction's market requirements and supermarkets has been further investigated. The auction and supermarkets were approached to join the network, but were not interested in cooperating[14]. Other adaptations to the envisaged approach did not seem necessary as indicated by the analysis of the match between the approach on the one hand and the perceived system barriers and opportunities on the other.

[14] Personal information project leader, Beerlings, May 26th 2008.

8.5 Conclusion

Although their contexts and the internal processes are completely different, the cases discussed show some similarities. The aim for all the innovation projects mentioned was to build a small network with a diversity of actors. The project teams' aims were quite ambitious in the sense that they addressed institutional barriers and the envisaged solutions needed the cooperation of and changed practices by several actors. This ambition was supposed to be served by the system analysis. In all cases the system analyses were based purely on the project team members' existing knowledge, that of the monitor/researcher and at times other actors in the innovation network in an indirect way (via interviews). Because none of the analyses was complemented by joint scientific-technical research, they may seem 'quick and dirty' to professional system analysts.

We would argue, however, that certain conclusions can be drawn: (1) in all cases the system analyses showed clearly whether the systemic barriers as perceived by (a number of) the participants matched the interventions for innovation; (2) the feedback on these perceived barriers given in two cases stimulated a reflection on the interventions in the project team, although not leading to adaptation of the interventions; and (3) in the single collective analysis, a discussion was stimulated on the character of the barriers mentioned in the greenhouse cultivation project. It can be stated that the analysis became more in-depth because of this discussion and systemic features were no longer considered as a 'given'. Moreover, it confirmed the direction chosen by the project team to be a fruitful one and suggested some slight adjustments.

The changes in the way the framework was used as a tool for analysis has not been just a matter of variation in the degree of interaction, but implies a drastic switch in methodology and its effects on the innovation process. In the first two periods it was an instrument to come to an inter-subjective idea of systemic barriers and opportunities mentioned in the interviews that were conducted by the researcher. The quality of the analyses in these periods depended, among other things, on the network analysis that provided the basis for the selection of the interviewees. In the last period it was used in a dialogue as a tool for scrutinising prevailing ideas about barriers to sustainable development and for considering the links between barriers and opportunities. This happened in a step-by-step, cyclical approach in which the definition of the barriers changed and became more in-depth because the interrelations were also considered. In one of these steps the framework was used to look for 'completeness' of the analysis by considering all kinds of systemic elements in the matrix one by one. The approach provided the project team with a sense of ownership of the analysis and its outcomes.

In this final, most collective analysis, the members of the project team were asked to draw on their experience and knowledge of the field and the role of the actors in the network. In contrast to the other cases, the team members did not need social-scientific competences they were neither educated in nor interested in to conduct interviews and read and analyse interview reports. It did however require them to reconsider and reformulate their initial ideas about systemic features and to question each other's perceptions in a critical but non-threatening manner. In this way system learning was stimulated.

As yet, we have not been able to experience the true potential of this instrument for collective system learning and challenging system barriers to sustainable development. To improve the quality of the system analysis as well as enhancing the opportunity for system learning, useful modifications to an analysis with the project team might include:

1. collective analyses with all participants once a network has been set up;
2. additional interviews with stakeholders who are not part of the project team or network that conducted the collective system analysis.

Initially, the IS framework seemed to have two limitations in comparison with other approaches of system analysis. Firstly, it is not in itself an instrument that clarifies the interrelations between system barriers. In fact, all factors and actors are fragmented when trying to translate parts of interviews into the cells of the matrix. In the last case however, the cyclical way of conducting the system analysis stimulated the project team to consider the connections between the barriers. Secondly, because it is not possible to project trends and developments into the matrix, it can be considered a static model. It would be interesting to seek combinations with approaches such as participatory scenario development in which virtuous and vicious feedback loops between all kinds of relevant developments and trends are projected in a figure (see e.g. Burt, 2007). In this way the apparent strengths of the *collective use of* the IS framework can be exploited, which is based in the experiences and perceptions of actors involved and the stimulus it provides to seek completeness by considering all types of system features and all kinds of actors involved, while its weakness can also be dealt with.

We contend that the IS framework may be useful as an instrument to stimulate system learning as it did in the case of greenhouse cultivation. It is obvious that no general conclusions on the use of the framework and its possible influence on ongoing innovation processes can be based on a single case, but we think it is sufficiently promising to justify continued experimentation with it in the directions suggested above.

Acknowledgements

We wish to thank Cees Leeuwis, Communication and Innovation Studies Wageningen UR, for his constructive feedback on the chapter and Frank Wijnands, Farming with a Future for his cooperation and feedback on the methodology.

References

Bergek, A., Jacobson, S., Carlsson, B., Lindmark, S. and Rickne, A., 2008. Analyzing the functional dynamics of technological innovation systems: A scheme of analysis. Research Policy 37: 407-429.

Burt, G., 2007. Why are we surprised at surprises? Integrating disruption theory and system analysis with the scenario methodology to help identify disruptions and discontinuities. Technological Forecasting and Social Change 74: 731-749.

Edquist, C. (Ed.), 1997. Systems of innovation, technologies, institutions and organisations. Pinter: London, UK.

Grin, J. and Van Staveren, A., 2007. Werken aan systeeminnovaties. Lessen uit de praktijk van InnovatieNetwerk. Van Gorcum: Assen, the Netherlands.

Hekkert, M.P., Suurs, R.A.A., Negro, S.O., Kuhlmann, S. and Smits, R.E.H.M., 2007. Functions of innovation systems: A new approach for analysing technological change. Technological Forecasting & Social Change 74: 413-432.

Jaeger, C.C. and Kasemir, B., 1999. Urban lifestyles, sustainability, and integrated environmental assessment. Final report. Darmstadt University of technology; Darmstadt, Germany.

Klein Woolthuis, R., Lankhuizen, M. and Gilsen, V., 2005a. A system failure framework for innovation policy design. Technovation 25(6): 609-619.

Klein Woolthuis, R., Van Mierlo, B., Leeuwis, C. and Smits, R., 2005b. Tussen actor en systeem. Een theoretische en empirische verkenning van leerprocessen en de rol van NIDO als systeeminstrument. Wageningen University, Communication and Innovation Studies: Wageningen, the Netherlands.

Leeuwis, C. [with contributions by A. Van den Ban], 2004. Communication for rural innovation. Rethinking agricultural extension. Blackwell Science: Oxford, UK.

Leeuwis, C., Smits, R., Grin, J., Klerkx, L., Van Mierlo, B. and Kuipers, A., 2006. Equivocations of the post privatization dynamics in agricultural innovation systems. In: Smits, R.E.H.M. (Ed.) The design of an innovation-enhancing environment. Transforum: Zoetermeer, the Netherlands, pp. 2-58.

Loeber, A., Van Mierlo, B., Grin, J. and Leeuwis, C., 2007. The practical value of theory: conceptualising learning in the pursuit of a sustainable development. In: Wals, A. (ed.) Social learning towards a sustainable world. Wageningen Academic Publishers: Wageningen, the Netherlands, pp. 83-98.

Lundvall, B.A. (Ed.), 1992. National systems of innovation. Towards a theory of innovation and interactive learning. Pinter: London, UK/New York, USA.

Negro, S., 2007. Dynamics of technological innovation systems. The case of biomass energy. PhD Thesis. Netherlands Geographical Studies, Copernicus Institute for Sustainable development and innovation: Utrecht, the Netherlands.

Nelson. R.R., 1993. National innovation systems: a comparative study. Oxford University Press: Oxford, UK.

Quist, J., and Vergragt, P., 2006. Past and future of backcasting: the shift to stakeholder participation and a proposal for a methodological framework. Futures 38: 1027-1045.

Rotmans, J., 2006. Tools for integrated sustainability assessment: a two-track approach. The Integrated Assessment Journal 6(4): 35-57.

Rotmans, J. and De Vries, B., 1997. Perspectives on global change. The TARGETS approach. University Press: Cambridge, UK.

Rotmans, J. and Van Asselt, M.B.A., 2001. Uncertainty management in integrated assessment modelling: Towards a pluralistic approach. Environmental Monitoring and Assessment 69 (2): 101-130.

Rotmans, J., Van der Brugge, R., Loorbach, D. and Dijk M., 2004. Zoeken naar duurzame systeeminnovaties. Het NIDO in transitieperspectief. ICIS: Maastricht, the Netherlands.

RIZA, 2005. Emissies van gewasbeschermingsmiddelen in de glastuinbouw [Emissions of crop protection chemicals used in Greenhouse cultivation]. RIZA: Lelystad, the Netherlands.

Van de Kerkhof, M., 2001. A survey on the methodology of participatory integrated assessment. Interim Report IR-01-014. International institute for applied systems analysis: Austria.

Van de Kerkhof, M., 2006. A dialogue approach to enhance learning for sustainability – A Dutch experiment with two participatory methods in the field of climate change. The Integrated Assessment Journal, Bridging Sciences and Policy 6(4): 7-34.

Van der Ploeg, J.D., 1999. De virtuele boer. Van Gorcum: Assen, the Netherlands.

Van der Sluijs, J.P., 2002. Integrated assessment. In: Munn, R.E. and Tolba, M.I (Eds.) Encyclopaedia of global environmental change, Volume 4, Responding to global environmental change. Wiley: London/ Chichester, UK, pp. 250-253.

Van Mierlo, B., Arkesteijn, M. and Leeuwis, C., 2007. Reflexieve procesmonitoring & telen met toekomst. De eerste fase. Praktijkonderzoek Plant en Omgeving: Lelystad, the Netherlands.

Van Mierlo, B., Arkesteijn, M., Elzen, B. and Leeuwis, C., 2009. Leren over reflexieve procesmonitoring. Lessen uit de tweede ontwikkelingsfase van een methodologie voor ondersteuning van projecten gericht op systeeminnovatie. WUR/ CIS: Wageningen, the Netherlands.

Vogelezang, J., Wals, A., Van Mierlo, B. and Wijnands, F., 2009. Learning networks in Dutch agriculture. In: Poppe, K.J, Termeer, C. and Slingerland, M. (Eds.) Transitions towards sustainable agriculture and food chains in peri-urban areas. Wageningen Academic Publishers: Wageningen, the Netherlands, pp. Chapter 6 in this Volume.

Chapter 9

Culture, innovation and governance in Europe: systems theories and the analysis of innovation in INTERREG programs[15]

Roel During, Kristof Van Assche and André van der Zande

Abstract

The relations between culture, innovation and governance are very complex. In this article we will present a theoretical approach for the analysis of these relations in their context within governance organisations in the European Union. The article supports the assumption that culture is an important factor when various communities (professional, institutional, regional, disciplinary) interact in EU-stimulated governance initiatives. These diverse cultural identities within the various groups have a large part to play in the rejection or invention of new repertoires of action. Little is known about the effect of European governance on the variety of local or regional governance practices, embedded in various national, institutional, professional cultures. Does EU-stimulated exchange between regions restrict or even eliminate repertoire, or does it favour innovation and increase variety? This crucial issue needs thorough further study. We will explore the complexity of European governance, using system theories and accountability within a social constructivist paradigm to explicate the interaction between actors, project-communities and societal structures on different policy and regional levels. Empirical data will be retrieved from the INTERREG practice, because this EU framework of programmes is based on both horizontal (between policy levels on the same level of hierarchy) and vertical policy interaction (along the axis of hierarchy between EC and local policy) with the explicit aim of stimulating regional development and innovation. As a nested case study (case within a case) we will use two INTERREG projects showing mechanisms of both closure and openness. One case deals with a project involved in the management of historical fortifications. This case study is an example of closed institutional cultures. We briefly touch on the second case study which shows cognitive openness in a network. Conclusions will be drawn regarding the role of culture in governance innovation, and the steps that are deemed necessary to improve our knowledge in this field.

Keywords: INTERREG, governance, innovation, culture, cognitive openness

[15] INTERREG is a Community Initiative which aims to stimulate interregional cooperation in the EU. It is financed by the European Regional Development Fund (ERDF). This INTERREG initiative is designed to strengthen economic and social cohesion throughout the EU, by fostering the balanced development of the continent through cross-border, transnational and interregional cooperation. Special emphasis has been placed on integrating remote regions and those which share external borders with the new Member States.

9.1 Introduction

9.1.1 A case of cooperation between regional actors: innovation or culture clash?

In several places in Europe different authorities are responsible for the restoration and redevelopment of historical forts and fortification zones. They all face similar problems. The European Union's INTERREG framework brought three of these authorities together in a project to exchange experiences and to formulate a European spatial development strategy for forts and fortification zones. The three partners were the cities of Utrecht (the Netherlands) and Mortsel (Belgium) and the Essex County Council (United Kingdom). The project was judged to be successful by the partners as they were more or less able to achieve some of their individual goals. However, they did not manage to take full advantage of the knowledge, skills and capacities that each of them brought to the project.

Regional cooperation is considered to make a significant contribution to a united Europe without internal borders. Within the INTERREG framework hundreds programmes and thousands of projects are executed causing a dynamic governance practice, involving all European regions. However, the success of many of these projects has been compromised by cultural misunderstandings and participating partners who were more preoccupied with their own situation. In this sense, the fact that the three partners in our case study failed to maximise the potential of the fortifications project was not unusual.

A project with the acronym CULTPLAN was set up to account for cultural differences in this broad and diverse European interregional practice and analyse their mechanisms and manifestations. It has been shown that cultural differences can be very intense and serious misunderstandings may occur within the context of cooperation. In some cases these misunderstandings became insurmountable obstacles and in others they were reframed to become a source of inspiration. These differences and their effects cannot be explained by simply characterising the partners in terms of cooperation and project features. CULTPLAN revealed that best practices occurred as a result of a co-production of innovative knowledge and practices that capitalised the mutual understanding between project partners representing different cultures. This requires a high level of (cognitive) openness by regional partners, which is something that is rarely witnessed. This raises the questions as to why some interactions become productive and others do not.

9.1.2 The questions at hand

Since the Lisbon Agenda, the European Union has put considerable emphasis on innovation. Almost all current programmes and projects are supposed to underwrite this objective. The Lisbon Agenda focuses primarily on economic growth in a free market situation without reference to cultural anchors and cultural boundaries. But this policy attitude can possibly lead to a loss of identity and decrease in cultural variety. Previous EU measures to stimulate governance seem to have focussed mainly on strengthening civil society, see e.g. the White Paper on European Governance (European Commission, 2001). The ultimate goal of those initiatives was to acquire a certain measure of control over a more democratic yet potentially unruly civil society, rather than innovation. Since the addition of the sustainability amendment initiated by the Gothenburg Agenda there is a widespread acknowledgement that culture is an important prerequisite or feature of innovations (World Bank, 1999; CULTPLAN, 2007; Loulanski, 2006). This 'innovation-by-culture' argument is clearly the rationale behind the Cultural Strategy of the EU (European Commission, 2007), and behind the shift from cohesion policy to innovation policy in the EU in programmes like INTERREG. This can easily be traced in recent publications by UNESCO (e.g. UNESCO, 1998) and OECD (2005).

This shift in focus, which attributes more significance to culture within innovation, raises fundamental questions. What exactly is culture and what is innovation? How can the relation between culture and innovation be described? Should we try to understand innovation by analysis within a specified cultural perspective? Or does innovation inherently mean that the cultural boundaries are relaxed so that new strategies and practices emerge? What is the effect of the cultural differences between regional actors? Can they be made productive?

In our analysis of an INTERREG project, we will try to answer those questions by using a combination of two social system theories. The social system theory proposed by Niklas Luhmann (1984) has been used earlier for governance studies (Schaap, 1997; Van Assche and Verschraegen, 2008; Van Twist, 1994). Itamar Even-Zohar's (1990) polysystem theory accounts for the origination of a new cultural political repertoire in social systems. Both theories facilitate the study of projects in a complex and contingent context, involving a network of relations between partners in a project, between projects, between projects and the governance system and the local governance practice.

By combining both systems theories in our analysis we believe this can prove fruitful in addressing the complexity of the INTERREG governance context. We will attempt to contribute to a better understanding of the relation between culture (in governance) and innovation in the next sections by analysing the role of culture in one European INTERREG project, using systems theoretical notions about culture, identity, governance and change. After comparing the theoretical considerations to INTERREG practice, we will formulate a number of conclusions concerning the potential of EU projects and policies to stimulate innovation in governance, bearing in mind the role that culture plays, and assessing the potential of systems theories for this study.

9.2 Description of INTERREG

The INTERREG Community Initiative, which was adopted in 1990, was intended to prepare border regions for a European Community without internal frontiers. INTERREG II ran from 1994–1999. INTERREG III was a EU initiative operating from 2000 to 2006 which aimed to stimulate interregional cooperation within Europe. It was financed under the European Regional Development Fund (ERDF). INTERREG III aimed to promote economic and social cohesion, a balanced and sustainable development of European territory and to foster territorial integration with candidate members and other neighbouring countries. Special emphasis was placed on integrating remote regions and those which shared external borders with the candidate countries. The idea was that national borders should not be a barrier to the balanced development and integration of the European territory as a whole. This phase of the INTERREG initiative was designed to strengthen economic and social cohesion throughout the EU, by fostering a balanced development of the continent through cross-border, transnational and interregional cooperation. As such, it aimed at innovation in governance, and had to deal with a variety of cultural differences.

The EU submitted a set of guidelines and principles as an implementation framework. In this framework INTERREG III was separated into three strands: A, B and C. Strands A and B were designed to provide concrete solutions for concrete problems in regions (Helanders, 2007). Strand A was aimed at local development of social facilities, economy and environment in cross border cooperation, Strand B aimed at trans-national spatial development strategies, linking cities and resource management in trans-national cooperation. Strand C was designed to focus on the exchange of experience in developing networks and to identify best practices in a large European Region (group of Member States). Altogether the strands aimed at improving the cooperation structures, leading to more and higher quality of joint projects and creating synergy between the exchange of best practices and the work in the mainstream structural funds programmes (EFRD). There was a specific programme structure within strands A and B. Strand A was the most decentralised and was composed of 64 programmes (INTERACT, 2005). Strand B contained 13 programmes, but strand C had no programme structure at all. The imposed management structure involved a Steering Committee, a Management Authority and a Payments Authority. Nevertheless, each strand had specific governance structures. The Steering Committee was decentralised and decisions on formulating, granting and executing projects were taken at a regional political level within strand A. Strand C was highly centralised, because the Management Authority was organised into four secretariats around Europe covering the one programme. Decisions concerning programme development and the selection of projects in strand B and C were taken by state representatives of the countries involved. Political responsibility was placed at ministerial level. Each INTERRREG project contained an international arena of actors cooperating in a context composed of European ideology and regulations and local or regional ambitions.

Due to the variety in management practices and political involvement a very complex and dynamic governance practice developed within INTERREG. In INTERREG III more than five thousand projects have been executed. There was cooperation between regions across Europe, between countries, regions and cities across Europe; there was competition and cooperation between projects, networks of partners and institutions. Cooperation mostly lead to networks of related institutions that developed mutual interests and exchanged and shared knowledge. Competition between projects partly took shape in a codification of practices. Almost every project promised to make an inventory of good and best practices. Being part of a best practice is an important prerequisite for new projects to succeed and to receive funding. There were successful teams that extended their cooperation from project to project and become really proficient insiders within the INTERREG community.

Identification and selection of good and best practices served as a vehicle for the EU to stimulate innovation in governance. Partner search meetings were organised by the EU and institutions used them to develop networks and search for new opportunities. Opinion leaders disseminated information about success stories that had lead to a huge return on investments or innovations, and this contributed to the codification of 'best practices'. One can therefore say that an INTERREG community developed, marked by shared assumptions, leading to shared interpretations of good and bad governance, of desirable innovation. Rhetorically at least, one shared assumption was a certain perspective on the EU. The INTERREG actors were supposed to share a common view on the development of Europe and this could be designated as a European cultural perspective. At least they had to express this view for their application to succeed. Whether this was an expression of a clever understanding of politically correct language or the applicant's heartfelt conviction isn't clear.

Participants in INTERREG represented a wide variety of interests and as a result projects had a wide range of strategic aims. Some projects tried to influence the European policy agenda and anchor their national problem solving methods in a new European approach. Other projects aimed at reinforcing or creating the identity of a region and at the same time tried to boost the regional economy. For some countries and institutions INTERREG served as a good platform to get acquainted with Europe, and this held especially true for the new accession countries. There was a wide range of strategic reasons for institutions and actors to get involved in INTERREG but even more important is the fact that there was a great variety in local and regional governance practices that intervened in this huge project fabric (Gablenz *et al.*, 2005). The local project practices and strategies were embedded in a broad variety of cultures (Swidler, 1986; Sanyal, 2005).

Potentially relevant cultures here are national, regional, local, political, institutional, disciplinary, ethnic and maybe other sorts of cultures. Every project bore the mark of it's own cultural mix, and only detailed empirical investigation will be able to identify the various ingredients. In our brief exposé on systems theories, we will deepen the understanding of 'culture' in the context of governance, and in our case study we will focus on one specific mix of cultural influences in a project. At this point in the reasoning, a simple representation of the relation between culture and governance is given in Figure 9.1.

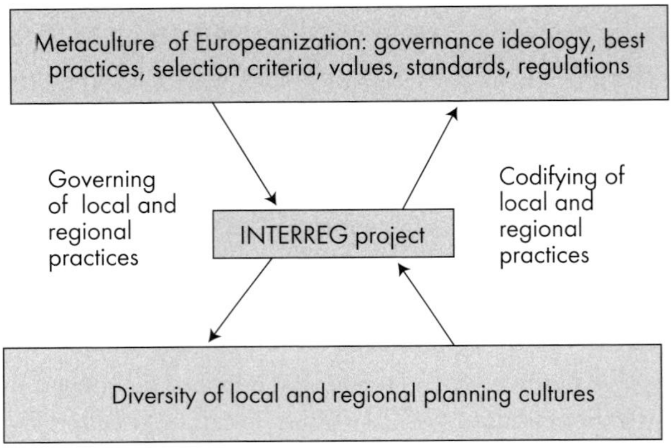

Figure 9.1. Governance perspective on interplay metacultures and local/regional planning cultures.

In DG REGIO[16], the organizational culture supports the idea of Europeanisation. An important feature of this culture is the idea that poor regions should be helped by wealthy regions. Prosperity should become more evenly distributed as a consequence of European programmes. Some indications have been found that this notion of Europeanisation has a cultural bias because of the dominance of western European planning traditions (During, 2008). A long list of success stories has been put on the DG REGIO's website in order to substantiate the idea that doing an INTERREG project is a piece of cake and that participating in a project sets a region apart as innovative. The INTERACT institute in Vienna that is affiliated to INTERREG published examples of good and best practices on their website (www.interact-eu.net), and the selection was based on the absence of administrative problems. This emphasis on administration is not an unusual one. The templates and formats for submitting INTERREG projects are primarily administration oriented. The content of a project only seems to be relevant at the moment of selection, but afterwards accountability for success is restricted to purely administrative criteria. This process is magnified because the application forms are so detailed that the result is built-in inflexibility. In this way, the whole idea of cultural diversity can easily be overwhelmed by the strong arm of bureaucracy!

9.3 CULTPLAN – a study of INTERREG cultures

CULTPLAN, an explorative study addressing the mechanisms of culture in INTERREG practice showed a complex picture (CULTPLAN, 2007). In CULTPLAN a number of projects have been analysed in depth, one of which, called *Crossing the lines*, will be discussed in this article, in relation to the influence of diverse cultures on governance and governance innovation. A second case called *FLAPP* will be briefly addressed because it shows different mechanisms

[16] Directorate General for Regional Policy of the European Commission.

of closure and openness. Before we enter the analysis of these cases and the systems theories used, it might be useful to present a few key findings of the broader CULTPLAN project.

9.3.1 Key findings

CULTPLAN found that cultural mechanisms affect the execution of programmes and projects in an ambivalent way. Actors referred both positively and negatively to cultural influences in hindsight, when a project was (almost) finished. Problems that came up time and again, related to a lack of understanding and lack of administrative synchronisation between regional partners. Positive aspects were found to be the opportunities to exchange practical knowledge and achieve a better understanding of the various ways actors from other parts of Europe succeed in organising the societal and political commitment to invest in regional development and to solve problems. The INTERREG practice showed that culture affected the cooperation between actors invisibly and unconsciously. These characteristics of culture are described by Gullestrup (2007). In the end, cross-cultural cooperation leads to a better recognition of cultural peculiarities in a partnership.

A deeper understanding of the role of culture has been achieved in CULTPLAN by analysing the cooperation and non-cooperation of actors in the projects. This in-depth analysis revealed some relevant aspects of the relation between culture and innovation. The capacity to innovate seems to be related to the ability to look at one's own cultural context through the eyes of the other partners, to reflect on one's own preconceptions and to have an open mind towards other cultural cognitive frames. In practice a copy and paste approach which just imitates other ideas and practices, does not lead to success. As mentioned before, we will explore the relation between innovation, governance and culture in more detail by discussing one of the CULTPLAN cases. But before doing so, we will explore the relation between culture and innovation in governance from a systems theoretical perspective, in order to get a better understanding of the determinants for success and failure.

9.4 Theoretical approach of culture and innovation in governance

9.4.1 Governance and government

Policy analysis in general is moving towards decentralised models, from government to governance (Bevir, 2003). While different concepts of governance abound, the overall understanding is that centralised steering mechanisms rarely work, and that governmental and non-governmental actors at several levels need to be taken into account in this analysis (Flyvbjerg, 2001; Arts *et al.*, 2000). Consequently, there is also a focus on multilevel governance (Gualini, 2004; Hooghe, 1996; Scharpf, 2000; Bache and Flinders, 2004) and a recognition of the importance of various sorts of networks (Termeer and Koppejan, 1997; Klijn and Teisman, 1992). These recent angles on policy studies facilitate a better and more detailed understanding of interactions, steering mechanisms and policy-formation in civil society.

More recent governance theories recognise that practices don't necessarily have to be seen as the result of deliberate governing. Sometimes practices occur spontaneously and governance has to deal with them (for an extensive description of this theme see Chapter 17 by Duineveld *et al.*, 2009 in this volume). One can observe a shift in policy analysis from concepts like 'preference' and 'institution' to new and more discursive and cultural concepts, where regulatory mechanisms are is considered to be the contingent products of diverse actions and political struggles informed by the beliefs of agents as they arise in the context of traditions (Sabatier and Henkins-Smith, 1988; Williams, 2004; Bang, 2004). These authors go beyond the traditional horizon of politics and administration, by focussing on the rapidly growing interest in empowering lay people such as consumers, customers and the public in general. In Newman's (2001) wake, Bang states that 'a new great narrative for connecting governance and direct democracy is taking shape outside the arena's of political parties and organised interests in representative democracy' (Bang, 2004: 158). New 'governmentalities'(Foucault) emerge, facilitating processes of self- and co-governance.

The functioning of INTERREG projects needs to be studied against the background of this broad move from government to governance (Hajer and Wagenaar, 2003; Pierre and Peters, 2000), and the projects themselves form the environments where policy networks can be reinforced or created.

9.4.2 Systems theory: Even-Zohar

The situation of practices originating in a governance context suffused with various cultures, fits concepts of repertoire and repertoire innovation used in the polysystem theory (Even-Zohar, 1990). Itamar Even-Zohar is commonly regarded as the father of the polysystem theory. The foundations of this theory can be traced back to Russian Formalism in the 1920's and later functionalism. It is related to Lotman's (1978) semiotic and cultural theories and involves the idea that semiotic phenomena should be studied as systems of sign-governed human patterns of communication rather than as conglomerates of disparate elements. The term polysystem is just an terminological convention addressing a system as dynamic, open and heterogeneous as opposed to the synchronistic approach to a system, which is seen as a closed set of relations. It involves the concept of a system centre and, surrounding the centre, a network of relations that are described by their processes and procedures. There are power relations between the centre of a polysystem and elements in systems of relations in the periphery, but there is no real control. Both Lotman (1978) and Even-Zohar hypothesise dynamism in the centre of an open system, resulting from ossification, adaptation and innovation.

In the polysystem theory the notion of 'system' is dynamic and heterogeneous. Systems are not uniform, they are stratified in a polysystem and a constant synchronic and diachronic interaction or struggle takes place between (semiotic) practices. Even-Zohar distinguishes canonised and non-canonised strata in the polysystem. Canonised strata can be understood as those (semiotic) practices which are accepted as legitimate by the dominant circles within a culture and preserved by the community to become part of its heritage. These practices may concern specific action models in a governance context.

According to this theory, the influence of 'culture' is something that is found in all levels and cannot be exclusively attributed to any one level of the polysystem and it determines to some degree which practices are to be canonised. The vitality of canonised practices depends on the constant dynamic tension with the non-canonised practices that are trying to replace them. This tension leads to adjustments and evolution of the canonised practices and it can be seen as a maintenance mechanism of the polysystem. As a rule, the centre of the whole polysystem can be defined as the most prestigious canonised repertoire. The concept, repertoire, is the aggregation of rules, principles, structures and actions (models) that govern practices. It is the group of actors (the elite) which governs the polysystem that ultimately determines the canonicity of a (cultural) model that evolves out of a certain repertoire. The focus of the elite may lie on simple-mindedness and conformism or on sophistication and eccentricity. Even-Zohar makes a distinction between static and dynamic canonicity. He sees dynamic canonicity as involving principles that can be applied by others and static as involving a finalised product which then serves as a standard. It is the dynamic canonicity that generates the canon and is contested by inferior practices.

Once canonicity has been established, the governing group either adheres to the properties that it has canonised or, if necessary, alters the repertoire of canonised properties in order to maintain control. The procedure of transfer between canonised and non-canonised repertoire governs the issue of conservatism and innovativeness. Referring to practices in literature, Even-Zohar explains that the ability to adopt the principles and models in the centre of the polysystem is vital to maintain one's position in the canon. If a writer sticks to his model then he will lose his position in spite of the quality of his previous writings. Understanding the transfer presupposes understanding the juxtaposition of primary and secondary elements, meaning innovativeness versus conservativeness in the repertoire. When a repertoire is established and all derivative models pertaining to it are constructed in full compliance with it, the repertoire can be designated as conservative. On the other hand, the formation and restructuring of a repertoire that includes new, less predictable elements can be seen as an innovative repertoire. When new primary models are transferred to the centre of the polysystem as a result of canonisation, they will, in due time, become secondary. At this point they may contribute to conservatism. The process of change from innovation to conservativeness is caused by simplification. A primary model is not necessarily more sophisticated than a secondary one but, as time goes by, a process of reduction takes place. Various kinds of ambiguity are reduced and complex relationships replaced by less complex. This process is unavoidable, as the more complex models, intimidating, demanding and loaded with information become more familiar and so on.

Procedures can be the precondition for transfer and/or they can be transformed as a result of the transfer. If the procedure is a precondition and canonised repertoire has the character of an action model, described as a set of detailed instructions, then conservativeness arises in the systems. Other action models in the repertoire may be constructed in full compliance with the canonised one and this will also lead to conservatism. In the polysystem theory, diversity of repertoire is as vital as dynamism in the polysystem centre in determining and governing the strata. Diversity is both a precondition for innovation and a result of an open culture where groups in power may be replaced or they may adopt new repertoires by using non-fixed procedures.

Polysystem theory can be applied in the analysis of INTERREG. INTERREG itself can be seen as the polysystem level, the strands and programmes as subsystems and the projects as systems. There is competition between local practices within a project trying to become best practices and trying to be part of a best practice guide in almost every project reporting opportunity. But the situation in INTERREG is far more complicated. No participant wants to qualify another partner's practice as bad because nobody wants to be associated with a bad practice, even though determining bad practices would be far more effective for their learning process. Normally every partner selects a best practice in his own governance context and all these best practices are put in the report. These practices are codified by extracting general principles to be adopted by other projects and actors. In addition, there is limited competition between projects to become best practices and the group in charge of transferring between non-canonised and good or even best practices can be found at the level of INTERREG programmes and within networks covering specific themes. There is in fact strong competition between regions and networks of cooperating institutes, like in the projects *Sharing solutions on structural funds* and *Adriatic action plan 2020* in the C-strand. The networks of institutes claim to establish a better strategic position than the regions can achieve individually, but the networks operate partly beyond political control. At the top level of the DG REGIO in Brussels no best practices are qualified, only success stories. Three or four levels of competition can be discerned; the project, the programme, the network and the EU. It can be observed that the power relations in INTERREG seem to be very diffuse and this results in a rather symbolic competition between projects. In this competition there seem to be no material arguments. This situation of content-oriented projects and administration-oriented governance is difficult to grasp as a single polysystem that produces its own canon. The criteria for success change in the vertical axis of the governance system.

A diversity of repertoire is present on project level because all partners represent their own local practices of governance. Innovations can sprout from both the interregional cooperation and the conflict between partners because they are unfamiliar with each other's governance practices and understanding of economic development in this process of Europeanisation. Even-Zohar (2007) sees complexity as a functional prerequisite for innovation at project level. Complexity leads to heterogeneity in repertoire and this is the substratum for innovations. According to the polysystem theory the innovativeness can be enhanced by the competition between projects in a situation of non-fixed procedures for transfer between the canonised and non-canonised repertoire. It is the culture in the centre of the polysystem that rules the scope of diversity that is tolerated and thereby the chances for innovation. In research practice, in the later analysis of our case study, the challenge will be to find a realistic delineation of the centre, the demarcation of canonical and non-canonical elements, and to determine the differences between fixed and non-fixed procedures. The issue is complicated by the fact that most fixed procedures are not entirely fixed, most canonical practices are not completely canonical and the centre is rarely one place or organisation. We will keep this in mind when we try to trace the mechanisms Even-Zohar describes within in our case study.

9.4.3 System theory: Luhmann

Niklas Luhmann's (1995) social systems theory gradually emerged in the course of the 80's and 90's, combining functionalist sociology, evolutionary biology, logics of form, post-structuralism and a reinterpreted phenomenology. A social system in his theory does not consist of people or actions, but of communications. People are called psychic systems and exist within the social systems environment. People and social systems are structurally coupled as the product of co-evolution, and they use the same medium, which is meaning. Luhmann's main categories in social systems are: (1) interactions (conversations), (2) organisations and (3) function systems (law, economy, art).

Society is the encompassing social system. People participate in various interactions, organisations and function systems; organisations participate in several function systems, and lodge many conversations. Systems have each other as environments, and the evolution of each system results from internal dynamics, and from adaptation to changing environments, interpreted through the codes of the system. Every observation of and adaptation to an environment, can only take place using the distinctions, the logic and the procedures of the observing system.

A project, as an organisation, is a social system and project practice necessarily involves many conversations. Projects in an INTERREG context have to deal with politics, economy (e.g. book-keeping), legal constraints and so forth. Decisions, actions or strategy can be seen as specific communications, with decisions driving the self- reproduction of the system.

Luhmann defines complexity to mean that the possible connections between elements in an observed system are too numerous to be grasped, and this, in turn, means that the elements themselves will change over time, making the observation even more imperfect. The elements of social systems, communications, change because social systems are *autopoietic*, which means that they continuously reproduce their own elements and structures using exclusively these same elements and structures. This implies that any environmental influence, any adaptation to environments, can only take place in an autopoietic way, using the elements and structures of the system. This is what Luhmann calls operational closure: interaction with environments is only possible when the environment has only indirect influence on the system. Every environment is interpreted within the system, using the structures and elements (including semantics) available to the system at that point in time. In that sense, social systems are self-referential.

Both Even-Zohar and Luhmann see complexity as having a positive function: environmental complexity functions as a resource for the observing system, and internal complexity will be constructed to deal with the environments that the system deems relevant. A higher degree of internal complexity equals a more refined model of external environments, and a more subtle understanding of possible adaptations. Part of this production of internal 'models' of the environment, is the development of adequate semantics, including words and ideas, to cover a certain aspect of an external reality. It is this level of semantics that was theorised by semiotics, e.g. the Russian structuralists that inspired Even-Zohar. Luhmann sees culture primarily as a matter of semantics, where different social systems might largely share similar semantics. The identity of a social system can consist in a specific semantic in the form and function of images of self and, not unimportantly, in the specific autopoiesis of that system – the product of its history of adaptations to certain environments. All these things form the identity of a social system. Culture also plays it's part in it. In the domain of culture, semantic practices can be discursively constructed and the games of hierarchy and marginality as outlined by Even-Zohar, find their place here.

Luhmann pays less attention to the concept of hierarchy than Even-Zohar, but leaves sufficient room to allow for a combination of both systems theories. In Luhmann's eyes, modern society is basically centre-less, meaning that there is no centre of society from which the rest can be observed and steered. This can be attributed to the move from government to governance theorised elsewhere – to the move from centralised government steering to steering in networks of diverse actors. Society as a whole does not have a centre, since the function systems are all operationally closed, but the principle of hierarchy (basically a remnant of an older organisational principle) is still present in various social systems. Some organisations have a clear centre, and are extremely hierarchical, but this will have to be observed empirically, it can never be assumed. Certain legal systems or political systems are very hierarchical, while others are not. This links up with our observations on Even-Zohar's notion of centre: if there is a centre, it is not necessarily an undivided one and it will need to be traced empirically. This will be a point of interest in the case study.

According to Luhmann, innovation can mean several things; change is taking place all the time otherwise a system would disintegrate very quickly. Some changes lead to structural changes, others do not. Some changes in the system are observed from within the system, others are not. Some are labelled as innovation, others are not. In organisation theory and policy studies, innovations are seen as positive and significant changes, leading to more success one way or another in changing and challenging environments. Product innovation in the economy seems to be the model. Luhmann allows us to paint a more subtle picture of organisational change, in adaptation to environments. In autopoietic systems, analyses of innovation, of success and the associated changing environments are all products of the system itself; the only 'objective' measure of success is the survival of the system, e.g. an organisation.

Social systems theory can assist in achieving a better understanding of the INTERREG governance situation. One can see the INTERREG community as a system, an organisation. External environments are manifold: the EU administration, the administrations of participating countries and regions, regional and national political systems, other EU-networks. Internal environments form the projects, each of them having each other as environments, as well as legal, political, scientific and economic function systems. Projects are organisations, otherwise seen as structured communication. Within the projects, interactions (conversations) take place. As for any other organisation, the basic elements for the reproduction of a project organisation are decisions, a special type of communication that allows an organisation to 'move on' while maintaining a self-image of unity and continuity.

The projects are not produced by the INTERREG system as elements. The INTERREG system functions as an environment in which projects (other systems) can be initiated but, once started, the projects become operationally closed. Luhmann is commonly misunderstood to mean that this entails full closure of the system. Nothing is less true: organisations, as social systems, need continuous observation of and adaptation to environments during their self-reproduction. The point is that everything will be interpreted using the elements and structures of the system, including self-images. (Questions like: Who are we? What is our organisational goal? What are our relevant environments?)

For example, if actors start communicating *outside* INTERREG, the INTERREG system interprets this information with the assistance of the concepts (regulations, information) in use and the response will reflect the way the issue is normally dealt with *within* INTERREG. This self-referentiality can in fact be observed in INTERREG. But it might be more interesting to show how the concept of social system fits various situations. An INTERREG project can be designated as a system of communication and environments can be found in both vertical and horizontal governance interdependencies. Actors often operate at different levels in different systems and bring their experiences from the one into another. It is worth reflecting on how this effects the system. Every actor plays a role in the system and in a part of the systems environment. For instance an actor can be a partner in a project and at the same time advisor in a regulative group and also a representative of a specific region or an associated institute or discipline. In all systems this actor is faced with different interpretative concepts of the outside world. Different forms of self-referential operations may occur and the actor may bring the information from one system as thematic information into another system. As said earlier, culture can be seen as the use of predefined concepts in communications based on common meanings that allow self referential operations; it operates on the level of semantics.

Given all this, innovation is not exclusive to a project; it can occur in any social system: an institution, a particular discipline, a project, etc. Innovations can originate anywhere anytime but the structures and elements of every system, including its culture, shape and limit the capacity for innovation.

9.4.4 Both theories and the understanding of culture and innovation

Both Even-Zohar and Luhmann hypothesise a degree of closure in a system to be necessary for it to be open towards the systems environment. The polysystem theory supposes that the hierarchical stratification in a system is constructed by the centre of the polysystem where power culminates. The idea of a centre of the polysystem and periphery seems very strong with reference to INTERREG governance. But in practice we can also observe that the formation of active networks can be interpreted as hierarchical layering originating from the periphery. So codification and canonisation seems to take place at the interface between bottom-up initiatives and top-down codification. This holds true especially for many projects that do not aim to codify their own practices, but aim to find and establish best practices in the repertoire of the network they participate in. The fact that projects sometimes succeed in influencing the European policy agenda can be seen as a special form of canonisation in accordance with the polysystem theory. These projects manage to alter the structure in the centre of the polysystem. In practice it could happen that projects generate innovations in other, larger systems with a self-description as centre. Conversely, self-described centres, as well as other systems observed to be at the same level, could generate an innovation at project level, an innovation that, in turn, can be observed by the centre and trigger reactions there, etc. *ad infinitum*. This is logical in Luhmann's perspective, since innovation can emerge anywhere anytime. If we limit Even-Zohar's concepts to the domain of semantics (one type of innovation), a contradiction can be avoided and both theories can be combined.

In our case study analysis, we will try to combine the ideas of both Even-Zohar and Luhmann in our investigation into culture and innovation in governance. Innovation can emerge in hierarchies and in centre-less structures and in the case of a centre-periphery structure, it can originate both in the centre and at the margin. Innovation can be a matter of changes in structures, changes in elements, or changes in both. It will only be an innovation when it is labelled as a significant change, in adaptation to a significant environment, with a significant degree of success. Every label of significance in this last sentence, will be applied bearing in mind the distinguishing characteristics of the observing systems, it will also be part of the ongoing autopoiesis and maintain the operational closure.

9.5 A case study: Fort restoration

The complexity of INTERREG may be best illustrated by analysing a specific project to see how culture and innovation relate to or even define one another. To this end we will look more closely at a project called *Crossing the lines*. The role of culture in this project has been analysed in CULTPLAN (2007). It was a fort restoration project with the objective of opening the relevant forts to a wider public. It was not a very complex project compared to other INTERREG projects, because the aims were quite clearly specified and, to a certain extent, the partners should have been able to understand each other's cultures. This project was a cooperative venture between three local or regional authorities in Belgium (Mortsel), the Netherlands (Utrecht) and the UK (county council, Essex). The aim of the project was threefold; to realise a restoration, to facilitate public access and to formulate a European spatial development strategy for forts and fortification zones. In Utrecht, the location of the lead partner, a choice had been made to restore *Fort aan de Klop* situated in the northern part of the city on the river Vecht. The Belgian partner wanted to develop the so-called *Fort 4*. This fort is part of the Brialmont-ring around Antwerp. In Essex the choice had been made to restore *Tilbury Fort* and *Jaywick Martello Tower*. The choice as to which forts would be restored, was made on the basis of local arguments.

In Utrecht, the city council wanted to restore the fort using European funding and they wanted the development to be economically sustainable by adding a camping ground and hotel accommodation to the site. The council in Mortsel also wanted to restore their fort, but they did not want to start restoration straightaway. They chose to develop an intense cultural programme first to ensure that the fort would win a place in the hearts of their citizens. The reason for choosing this strategy, according to the project manager, was because they believed that public appreciation of the value of military and fortification monuments in Flanders was much lower than in the Netherlands or in England. In England the County Council wanted to use European funds to restore Jaywick Tower, thereby giving a much needed boost to the local community. This community, located in an industrial and marginalised part of greater London, is impoverished and facing social problems. By collecting personal stories about the Tower from within the local community, the County Council attempted to generate publicity and at the same time boost public interest in the monument. They developed a museum in the tower, where these stories can be heard, using interactive technology. A second goal for the County was to establish a European network foundation for fortification zones. In their county they saw their role primarily as a spider in the web of the heritage institutions they cooperated with on a regular basis.

In this project great differences could be seen in the governance and working cultures of the partners. The Dutch partner represented a very strong regionally-oriented planning culture. *Fort aan de Klop* is part of the New Dutch Water Defence Line, and this is an immense Dutch Government planning project covering almost the whole eastern part of the Randstad (city conglomeration along part of the west coast of the country). In their institutional culture the ultimate economic sustainability of a project is very important and the institute normally calculates the economic situation ten years ahead. The maintenance of monuments is a municipal task. Serious problems were encountered in the implementation phase, because the residents in the vicinity of the fort didn't want any disturbance in their quiet neighbourhood. They had not been involved properly at the beginning of the project and this resulted in much opposition. One of the original project objectives, to develop a European spatial development strategy for historic defence lines or fortifications, was removed from the agenda one-sidedly by the lead partner. The Dutch partner just didn't see the value of that goal. For the Belgians this was rather disappointing, because they themselves didn't have an integrated approach for the maintenance and restoration of the Brialmont ring with the consequence that all forts were treated as similar entities under generic regulations. There was no specialisation or differentiation between the forts, based on a spatial strategy.

The Belgian partners had a very limited financial mandate; they had to get official approval for every expense incurred that exceeded €50. At the same time they had almost total freedom to do anything they wanted with the fort. They were impressed by the financial management of the Dutch partner and successfully tried to implement some small elements of the Dutch approach in their own organisation. The way the Belgian partner developed a diverse and attractive cultural and social programme at the location of the fort was remarkable. As a result, the fort became quite well known during the execution of the project and many television programmes were recorded at the site. They wanted the fort to resemble a park, because this attracts visitors. The idea of a park was stronger than the idea of a monument, so trees were allowed to grow almost anywhere at the site. The biggest irritation for the project management was caused by the tenants of a small building on the site which had been rented out to clubs with strong local political contacts. It wasn't possible to get rid of them to take full advantage of the location.

The English partner didn't have any mandate to manage the monument site because it was under the jurisdiction of the English Heritage Trust. Ideas to develop a sustainable energy situation with the aid of solar panels could not be realised because of an intervention by the English Heritage Trust. The Trust wouldn't allow solar panels that were visible from the outside, no trees were allowed in or alongside the monument site and they prescribed lawns for the surroundings. Their idea was that the site should become a museum for the benefit of local inhabitants.

There were several instances where the collaboration between the participating partners fell apart. As mentioned before the project aim to establish a European foundation for fortification zones had been put aside by the lead partner, much to the annoyance of the English and Belgian partners. The Dutch abandoned this goal because there was a foundation on the European scene with a very similar goal already, an organisation called Europa Nostra. The English partner has been assigned the task of developing a website and asked for photos of all sites. The English removed the trees from the photos of the Belgian location because they considered them inappropriate (see Figure 9.2). This annoyed the Belgian partner because they valued the park association. The English partner had an interesting strategy to involve the local community. They investigated every aspect of the history of the site, including the oral history. Especially this last historical category is important because it links the monument to the social memory of the community. The English had achieved very good

Figure 9.2. Photo of the monument site in Mortsel.

results using this working method. The English partner offered to do a similar investigation in Utrecht and Mortsel, but the Dutch and the Belgians considered this too sophisticated and rejected the offer.

The project was deemed successful by all partners in the sense that every partner managed in a way to achieve some of his individual goals. But they did not manage to take (full) advantage of the knowledge, skills and capacities of the other partners. It is interesting to see that the problems that one actor encountered, almost perfectly matched the strengths and successful past experiences of a partner from another country. Overcoming one's weaknesses by using the other partners' strengths as a model, would certainly be an innovation of practices. However, in this project and in almost every other project analysed in CULTPLAN, this did not occur. As a rule, the partners were more preoccupied with their own situation and problems and did not have the skills to step back from the cultural cognitive framework that prescribed their modus operandi. This can be illustrated with some statements by the Dutch project leader qualifying his English partner as unprofessional because his network orientation lead to too many overseas visits, high costs and an untidy administration.

9.6 Systems analysis of the case

There was no question of any impulse whatsoever from the European Union itself to set up this project. It was wholly the result of a bottom-up initiative; several organisations decided to work together and formulate a common theme. Originally the project had been submitted by the Dutch city of Den Helder, but the application failed because of a lack of synergy between partners. Then the city of Utrecht, which became involved in rewriting the application, took over the initiative at the request of the other partners. In this respect, it seems to illustrate the broad move from government to governance mentioned earlier. In this case, it is interesting to note the mix of project goals: from modest, to quite ambitious – the development of a European foundation of institutes involved in fortification zones. This last can be qualified as an institutional innovation at the centre.

Another of the project's ambitions had been to disseminate their knowledge regarding heritage tourism and restoration techniques to all European institutions facing similar problems. This can be interpreted as an ambition to establish a network of excellence that sets standards for a codified spatial developmental approach, a new repertoire of thinking and acting. Both ambitions can be seen as intended synchronic cooperation and competition as described in the polysystem theory. The project initially wanted to relate to similar projects, form a network and move to the centre of the INTERREG polysystem by submitting a powerful planning instrument that even might influence the policy agenda of the EU. But local practice, local demands, and internal competition undermined this ambition.

In the end the project didn't pursue the parts of the project that aimed towards moving it into the centre of INTERREG and even the EU. This happened because the lead partner from the Netherlands did not consider this objective to be important. (They had the role of lead partner as a result of a *de facto* hierarchy between the partners in the project.) The Dutch decision was a good example of self-referentiality. In their institutional culture, they only considered a project to be successful when completed in a economically sustainable way, which means that exploitation should be viable for an entrepreneur. As a result, they mainly adopted projects that could be executed practically and financially in their traditional way (element of autopoiesis). Other project orientations were considered unprofessional. These were stigmatised as deviant approaches, not meeting the standards of the lead partner, illustrating the autopoietic character of the organisation.

The only person who seemed to learn from the interaction was the Belgian project manager. Because he was open to change, the project lead to minor changes within his organisation that were inspired by his Dutch partner's good accounting practices. Innovations in the project did not originate as a result of the interaction of the project partners. No new repertoire developed in this case. The conservative relations between the project and its institutional and societal environment acted as a constraint to innovations. The new repertoire builders in the project were too far from the polysystem centre, being both the

INTERREG polysystem (Brussels) and their own national polysystem centres (Member States' capitals and disciplinary science centres).

Relating this back to the main issue of innovation, our case sheds an interesting light on the labelling of changes as innovations. The changes in the broader EU policy environment proposed by two partners in the project, were considered irrelevant, and were not regarded as an innovation by the lead partner. We refer to those innovations initiated by two of the project partners, that transcended the project itself. Presumably, the larger innovation would have promoted the goals of those project partners (English and Belgian) in the long run, as well as the goals they thought to be in the interest of all EU citizens. Within this single project, one of the partners clearly did not share this perspective on the common good, on the responsibilities of the participating organisations, on the project boundaries, on the kind of innovation needed.

One could observe that the only partner where the cooperation led to internal innovation, was an organisation that had a fairly large degree of autonomy in its functioning – despite the modest budget. This Belgian partner's organisation allowed itself to change its procedures, while respecting its own local goals and those of the other partners. This result was made possible because their organisational culture highly valued flexibility and it didn't have rigid accountancy procedures, nor a strong bureaucratic identity.

The Belgian partner understood the increased potential for self-innovation, but they assumed that the relevant boundaries for this were political. Local politics and its clientele, was seen to be the limiting factor for the envisaged innovation. The multifunctional land-use of the project site, conjured up by the Belgian partner, did not directly derive from the project partners, nor was it directly the result of their interaction within the project. The rationale behind this designation was a presupposed lack of appreciation of historical military objects by Mortsel's inhabitants. The Belgian partners interpreted the innovation, to some extent, in light of their assessment of local preferences and this lead them to designate their monumental site as a park. This lead to different recreational options and a different conservation strategy. This interpretation of the innovation stood in sharp contrast to that of the English partner. The English partners were much more focussed on heritage policy as they were, in turn, under scrutiny of the English Heritage Trust, which represented an even more conservative and mono-functional heritage approach. The culture of the English Heritage Trust, and to a lesser extent the English project partner, is imbued with heritage disciplines (art history, conservation, archaeology). This tradition is not shared by the Belgian partner. What they saw as a pragmatic innovation embraced by the local community, leading to an optimal use of the site, was perceived by the English partner as heresy. The self-description of the English partner, led to a semantic that did not allow for the procedures and contents of their Belgian partner; all those things were seen as foreign, alien to the goals supposed to be common. The whole idea of sharing a project goal and working together towards that goal internally and in the field of EU policy, gradually eroded away because of those unspoken differences in basic observational distinctions. Once those differences became clear, (for example, when the English partner manipulated the photos of the Belgian site) what belief there was in a shared goal and what little trust there was between them, evaporated. Luhmann believes that it is an inherent feature of communication that it pre-empts double contingency situations. He means that any communication implies a minimum of trust, of belief that the other will more or less understand what I mean and that his response will be more or less delineated by my horizon of expectations. Social systems allow actors to overcome this situation of double contingency, by narrowing down the field of selections, by structuring the mutual horizons of expectation. Social systems prevent double contingency situations by narrowing down the field of selections and by structuring the mutual horizons of expectation. Once this minimal trust breaks down, communication breaks down and the social system disintegrates. According to Luhmann, many organisations cease to exist as such, long before they are officially buried by the internal and external environments. In the case of the forts project, the drive towards innovation stretched the observational capacities of the partners in several cases to the extreme. The rigid procedural identifications of the Dutch partner undermined the originally shared goal and caused a measure of alienation between the partners and it reduced the vitality of the project to one of mere survival of the organisation. Despite the bottom-up origin of the project, a project culture barely developed and a shared identity did not evolve either.

This is not as tragic as it may seem; one can even say that this case illustrates the risk of innovation in general. Innovation is always limited by the cultures of the participating systems, just as it is fuelled by them. Innovation is inherently risky because it aims at significant change in the autopoiesis of the participating systems. Within INTERREG, and certainly in the case of this Forts project, the episodic and varied nature of the projects functions as a testing ground for innovation. Because the EU funds the project organisation, and precisely by virtue of their temporary nature, risks are more easily taken by the participating organisations. Centralised imposition of 'best practices' attempts to minimise the risks involved, but will necessarily also diminish the drive toward innovation. Pushing people to follow standardised procedures will naturally lead to risk-avoidance behaviour.

9.7 Cognitive openness

The mechanisms of cognitive closure in the *Crossing the lines* example can be found in all INTERREG projects analysed in CULTPLAN. Nevertheless openness towards the unfamiliar repertoire between partners in a partnership still occurs in many projects. A project in the C-strand concerning Flood Awareness (FLAPP) can serve as an interesting example where openness has been observed (CULTPLAN, 2007). In this project 37 partners exchanged practical and theoretical knowledge concerning the best ways to avoid flooding and to deal with the effects of flooding. The project aimed at both transferring knowledge about successful flooding approaches in a network across Europe and influencing the European flooding policy. The idea was to submit a flooding guideline that would be applicable in all European regions. Even though each region had it's specific interest area, the western European countries strove to anchor their flooding policy principles to this guideline. The German regions were primarily focussed on risk management and risk mapping, other regions were in favour of improving the technical means to control water discharges and the Dutch lead partner was focussed on organising more space for rivers. During the project the EU submitted a flooding directive, so the central aim of the project was rendered obsolete. Despite this setback, the project partners continued to organise excursions to exchange practices concerning flood management. Partners from the new member states were very eager to take part in every excursion and event organised. The Hungarian partners explained in an interview that they had spent many years studying literature about flood and water management in West European countries and now they were in a position to go abroad and see for themselves. They travelled with seven colleages by van through Europe to take part in a project excursion in Ireland. The people from the Hungarian institutions that were involved were highly esteemed in their institutional culture because of their language skills and their international experience. One of them admitted to being the first employee ever to work on a 'project', a phenomenon not well understood in Hungary. This meant in practice that some workdays were short and others were long, depending on the amount of work to be done for FLAPP. This interviewee said that, other than her boss, she was the only person that could speak English in an institute numbering 200 employees. It appears that the first step has been taken to include 'projects' within the repertoire of that institution.

The Lithuanian partners, representing local council, declared themselves to be a newcomer to the European democratic spectrum. They were eager to learn about the different governance practices of all the partners in FLAPP. The issue of flooding as such was not of prime interest to them, but it was interesting because it required interaction between government and the general public. In answering a question as to which partners were most interesting for them, they mentioned Ireland and explained this by referring to their involvement with school classes doing water education projects. A governance practice including youngsters was their ideal at that moment. They felt there was a long way to go, because of the long history of mistrust of the Lithuanian government by the people. The issue of flooding was seen as a good opportunity to restore or build confidence. The Lithuanian partner was in search of a new governance repertoire without calling it innovations as such.

The FLAPP case shows a specific situation where competing practices assist each other to achieve their goals, because a better recognition of a specific practice in the EU guideline would lead to a better position for claiming structural funds in the execution of a governmental policy. Within this network several power centres can be observed and a related transfer of repertoire between West and East. Cognitive openness can be seen as the result of a strategy to move towards the centre of the polysystem of the EU. In short, say the right words and you will stand to gain something.

9.8 Conclusion: back to the main questions....

INTERREG can be seen as a network, or as an example of governance which is distinct from government. Top down ideology and regulations meet thousands of bottom-up projects involving even more institutions. The complexity of this governance network leads to questions of enlargement or narrowing down of the variety of local governance practices. Are the regions becoming standardised in their governmental approach to regional development? Does the cultural variety and dynamics in projects lead to institutional innovations? Or do innovations lead to a more standardised culture? The mutual relationship between culture and innovation should be clarified in order to achieve a better understanding of governance processes. Social system theories have been used in an attempt to shed a light on this relationship between culture and innovation. The results are promising but not yet satisfying. On a theoretical level the constraints and mechanisms of innovation seem to lie both on the system level as on the level of localised repertoire. The way practices are codified and the flexibility of the institutions in the centre of the power structures are essential. Too much emphasis on codifying (for example, formulating a set of detailed instructions) seems to be counter productive. Too much emphasis on the codification procedure is also a threat to innovation. The inner circle of INTERREG representing the ideology of Europeanisation is flexible when assessing the governance content of good and best practices, as the criteria differ when referring to accountancy or to regional exposure (CULTPLAN, 2007; Helander, 2007). Accountability and good administrative procedures, however are the more dominant policy goals. This has come about as a result of the issues that lead to the fall of the Santer EU Commission.

On a local level the processing of information by the institutions involved seems to be crucial. The culture of organisations directly and indirectly involved determine the capacity to adopt and develop new repertoire of governance. In practice, the complexity resulting from the international context, the national institutions and the societal implementation context, leads to cognitive closure within institutions. Actors on project level, can only partially understand the variety of environments that are supposed to be relevant, according to the original INTERREG goals. Complexity and risk are reduced by producing concepts of good practices, concepts that tend to become fixed, ossified, losing their adaptive capacity. Evidence has been found that culture determines the capacity for innovation. To put it more strongly, *innovations require (institutional) cultures to open up.* It is important to know the conditions and mechanisms of openness to be able to achieve this. Perhaps it is not cooperation but conflict and frustration that is the real key to this door. This process should be investigated in detail with more case study analysis. It is clear that the actors' institutional contexts in local or regional projects is a relevant factor, but this needs to be investigated further. Understanding the specific influences of various forms of culture in a governance network involves a thorough investigation into cultural drivers and constraints. Cultures either propel projects or hold them back. By analysing the enormous number of projects with this in mind, one should come to a better understanding of cultural dynamics, as shown by

CULTPLAN. This awareness of one's culture can be a first step in the process of opening up and achieving innovations.

Despite themselves, failed projects can also produce systems innovations. A combination of governance theories and social system theories can supply a better understanding of practices like INTERREG where bottom-up ambitions meet top-down ideologies. More empirical investigations are nevertheless needed, to map the variety of governance mechanisms affected by culture, whether it be the culture of the organisations directly involved, cultures of broader embedding institutions, regional and national cultures, professional and disciplinary cultures or otherwise. Every case study will show a different pattern of relevance, different cultures affecting the governance structure, strategies and results. The interplay of the various cultures will structure and delineate the potential for innovation, and it will co-determine the labelling of innovation, as well as the measurement of success. Even-Zohar, Luhmann, and our first case-study already clarify those essentials.

Just as transition too often is portrayed as a process that can be clarified by science, it is assumed too often that innovation is one single process, a process with characteristics independent of the observing system. Luhmann, Even-Zohar and INTERREG tell us very clearly that innovation is essentially an adaptation to a multitude of environments and that innovation is always a self-description of a system, using the semantics of that system, restricted by the autopoietic identity of that system, bound by the culture produced in the system. Given these features, innovation is always risky, and temporary organisations, project organisations like in the INTERREG network, can form fertile testing grounds for such high-risk endeavours.

References

Arts, B. and Van Tatenhove, J. and Leroy, P. 2000. Policy arrangements. In: J. Van Tatenhove, B. Arts, and P. Leroy (Eds.) Political modernisation and the environment: the renewal of environmental policy arrangements. Kluwer: Dordrecht, the Netherlands.

Bache, I. and Flinders, M., 2004. Multi-level governance. Oxford University Press: Oxford, UK.

Bang, H.P., 2004. Culture governance: governing self-reflexive modernity. Public Administration 82(1): 157-190.

Bevir, M., 2003. A decentered theory of governance. In: H.P. Bang (ed.) Culture governance as political and social communication. Manchester University Press: Manchester, UK.

CULTPLAN, 2007. Cultural differences in European cooperation. Learning from INTERREG practice. Publication of the INTERACT project CULTPLAN. Main report; appendix.

Duineveld, M., Beunen, R., Van Assche, K., During, R. and Van Ark, R., 2008. It holds a promise. The relation between description and prescription in transition-research. In: Poppe, K.J, Termeer, C. and Slingerland, M. (Eds.) Transitions towards sustainable agriculture and food chains in peri-urban areas. Wageningen Academic Publishers: Wageningen, the Netherlands, Chapter 17 in this Volume.

During, R., 2008. Nederlands erfgoed in Europees perspectief. In: During, R., Eerden, M., Kolen, J., Luiten, E. and Zande, A.N. (Eds.) Op historische gronden. Projectburo Belvedere: Utrecht, the Netherlands, pp 105-122.

European Commission, 2001. Commission of the European Communities. European Governance; a white paper. Brussels, COM(2001) 428 final, 2001.

European commission, 2007. Communication for a European agenda for culture in a globalizing world. Brussels, 10 May 2007, COM 242.

Even-Zohar, I., 1990. Polysystem studies. Poetics Today, International Journal for Theory and Analysis of Literature and Communication 11 (1): 1-268.

Even-Zohar, I., 2007. The complexity of culture. Paper presented to the international conference 'Cultural variety in Europe: policy and practice'. Amsterdam, November 23, 2007.

Flyvbjerg, B., 2001. Making social sciences matter. Why social inquiry fails and how it can succeed again. University Press: Cambridge, UK.

Von der Gablenz, O., Plesu, A., Mak, G., Nauman, M., Pomian, K., Licht, S., Abu Zayd, N., Van Veen, M. and Klaic, D., 2005. Europe as a cultural project. Final report of the Reflection Group of the European Foundation (2002-2004). European Cultural Foundation.

Gualini, E., 2004. Multilevel governance and institutional change: the Europeanization of regional policy in Italy. Ashgate: Aldershot, UK.

Gullestrup, H. 2007. Cultural analysis: towards cross cultural understanding. The University of Aalborg, Centre for International Studies: Aalborg, Denmark.

Hajer, M.A. and Wagenaar, H., 2003. Deliberative policy analysis; understanding governance in the network society. Cambridge University Press: Cambridge, UK.

Helander, E., 2007. Reflections on INTERREG III and INTERREG IV. Presentation at the international conference 'Cultural variety in Europe: policy and practice'. Amsterdam, November 23, 2007. See www.cultplan.org.

Hooghe, L. (Ed.), 1996. Cohesion policy and European integration: building multilevel governance. Clarendon Press: Oxford, UK.

INTERACT, 2005. Cross-border cooperation in Europe. An overview of INTERREG IIIA programmes 2000-2006. Leaflet, developed by the INTERACT Point for Qualification and Transfer, Denmark.

Klijn, E.H. and Teisman, G., 1991. Effective policy making in a multi-actor setting: networks and steering. In: In 't Veld, R.J., Termeer, C.J.A.M., Schaap, L. and Van Twist, M.J.W. (Eds.) Autopoiesis and configuration theory: new approaches to societal steering. Kluwer Academic Publishers: Dordrecht, the Netherlands, pp. 99-111.

Luhmann, N., 1995 [1984]. Social systems. Stanford University Press: Stanford, USA.

Lotman, J., 1978. On the semiotic mechanisms of culture. New Literary History 9(2): 211-232.

Loulanski, T., 2006. Cultural heritage in socio-economic development: local and global perspectives. Report.

Newman, J., 2001. Modernising governance. New labour, policy and society. Sage: London, UK.

Organisation for Economic Co-operation and Development (OECD). 2005. Culture and local development: local economic and employment development. OECD: Paris, France.

Pierre, J. and Peters, B.G., 2000. Governance, politics, and the state. Macmillan: Basingstoke, UK.

Sabatier, P.A. and Jenkins-Smith, H., 1988. Policy change and learning: an advocacy coalition approach. Westview Press: Boulder, USA.

Sandercock, L., 1998. Making the invisible visible: a multicultural planning history. Berkely: Berkely, USA.

Scharpf, F.W., 2000. Toward a theory of multilevel governing in Europe. MPIfG Working Paper 00/5, Köln, Max Planck-Institut für Gesellschaftsforschung, Germany.

Sanyal, B. (Ed.), 2005. Comparative planning cultures. Routledge: New York, USA. 415 p.

Schaap, L., 1997. Autopoiese, sturing en stadsprovincie. PhD thesis. Rotterdam, the Netherlands.

Swidler, L., 1986. Culture in action: symbols and strategies. American Sociological Review 51: 273-286.

Termeer, C.J.A.M. and Koppenjan, J.F.M., 1997. Managing prceptions in networks. In: W.J.K. Kickert, E.H. Klijn and J.F.M. Koppenjan (Eds.) Managing complex networks. Strategies for the public sector. Sage: London, UK, pp. 14-34.

UNESCO, 1998. The power of culture. Intergovernmental Conference on Cultural Policies for Development, UNESCO: Stockholm, Sweden.

Van Twist, M., 1994. Verbale vernieuwing; aantekeningen over de kunst van het besturen. PhD thesis. Rotterdam, the Netherlands.

Van Assche, K. and Verschraegen, G., 2008. The limits of planning. Niklas Luhmann's social systems theory and the analysis of planning and planning ambitions. Planning Theory 7(3): 263-283.

Williams, B., 2004. Formal structures and social reality. In: Boland, R.J. and F. Collopy (Eds.) Managing as designing. Stanford University Press: Stanford, USA.

World Bank, 1999. Culture counts: financing resources and the economics of culture and sustainable development. Culture Counts, Florence, Italy, World Bank, Washington, USA.

Chapter 10

The Wageningen innovation assessment toolkit: how to improve the potential of transition projects?

Frances Fortuin and Onno Omta

Abstract

In this chapter we will discuss how the experience gained with innovation management in high tech industries can be applied to transition projects in the agri-food sector, using the *Wageningen innovation assessment toolkit* (WIAT). Using Company X as an example, we discuss how the use of WIAT can has helped companies identify the drivers and barriers to innovation and the potential pitfalls for innovation projects at an early stage when correction is still possible. In order to realise transitions in the agri-food sector, innovation networks have to deal with a large number of transition partners with potentially conflicting ideas and insights. In this chapter we propose that WIAT will prove to be even more effective in the effort to gain insight into potential success and failure factors in transition processes in this sector than it has proved to be for in-company innovation projects.

Keywords: WIAT, transitions, agri-food sector, project assessment

10.1 Introduction

'It was such a good project, why did it flop in the market?' Company X is a large multinational prospector food processing company. It is one of the largest multinational agri-food companies in Europe with an annual sales volume between US$ 500 million and US$ 1 billion. It is a primary produce processing company, with a strong technology base. It pursues a prospector business strategy: it aims at staying ahead of competition by introducing new products and implementing new processes faster than its competitors. It aims at bringing products to the market in innovative ways, involving not only new production processes, but also introducing new products that are geared to new ways of food preparation at home. But Company X has a problem. Despite the fact that it can build on a long history of technological superiority, its current innovation projects somehow seem to miss their target. Recently a number of very promising projects had to be cancelled before completion, or led to product introductions that did not bring the market success the company had hoped for. It was clear to top innovation management that 'something was wrong', and that 'something had to be done to make the company more innovative', but nobody knew exactly what.

10.2 Breaking the mould

Following on from the definition of the famous Austrian economist, Schumpeter (1934), transition to more sustainable forms of agriculture can clearly be classified as a form of innovation. Schumpeter defines innovation as ' a process of creative destruction" ... 'breaking old rules to establish new ones'. In order to transform the current system of agriculture into a more sustainable one, a whole range of innovation projects is needed to develop new processes, products and services that combine sustainable characteristics with a sound commercial basis.

Managing the innovation process, however, is a high risk endeavor because of the inherent uncertainty of every innovation process in itself. From innovation literature it is known, that only a very limited number of innovation projects will turn out to be successful (Cooper, 1999). So if we want to realise the transition to more sustainable forms of agriculture, we must learn how to manage the innovation process effectively, thereby raising the chances of success for innovation projects. In this chapter we want to discuss how the experience gained with the management of innovation in high tech industries such as the pharmaceutical, ICT and aviation industries can be applied to innovation projects in the agri-food sector, using the Wageningen innovation assessment toolkit (WIAT).

Based on data from over 50 agri-food prospector companies around the world, WIAT provides a methodology that enables agri-food companies to compare the self assessments of the drivers and barriers to innovation at the company level and the critical success and failure factors of innovation at the innovation project level with comparable companies in their own sector. We will first demonstrate how WIAT works in practice using the Company X case. As we mentioned above, Company X is a large multinational prospector food processing company, that aims to launch innovative products into the market in innovative ways, involving not only new production processes, but also introducing products that are geared to new ways of food preparation at home. By analysing this company's innovation projects and innovation processes, we show how we have been able to help it to understand and improve its suboptimal innovation performance. We then elaborate on the theoretical foundation of the WIAT toolkit and discuss how it can be used in transition projects that typically go beyond the individual company – the so called institutional or system innovations. The distinguishing characteristic of transition projects is that they are complex, because they often involve many actors such as the business sector, the government, non-governmental organisations, and of course consumers, and they often require the balancing of ecological, physical, spatial and socio-economic values (Omta, 2002).

10.3 The case of Company X

'To know or not to know – that is the question!' The central question facing Company X was: how come that they, with their superior technologies, did not achieve the market success

that they should expect from their innovations? It was their objective to raise the success rate of individual innovation projects, and improve the innovative climate of the company as a whole. Company X's core problem was that they could not find a viable way to reach this objective. Whenever they discussed this problem internally they came to the same conclusion: they could see that projects went wrong, but were unable to identify the underlying causes of these failures, let alone that they could identify problems before a project turned into a failure. To help them answer this question, they hired expertise in from Wageningen UR. The lens used to look at Company X's problem was one that combined the insights gained in the management of innovation in technology-based industries with the tacit knowledge available within individual project team members, as revealed by the WIAT toolkit. By using this lens, Company X learned to identify the underlying factors for success and failure and pick up early warning signals. They could then use this knowledge to redirect their innovation processes.

10.4 The theory behind the Wageningen innovation assessment toolkit

The previous decades have produced a number of studies that identify characteristics and factors leading to innovation success as well as failure. The prominent focus of these studies was to open the black box of innovation and provide in-depth understanding of how products are actually developed within companies (Brown and Eisenhardt, 1995; Tidd *et al.*, 2001). Some of these studies explicitly compared successful with failed projects (the so-called dyadic studies). One of the best known is the SAPPHO study (Rothwell, 1972; Rothwell *et al.*, 1974) in which 43, mainly British, case studies were compared in pairs. Another study was the Stanford project, which was mainly directed at Californian companies (Maidique and Zirger, 1984; Zirger and Maidique, 1990). Perhaps the best known is the NewProd project (e.g. Cooper, 1979; Cooper, 1992), which was a large study of pairs of product successes and failures. There have been some extensive reviews of these studies (e.g. Brown and Eisenhardt, 1995; Montoya-Weiss and Calantone, 1994; Ernst, 2002; Hart *et al.*, 2003). From the reviews we list the central factors that determine either success or failure:

- Product superiority: the product uniqueness and superiority from the customer's perspective.
- Proficiency of marketing and technological activities: 'up-front' activities such as initial screening, preliminary market and technical assessment, detailed market study and/or marketing research, and business/financial analysis.
- Protocol: clear definitions of the target market; the customers' needs, wants, and preferences; the product concept; and the product specifications and requirements.
- Market potential: market need, growth and size.
- Organisational relations, cross-functional integration, team communication and co-operation.

It is important to realise that most of these insights are based on research in high-tech industries, such as the computer, biotech, or pharmaceutical industry, where other industries, for instance, supplier dominated industries such as the agri-food industry remained largely unexplored. An exception is the study by Pannekoek *et al.* (2005) of 74 entrepreneurial innovation projects in Dutch greenhouse horticulture. The authors identified product superiority, and cooperation with supply chain partners as the most important success factors for entrepreneurial innovation. Other studies indicated that successful innovating agri-food companies have a strong market orientation (Batterink *et al.*, 2006), and that economic considerations and insufficient innovation competencies are the main barriers to innovation in this sector (Batterink *et al.*, 2006; Garcia Martinez and Briz, 2000). Costa and Jongen (2006) list major barriers to agri-food innovation as being (a) a lack of concrete guidelines for the effective implementation of consumer oriented food development, (b) the sequential approach of the innovation process and (c) the lack of intra- and inter-organisational coordination or integration of R&D and Marketing's activities and know-how.

WIAT tailored the insights derived from the studies discussed in the previous section to the needs of the agri-food sector. This tool adds to the diagnostic value of existing tools by effectively utilising the rich tacit knowledge of the members of innovation project teams that come from such diverse backgrounds as food science, marketing, engineering and sales. It deepens these insights by linking information about critical success and failure factors of individual innovation projects to the drivers of and barriers to innovation present in the company as a whole. WIAT uses the following constructs to assess the chances for success and failure of individual innovation projects:

- Two constructs at the company level:
  - project-company fit;
  - project resources.

  The project-company fit indicates that an innovation project should fit with the company's strategy, if a project is not in line with the company's strategy, the project leader should ask why this project was initiated in the first place (Hollander, 2002; Fortuin, 2007). The project resources construct indicates that the success of an innovation project relies on the financial and human resources devoted to it, as well as the technical, managerial and marketing skills of the team members and the company at large.
- One construct at the team level:
  - team communication.

  Project team members are key to every innovation project, of course. Without them, the best innovation process cannot develop a new product or process successfully. In effect, the project team members all have to share the same vision and cooperate with one another. This necessitates good technical and communication skills as well as sufficient decision making authority.
- Two constructs at the product/process level:
  - product superiority;
  - product aspects.

  Product superiority indicates that a new product should possess distinctive features compared to competitors' products in order to be successful in the market, such as a higher quality, or unique features. It is assumed that if a new product has a clear economic advantage and meets certain customer demands, it will have a higher probability of achieving success in the market. Furthermore, there are specific product aspects that define the innovation process, such as high product innovativeness and/or technological complexity.

- Three constructs define the market:
  - market competition;
  - market volume;
  - environment.

  A new product is developed for a certain market, where volume, size, potential value, and growth of the market determine if it is possible to sell a product in the predicted volumes. However the new product has to compete with other products or substitute products in the market. The constructs define the level of competition and the market attractiveness as well as the level of hostility of the (institutional) environment.
- Three different time-dependent constructs are used to measure performance:
  - project;
  - product;
  - future performance.

  Project performance refers to whether the project is within planning, budget, and to what extent the original project objectives are fulfilled. Product performance refers to benefits for end-users and if the project is expected to earn money for the company, and future performance refers to possible spin-off products or processes and its potential to improve customer loyalty. The constructs and the individual items are listed in the appendix.

In Fortuin *et al.* (2007) we described how WIAT was applied in 12 prospector agri-food companies in the Netherlands and France. All investigated companies were large, multinational prospector agri-food companies, with annual sales ranging between US$ 100 million up to over US$ 1 billion. All these companies allocate resources to innovation on a structural basis and have a central R&D department where innovation projects are carried out by multidisciplinary teams. By comparing the average assessment of the successful projects (11 projects, 35 respondents) with the failed projects (6 projects, 30 respondents), the key success factors were determined. All construct scores for successful projects proved to be higher than for failed projects, except for the construct 'project-company fit', which is more an indicator for the radicalness of the project analysed. The greatest difference between successful and failed projects was found on the constructs 'product superiority' and 'project performance', ($P<0.01$), directly followed by the factors 'team communication' and 'expected market volume' ($P<0.05$). The factor 'product resources' had a P-value<0.1. When looking at the level of individual statements, the most striking finding was that successful projects scored higher for all five market related statements. Team members from successful projects were clearly more certain about the market features. This implies that the teams of successful projects were better informed about the market characteristics (e.g. through dedicated market research) then teams from unsuccessful projects. Team members from successful projects also proved to be more confident about the product development process. Overall these findings were well in line with the outcomes of similar studies in other industries.

10.5 The Company X case revisited

10.5.1 What did we do?

When applying WIAT in Company X we started out by analysing its overall innovative climate by assessing the perceptions of three top (innovation) managers and comparing their judgment of the situation to an American Management Association (AMA) database, consisting of 1,396 executives of leading innovative companies in North America and Europe. Next we asked the CTO to select a number of projects for analysis: a number of past projects (clearly successful as well as clearly unsuccessful ones), and a number of running projects. The clearly successful projects were defined as projects that not only were a success in terms of engineering/technological accomplishment, but also performed well after market introduction and generated substantial sales for the company. The unsuccessful projects were projects that were either stopped before project completion or market introduction, or proved to be a failure in the market. Then, these projects and a number of running projects were evaluated by 3 to 5 team members using the WIAT project tool. They measured how well their project performed by assessing the following factors:

- project-company fit;
- project resources;
- team communication;
- product superiority;
- product aspects (level of innovativeness and product complexity);
- market competition;
- market volume;
- environment;
- performance.

At the same time they had to give an indication of how certain they were about their answer. This implied that they had to provide an assessment between 1 to 10 (1 = I totally disagree with this statement and 10 = I totally agree with this statement) for 55 statements, and the level of certainty (1 = I am completely uncertain about my assessment of this statement, and 10 = I am completely certain about my assessment of this statement). Based on the team's response score, the optimism within the team regarding the measured factors was determined. Based on the team's certainty score, we determined the confidence they had in their answers.

10.5.2 What did we find?

The overall innovative climate of Company X at first glance seemed to be good when compared with the figures of the AMA database.

The three top managers recognised the importance of innovation, and had a clear understanding of what innovation means. Figure 10.1 shows that Company X's managers are as convinced as, or even more convinced than, the average AMA executive about their company's culture of risk-tolerance, diversity, ability to select the right ideas, to provide the appropriate resources, and to find the right balance between incremental improvements and breakthrough discoveries. This is not surprising because Company X really is at the top of its industry, while the AMA-results are based on the assessment of companies with an average performance. However, in assessing the customer focus, organisational communication and teamwork and collaboration with other departments, Company X's managers scored their company significantly lower than the assessment done by the AMA executives. Even more surprising was the finding that all three mentioned that Company X had a lack of clear goals and priorities, when it comes to innovation. Although Company X considered itself a prospector company, this assessment clearly pointed out that the implementation of their innovation strategy could be improved.

10.5.3 Analysis of the innovation projects

Figures 10.2 and 10.3 present the assessments of two current innovation projects at Company X: one that seems to be a potentially successful one and one that appears to be a potential failure. The figures are presented the way they were shown to the project teams. The zero line in the figures represents the mean construct score of successful agri-food companies.

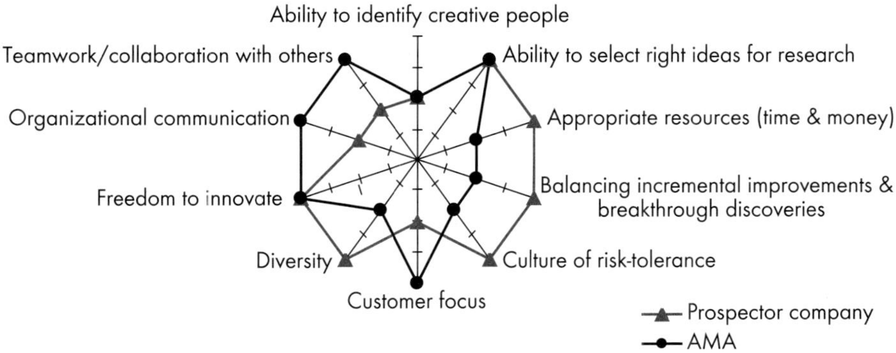

Figure 10.1. Innovation drivers and barriers of Company X compared with the AMA average.

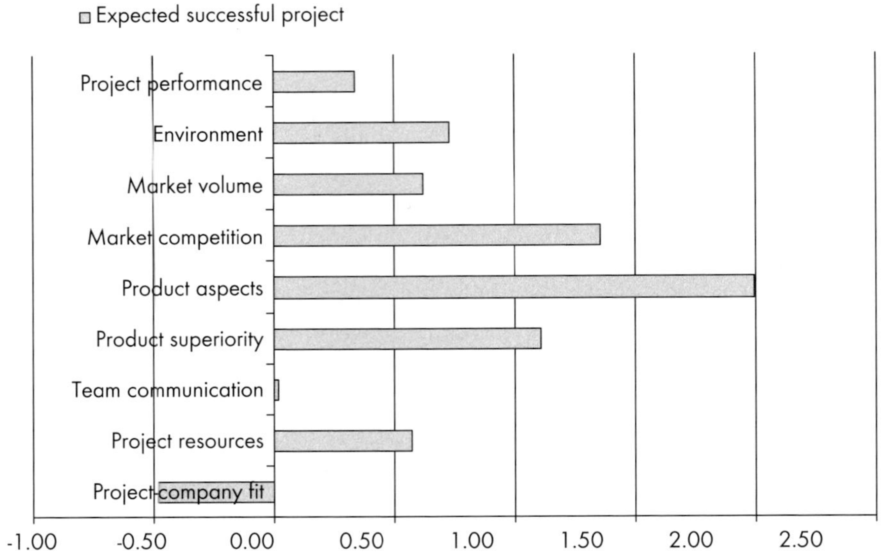

Figure 10.2. A potentially successful innovation project compared to successful projects. The zero line in the figure represents the mean factor score of successful agri-food projects.

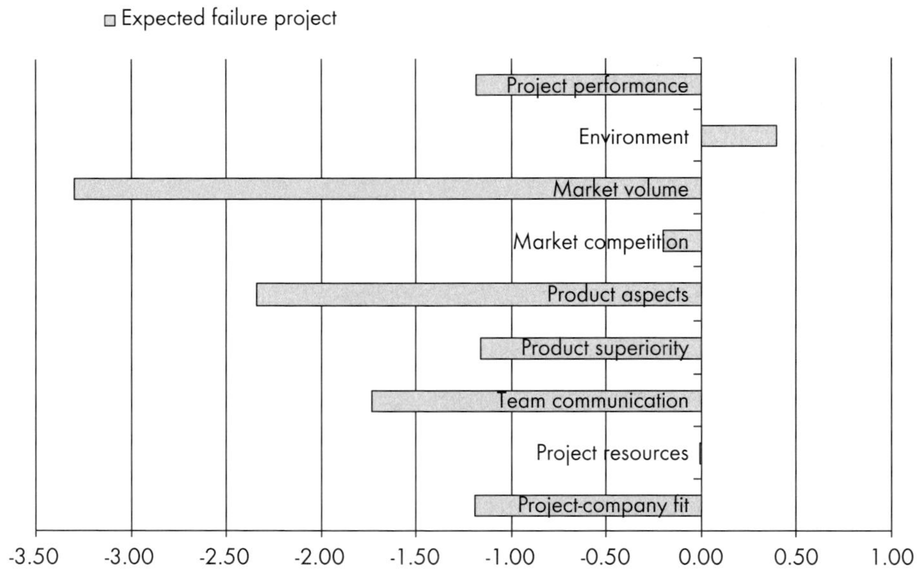

Figure 10.3. A potentially failing project compared to successful projects. The zero line in the figure represents the mean factor score of successful agri-food projects.

Figure 10.2 presents the results for a potentially successful project. The scores of this project are high compared to successful projects for almost all constructs. Only for project-company fit the score is lower, indicating that this project concerns a product that is relatively unknown to the company (a more radical innovation). It is important to note that the high score for 'market competition' is a potential weakness, as this construct represents the expected level of competition this product will face on the market. The team factor is perfectly in line with successful projects and the team evaluates the product as superior to competing products. This is important because it is the most important success factor. In addition, the high score for 'product aspects' indicates that this project concerns a relatively innovative product, with highly advanced technologies. As this is a comparison with successful projects, the prospects for this project look extremely good. If we had compared the assessment with failed projects, the figure would even have looked better.

Figure 10.3 presents the results of a project that is a potential failure. Most notable is the low score for market volume. Apparently, the team is not very positive about its market potential (below 5 on a 10-point scale!). Moreover, for key success factors such as product superiority and team communication the scores are low. The management should clearly ask itself whether they should continue this project.

10.5.4 What was the effect?

When these results were discussed with the team members, project leaders and the top innovation managers of Company X, the effect was tremendous. Their vague feelings that 'something was wrong' in a number of projects, and that 'something had to be done to make the company more innovative' were replaced with relevant and accurate information that could be addressed with targeted management measures. Moreover, comparing Company X's current projects with the ex-post insights of its own clearly successful and failed projects, further enhanced the *ex-ante* predictive value of the findings. WIAT provided important diagnostic clues that not only now help Company X in the go/no go decisions for current projects, but even more importantly, delivered critical information on strong and weak points of the projects as they are being conducted. As such, it enabled managers to interfere at a moment that this was still possible.

Using WIAT now enables Company X to use its limited resources for the most promising projects and to effectively steer these projects past pitfalls and threats. By connecting the feedback on project level with information on the drivers and barriers to innovation present in the culture of the company as a whole, the instrument deepened the understanding of its managers of what underlies its overall innovation success. A critical element that came out of the evaluation for the company as a whole was that the company lacked a well-structured marketing function. Innovation team members based their decisions to a large degree on personal assumptions about their customers, and not on thorough market research, while the final consumer was completely overlooked. As a result of this study, the company reorganised its R&D and marketing functions world-wide.

10.6 Lessons learned

10.6.1 In the case of Company X

A number of conclusions can be drawn from this case study. First the Company X example proved that revealing the tacit knowledge of the project team members by use of WIAT acted as a very powerful tool in getting all relevant information out on the table. One may ask, why the team members did not share their insights on the strong and weak points of their projects before they were asked to do so for the WIAT evaluation. We think the reason for this lay in the fact that every team member was an expert in his or her own field, and not in other fields. None of them felt qualified to address weak points in the project that fell outside their own particular field of expertise, in order not to challenge the expertise of their colleagues. In some cases this meant, that even if the majority of the team members had serious doubts about the feasibility of a project in an early stage, no one dared to say so. In the WIAT project questionnaire they were asked to give their opinion on all factors critical for project success, ranging from the typical technical performance criteria to the market potential and customer related aspects. The fact that they could add to this assessment an indication of how certain they were about their judgment enabled them to give their opinion on subjects that did not belong to their typical field of expertise. When these results were discussed, the collective knowledge of the team that until this point had been largely hidden, was revealed and many team members were surprised to find out that they were not the only ones with doubts or concerns on a number of critical factors. Next it proved that the critical success factors used in WIAT aligned well with critical aspects in the projects and the overall innovative climate of Company X. This meant that Company X confirmed once more the findings in other agri-food companies that factors like customer focus, product superiority, team communication and market information have become as critical for the agri-food sector as they are in other industries (Fortuin *et al.*, 2007).

10.6.2 For transition projects in general

We mentioned in the introduction that transition processes are very complex, because they involve many actors from different backgrounds: from the business sector, the government, non-governmental organisations, and of course consumers and they often require the balancing of ecological, physical, spatial and socio-economic values as well. Based on earlier in-depth studies of four sectors in the Dutch agri-food industry, Omta and Folstar (2005) indicate, that while some have improved (e.g. cut flowers and vegetables), others have lagged behind (beef and pork). They concluded that an important failure factor was the – sometimes very high – number of actors (that had to be) involved in the transition process, each with their own, sometimes conflicting, ideas and interests. Gaining insight into the ideas and interests of the transition partners is of paramount importance for the successful implementation of transition projects. To this end, we believe that WIAT is even better able to gain insight into the potential success and failure factors in an early stage of a transition process than it has proved to be for in-company innovation projects. The results of the WIAT database clearly indicate that where in the past the Dutch agri-food sector could flourish by counting on their technological expertise as driver for commercial success, the sector nowadays needs to take into account a whole range of new factors, including market and product related up-front activities as well as to aspects of sustainability, in order to keep its license-to-produce.

References

Batterink, M.H., Wubben, E.F.M. and Omta, S.W.F., 2006. Factors related to innovative output in the Dutch agri-food industry. Journal on Chain and Network Science 6 (1): 31-45.

Brown, S.L. and Eisenhardt, K.M., 1995. Product development: past research, present findings, And future directions. Academy of Management Review 20 (2): 343-378.

Cooper, R.G., 1979. Dimensions of industrial new product success and failure. Journal of Marketing 43(3): 93-103.

Cooper, R.G., 1992. The Newprod system: the industry experience. Journal of Product Innovation Management 9 (2): 113-127.

Cooper, R.G., 1999. From experience: the invisible success factors in product innovation. Journal of Product Innovation Management 16 (2): 115-133.

Costa, A.I.A. and Jongen, W.M.F., 2006. New insights into consumer-led food product development. Trends in Food Science & Technology 17 (8): 457.

Ernst, H., 2002. Success factors of new product development: a review of the empirical literature. International Journal of Management Reviews 4 (1): 1-40.

Fortuin, F.T.J.M., 2007. Strategic alignment of innovation to business: balancing the exploration and exploitation function. Innovation and Sustainability Series – Volume 2. Wageningen Scientific Publishers: Wageningen, the Netherlands.

Fortuin, F.T.J.M., Batterink, M.H. and Omta, S.W.F., 2007. Key success factors of innovation in multinational agri-food prospector companies. International Food & Agribusiness Management Review 10 (4): 1-24.

Garcia Martinez, M. and Briz, J., 2000. Innovation in the Spanish food & drink industry. The International Food and Agribusiness Management Review 3 (2): 155.

Hart, S., Hultink, E.J., Tzokas, N. and Commandeur, H.R., 2003. Industrial companies' evaluation criteria in new product development gates. Journal of Product Innovation Management 20 (1): 22-36.

Hollander, J., 2002. Improving performance in business development: Genesis, a tool for product development teams. PhD thesis, Groningen University: Grongingen, the Netherlands.

Maidique, M.A. and Zirger, B.J., 1984. A study of success and failure In product innovation: the case of the United States electronics industry. IEEE Transactions on Engineering Management 31 (4): 192-203.

Montoya-Weiss, M.M. and Calantone, R., 1994. Determinants of new product performance: a review and meta-analysis. Journal on Product Innovation Management 11 (5): 397-417.

Omta, S.W.F., 2002. Innovation in chains and networks. Journal on Chain and Network Science 2: 73-79.

Omta, S.W.F. and Folstar, P., 2005. Integration of innovation in the corporate strategy of agri-food companies. In: Innovation in agri-food systems (2nd ed.). Wageningen Academic Publishers: Wageningen, the Netherlands, pp. 223-246.

Pannekoek, L., Van Kooten, O., Kemp, R.G.M. and Omta, S.W.F., 2005. Entrepreneurial innovation in chains and networks in Dutch greenhouse horticulture. Journal on Chain and Network Science 5 (1): 39-50.

Rothwell, R., 1972. Factors for success in industrial innovations, from project SAPPHO: a comparative study of success and failure in industrial innovation. SPRU: Brighton, UK.

Rothwell, R., Freeman, C., Horlsey, A., Jervis, V.T.P., Robertson, A.B. and Townsend, J., 1974. Sappho updated: project Sappho phase Ii. Research Policy 3 (3): 258-291.

Schumpeter, J.A., 1934. The theory of economic development. Harvard University Press: Cambridge, USA.

Tidd, J., Bessant, J. and Pavitt, K., 2001. Managing innovation: integrating technological, market, and organizational change. John Wiley & Sons: New York, USA.

Zirger, B.J. and Maidique, M.A., 1990. A model of new product development: an empirical test. Management Science 36 (7): 867-883.

Chapter 11

Two complementary transition pathways: supporting strategies for innovation towards sustainable development in Dutch agriculture

Frank Wijnands and José Vogelezang[17]

Abstract

Agriculture in the Netherlands is facing a huge challenge. It has to evolve into a viable sector that matches the requirements of both the market and society in a sustainable way. The plan for this transition requires substantial changes at system level, since the current rate of progress is too slow. The greatest challenge for the forthcoming period is to link the innovative capacity of current stakeholders in agriculture in new settings and in varying coalitions to the long term goal of sustainability. Two complementary pathways to support transition processes in the field were identified. The first pathway begins with a target vision for the future and works back to current practice in the field. The second pathway runs in the opposite direction: working from current practice in the field towards the future. Both pathways are further elaborated on in this chapter and are illustrated by key projects. Finally the connection between the pathways is discussed, taking a look at the merits and pitfalls of a dual approach.

Keywords: sustainability, transition pathways, system innovations, transition points

11.1 Challenges for Dutch agriculture

The Netherlands struggles with a number of complex problems whereby socio-economic, cultural and technical aspects are fundamentally interwoven. Problems such as the loss of landscape quality, biodiversity, the ongoing crisis in animal health control, the depletion of natural resources and the competing claims on land use. Moreover society and agriculture estranged from each other in the last decennia by a strict division of functions in land management The National Plan for the Environment in 2001 (VROM, 2001) proposes that Dutch agriculture should be sustainable by the year 2030 within social, economic and technical preconditions. This plan can only be realised if the current supply-driven agricultural sector becomes a sustainable sector that conforms to the community's desire for quality food production, as well as an attractive countryside.

---

[17] Frank Wijnands and José Vogelezang developed the model of the two transition pathways.

To achieve this, drastic changes are needed that transcend individual organisations. These are known as *system innovations* which transform the relationship between vested interests, organisations and working methods (Rotmans, 2003). In order to get to grips with this complexity, Geels and Kemp (2000; see also Geels, 2002) developed a theoretical model whereby they highlight three levels: macro (global trends), meso (coherent system of dominant practices, rules and protocols) and micro (product and process innovations). Working at sustainability *involves* many different players because the level of sustainability in agriculture is *determined* by many different stakeholders. They often have conflicting interests but are increasingly forced to co-operate, form coalitions and alliances to realise their goals. These days each of them must face the question of how to envisage their future bearing in mind both social accountability and sustainability.

There are two autonomous trends to be seen in the course currently being followed by businesses and production chains. The first is up-scaling and increasing efficiency, with the emphasis on high-tech applications, internationally orientated and market driven chains; integral quality control and efficient logistics. The second trend is developing at the same time. This is a more regionally based agriculture and horticulture with the emphasis on small scale, locally oriented chains and diversification beyond the production of food and resources. These include providing health care and welfare services, such as therapy, recreation, education, etc. and the management of collectively owned resources such as water or nature and landscape values; or the manufacturing of regional products eventually selling them from home.

What effect do these autonomous trends have on the future for the Netherlands? Is it desirable to modify these dominant development trends or even to reverse them? These are urgent questions because the demands on the Dutch countryside are growing quickly and are very diverse; from the need for large scale water catchment or nature parks, to suburban growth and infrastructure. Are other desirable, realisable future visions possible? These questions are, of course, interwoven with the many questions surrounding sustainability and social accountability. The new avenues that have to be explored to achieve climate neutral, energy efficient and sustainable agricultural systems, sustainable water management, etc. will have consequences for the type of farms and the context in which they will be active in the nearby future.

The challenge for the current and the coming decennium is to get enough momentum and focus into this quest for new avenues of development to promote sustainability and social accountability. In our opinion the focus will have to come from a long term agenda for sustainable system innovations that is supported by stakeholders as a group. The momentum has to originate out of the current initiatives and the innovative drive of today's stakeholders and to bind them in new alliances coupled to this long term agenda. This chapter describes a strategy to achieve this.

11.2 The two transition pathways

The answer lies in the model of the two transition pathways and its constituting parts. This model was developed by Wageningen UR as framework for a group of research programmes directed at system innovation[18]. These research programmes encompass both the arable and horticultural sectors and include organic farming and multifunctional agriculture. The model of the two transition pathways is a methodical way of designing, implementing and facilitating innovation projects and processes in practice (see Figure 11.1).

The pathway from Future to Practice (F to P) runs from the target visions of the future to current practice in the field. The first step in this process is to make an inventory of desired and supported visions of the future among stakeholders as a source of inspiration for feasible directions for development. Obstacles that hold up progress and that appear to be resolvable (transition points) are sought out via backcasting. Innovation projects are then developed to tackle these transition points. This opens the perspective that the route to the future actually becomes passable in the middle to long term.

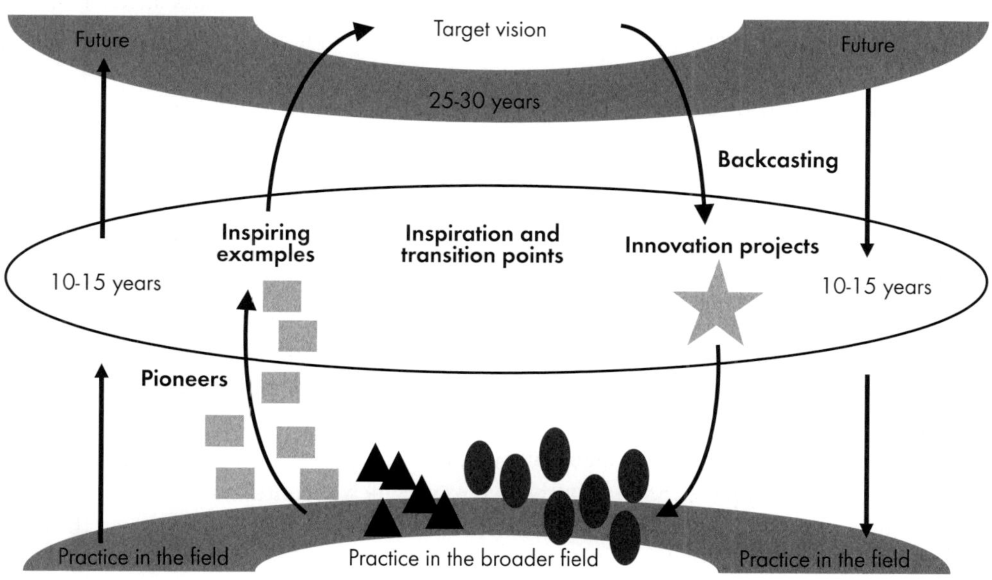

Figure 11.1. Schematic illustration of the two transition pathways: The first pathway (right) from Future to Practice (F to P) runs from the target visions of the future to current practice in the field The second pathway (left) runs in the opposite direction: working from current practice in the field towards the envisaged future (P to F).

[18] Research was financed by the Dutch Ministry of Agriculture, Nature and Food Quality (LNV).

The second pathway Practice to Future (P to F) runs in the opposite direction: from current practice in the field towards an envisaged future. This pathway puts all its money on innovation. Pioneer's innovations are used as source of inspiration in this pathway. When innovations come to a standstill or parties can't agree, assistance is offered to keep the innovation process moving.

By working on two fronts at once, the innovation process can be enhanced and its implementation in the field facilitated, which is essential for the development of businesses for the future. The pathways we have outlined form the basis of the intervention logic for a large series of projects that are being carried out under the flag of the aforementioned system innovation programmes.

11.3 From the future to current practice (F to P)

The way that the explicit visions of the future are arrived at, will be described here at length. Using one particular example for arable cultivation we will show how that translated into an innovation project with an implementation horizon of the mid to long term (5-15 years). Similar routes have been followed by protected horticulture (Poot, 2004) and multifunctional agriculture (Kommers and Hopster, 2004).

11.3.1 Visions for the future

The project 'Toekomstverkenningen' [*Exploring the future*] started in 2003. It's objective: to develop a number of inspirational visions of the future for a sustainable, socially acceptable agriculture by 2030. The idea was to outline the steps that are needed to be taken now in order to reach that goal. The methodical framework for the working strategy was formed with a combination of the Sustainable Technology Development (STD: Aarts, 1998) and the Interactive Technology Assessment methods (Grin en Van de Graaf, 1996). The project began with a strategic problem orientation around the target visions of the future. Initially, 50 stakeholders were interviewed about their individual visions of the future and their background motives. From this round of interviews it was apparent that the stakeholders saw future development primarily happen in the two ways which we outlined in the introduction. On the one hand large scale businesses which produce for the global market and on the other hand the regionalisation and diversification in function by small scale businesses.

These two visions for the future may appear to preclude one another, but in the workshops that followed the initial interviews, it became apparent that a combination of both lines of thought was an important source of inspiration for new visions of the future. This combination envisages business systems that implement large scale, high-tech production a in small scale and multi-functional environment. In other words: how do you combine an efficient and effective, economically viable and competitive agriculture and horticulture with small scale production systems that are more accessible to the people in the towns and cities – and which can fulfil more than one role?

Two visions resulted from discussions in the ensuing workshops: one for built-up areas (urban agriculture) and one for the countryside (regional clustering) (see Box 11.1). These visions have been made concrete in artist's impressions.

Box 11.1. Target visions of the future for outdoor cultivation (integrated and organic agricultural systems).

A. Urban agriculture: the ring model

Agriculture integrates in rings of diminishing urban density round the city with a changing mix of functions and production of food and horticultural products for urban dwellers.

- Close to the city, woven into the outskirts or as an integral part of newly built areas, agriculture is devoted to creating a rural experience for the city dweller (food origin, feel and taste). Agriculture in combination with health care, day-care for children, education, pick and pay orchards, meeting centres under glass, architecturally interesting glasshouses, animal encounter zones, etc.
- In the outer rings of the model there's room for specialisation, production of fresh produce for the city and production for other markets.

This model links production with consumption, thereby boosting the social accountability within the food production and rural sectors.

Artist's impression of the vision for the future urban agriculture with four rings of diminishing density of urban development.

B. Regional clustering model for rural areas

In this model, agriculture is multifunctional and combines two different business strategies: in a small part of the business, innovative, high-tech, high profitable crops are cultivated (cosmetics, functional foods, bio-based resources, etc.). The largest area of the business is utilised for economically less viable functions such as extensive agriculture, park and water management and recreation. These large scale businesses combine cultivation of innovative products with providing social services. In this model businesses work together in clusters to form regional entities. Thereby, specialisation in tasks is possible and enhances the professionalism of the total cluster. At the same time small scale landscapes can be maintained in a viable way without compromising the environment or natural resources.

Artist's impression of the vision for the future regional clustering model for rural areas.

These visions of the future are not a blueprint for the future but rather an indicator for the direction that plans could take for viable business, production chain or regional development, making agriculture more sustainable (Planet), making it more accessible to the community (People) and improving its economic viability (Profit). Whether horticulture or agriculture realises these visions of the future is not certain. These visions provide a tangible direction for development which has a broad support base, without which there is no chance that this development line can be pursued. Formulating visions of the future is not the same as extrapolating trends: the future of agriculture is envisaged based on what people see as its role and the desirable features that it should have in the Netherlands in 2030. This produces radically new ideas as well as solutions for current system limitations (Aarts, 1998).

11.3.2 Transition points and innovation projects

The most important transition points are analysed via backcasting as the first step on the route from the target vision of the future to current practice. Transition points are the obstacles that stand in the way of realising the vision for the future. Three types of transition points have been defined through Jansen and Vergragt (1993), shortened as CST: Cultural, Structural and Technological (CST as part of the STD approach). The most important transition points that emerged in interviews with the stakeholders are summarised in Table 11.1. In order to surmount these obstacles, action needs to be undertaken in different areas. Behavioural patterns determined by our standards and values – our choices – fall under 'culture'. The

Table 11.1. Transition points grouped in themes and possible avenues towards a solution. Transition points are obstacles that have to be overcome to free up the route to the desired vision of the future.

Theme	Transition crossroad	Possible avenues towards a solution
Ecology, the environment & technology	current businesses don't meet environmental quality demands to much dependence on non-sustainable energy sources traditional crops with low profits	minimal emissions into the environment and sustainable soil management new affordable forms of sustainable energy new, high value crops
Planology	planology does not accomodate multifunctional agriculture lay-out and management of farms not adjusted to needs of urban dweller's	development planology (dynamic and interactive) creating experience added value for urban dwellers by new services and improved design of farms
Economy and business	lack of inspiring new arrangements for collaboration (businesses and partners) insufficient knowledge and skills in non-production related challenges no adequate financiering of collective resources such as nature and landscape	new arrangements for collaborations knowledge and skills for new types of businesses new payment mechanisms for a diversity of services and for urban agriculture,
Market and logistics	'lack'of tuning production to consumers demands	short, efficient lines between production and consumption demand driven production processes intelligent distribution systems

changes in methods of payment for the new services and functions in agriculture; new laws and regulations and the changed roles of organisations in the innovation process fall under 'structure'. 'Technological' transition points are mainly dealt with by breaking new ground inventing new methods and techniques, because existing technology and mechanisms no longer suffice. In the first instance, this means proving that the new concepts do work in practice [proof of principle]. Research into this is often risky, experimental work with a time horizon of five to 15 years. If the new concepts work, then there is still a lot of fine-tuning necessary to transform that into economically viable applications. Simultaneous changes in diverse areas are often needed to allow all the individual innovations to be fully implemented.

We have defined innovation projects as those which are designed to find new routes for specified sets of transition points (Figure 11.1). Researchers and stakeholders work closely together in this search. The solutions that are found assist in surmounting the obstacles standing in the way of system innovation and have an inspirational and promotional role in the field.

Dutch system research has been active for 25 years in an effort to sever the direct relationship between production and pollution. They have achieved great progress but the link between production and pollution is still a real obstacle to the realisation of this vision for the future. Completely new methods and technologies need to be invented to solve this problem. Three innovation projects were designed to address these more technical issues which require a novel approach to systems. These projects began in 2004/2005 and one of them will be discussed here – the innovation project 'De smaak van morgen' [*A taste of tomorrow*] (Box 11.2). This project has a strong and explicit link to both visions for the future, but primarily with suburban agriculture. It is working to realise peripheral urban agriculture in co-operation with stakeholders in the city of Almere.

11.4 From practice to the future (P to F)

The second pathway runs from current practice in the field towards the future, by seeking co-operation between different groups of entrepreneurs and businesses in practice. The objective is to sharpen the focus on promising innovations that contribute to the desired goals of sustainability and social accountability. Also here in this pathway, the binding factor is a communally supported vision for the future. The innovators are the main players in this pathway. They form the vanguard in the agri-business as they search for their route to the future and they are well-placed to point out the transition points. To realise more impetus, it is necessary to intensify the innovative drive found in the field and facilitate the spread of knowledge between the various groups. The strategy in this pathway is characterised by the creation of and support given to networks as described in Chapter 6: *Learning in networks* (Vogelezang *et al.*, 2009, this volume). Networks are defined as temporary or semi-structural cooperatives of primary producers, other agribusinesses and stakeholders. The coalitions can be fluid. There are roughly three types of networks: (1) networks exclusive to pioneering

Box 11.2. Innovation project 'A taste of tomorrow'.

This innovation project has a clear link with both visions of the future, but especially with urban agriculture. Agriculture near cities has to fulfil a number of technical and environmental criteria and needs to be attractive for town dwellers and consumers in general. Fulfilling the functions that an agricultural business can have in the city (outskirts), requires the crop choice and cultivation to be adaptated to this situation. Experimental research is being carried out on two experimental farms operated by Wageningen UR: research into annual open field crops (30 hectares) at the Professor Broekemahoeve [*farm*] in Lelystad and research into fruit cultivation on an experimental farm in Randwijk. On both locations integrated and organic farming systems are operated next to each other. At both locations new methods and technologies (proof of principle) are developed. The main challenge for the integrated system lies in finding alternatives for pesticide use and/or emissions because of the nil-emissions objective for the urban environment. The organic system looks for better control of quality loss due to pests and diseases and improved non-manual weed control. For both systems the same search routes are functional: namely integral soil management, utilisation of functional agro-biodiversity (in crops and environment), pre sowing/planting and post harvest treatments as part of the integral chain management and precision agriculture and automation. Moreover, attention us given to 'down scaling' of mechanisation and precision farming, to facilitate efficient work routines even in the small scale peri-urban agricultural farms. New crops that are attractive to urban dwellers are also being developed.

In 2004 *A taste of tomorrow* began with exploring the possibilities of and the establishment of a network of vested interest parties around urban agriculture in Almere. In the following three years a large number of activities have been carried out (workshops, events, interviews, surveys, publications) which have contributed to the building of a large network dealing with urban agriculture. Almere's ambition is to lay the first stone for an urban agricultural city quarter, *Agromere*, in 2010.

entrepreneurs (2) networks aimed at facilitating the spread of knowledge through to a broad group of entrepreneurs, the 'early adopters' and (3) networks from the field which are formed in response to a communal quest for change or innovation.

11.4.1 Network of innovators

Businesses out in the field react in different ways to developments in the market and their region. Innovators are interested in change and often have eyes open and ears to the ground in order to become aware of any opportunities that developments in the market and society offer them. They develop a broad range of innovative, promising working methods from the desired perspective for their future business, and realise new coalitions between businesses which have never worked together before. Not only that, but these businesses often seek their inspiration from sectors outside of agriculture and translate this inspiration into workable ideas for their own terrain. This co-operation between innovators is formalised in so-called innovation networks. Within these networks of pioneers, our task is to help intensify promising lines of innovation and facilitate their availability to others. Innovations that straddle at least two of the three 'Ps' are in this way the most interesting, because they have found ways to innovate on more fronts at once. The task of working with innovation networks can be split into three steps. The first is the selection of a coherent, promising group of entrepreneurs based on in-depth interviews and the identification of innovations with potential. The second step is formulating with them a common perspective on the future for a specific line of innovation, bearing in mind that a balanced strategy is needed for the three 'ps' of sustainability.

In this way it is soon obvious what the obstacles are which prevent the intensifying and up-scaling of the innovation. The challenge in step three is to tackle the obstacle in such a way that the innovation can transcend the niche and enter into a fully-fledged system. Exploiting opportunities and overcoming obstacles depends on cooperation between many vested interests. By making the 'breakthrough' plan widely known, the interested parties can be united for activities they can do together. Follow-up activities can differ widely, from technical developments to changes in the law and regulations, depending on the type of opportunity or obstacle. 'Waardewerken' [*Value works*] is an example of an innovation network that has been operational for some time in multifunctional agriculture (see Box 11.3).

Box 11.3. Value works.

The innovation network *Value works*, consisting of 18 participants, originated in 2004 as a result of a survey amongst innovative businesses in multifunctional agriculture. It is a broad-based group involving diverse businesses fulfilling functions outside of primary production. These include: education, recreation, selling from home, health, natural resources and landscape management, water catchment, energy production, etc. In the first workshop with this group, the central issue was to establish a rationale for collaboration and to determine a common goal for the group. The workshop revealed that what these businesses had in common was their search for recognition, appreciation and inspiration, as well as their desire to come to a communal vision on multi-functionality. Since 2004 this group has met several times per year and is largely self-governed. It is mostly the entrepreneurs themselves that determine the shape and content of a project and the activities. The role of the researchers is to facilitate, analyse and reflect. Together with the group, bilaterally and in fluid coalitions, they work on issues which the group have put on the agenda.

11.4.2 Other networks

Innovations that are realised by pioneers or groups of pioneering entrepreneurs still have to find their way to the field by way of 'early adopters'. This up-scaling of the innovation requires obstacles to be removed that arise around its introduction, application and promotion in the field. The question of how to reach the potential users and how to maintain the momentum has been tested in empirical experiments such as *SynErgie*; a network for low energy glasshouses.

Impulses for innovation are not always visible. Sometimes people in different places are working on an innovation at the same time. Because there is no contact between them there is no 'critical mass'. The development can suddenly gather momentum when businesses are brought together. The government has recently made funds available to finance a programme in the animal husbandry sector so that demand-driven networks can be formed or their formation facilitated (see Chapter 7: *Networks with free actors*, this Volume). The experience and knowledge gained in these networks is promoted in the field in many different ways (magazines, demonstrations, field days, lectures, workshops, study groups, etc.) often together with stakeholders. Communication proved to be crucial to the success of the programme – especially the role of internet in bringing farmers together.

A larger number of networks are active in 'Practice in the broader field' at the base of Figure 11.1, varying from study groups operating under the auspices of the process industry or rural organisations, to larger scale networks specifically devoted to certain topics such as sustainable crop protection or organic farming. In these networks, farmers work together with researchers, extension services and other stakeholders. The whole entourage around the agricultural entrepreneur has to innovate as well. *Farming with a future* is an example of a network where the broad implementation of more sustainable crop protection and nutrition management in the arable and horticultural sectors is adressed (see Box 6.4, Chapter 6).

11.5 Lessons learned: theory and practice

We have been working with the two transition pathways model and related projects for a number of years now. In this paragraph we reflect on our experiences using primarily the examples presented in the boxes as reference. In our assessment of the results and the value of the projects for the desired innovation, we have leaned on the commonly used criteria for monitoring and evaluation processes: output, outcome and impact. *Output* refers to directly measurable results, such as: products, new networks, new initiatives and unexpected results (spin-offs). *Outcome* refers to results based on the output such as changes in mindset and behaviour of stakeholders, organisations and networks. *Impact* finally refers to the resulting concrete changes in practice as a result of the doings of the various stakeholders. Are the results used by the target groups or other stakeholders. Is the project contributing to the agenda of change.

11.5.1 Visions of the future

One year after the 'Toekomstverkenningen' [*Exploring the future)* project had been concluded, the stakeholders that had taken part were asked to review the experience (De Wolf *et al.*, 2006). The original project consisted of three phases: (1) the exploratory interview phase; (2) the design studio and the workshops; (3) and the development of innovation projects to realise the target visions of the future. Stakeholders proved to be enthusiastic about the process: in their opinion it stimulated creativity and future-oriented thinking and the interaction between them generated a shared vision. Stakeholders also saw the process as a means to influence the research goals. The target visions of the future were seen as innovative and stimulating. The concrete realisation of follow-up activities was valued, whereby the innovation project was seen as a bridge between current practice and future promise. There was also some criticism, especially when stakeholders were not involved in specific phases of the project. In this way the choices and decisions were not always transparent or understood by those stakeholders. They also sounded a warning that the group should ensure that the visions of the future kept pace with changing reality. This *Exploring the future* project has brought the Wageningen UR researchers in contact with more stakeholders more intensively than 'normal'. This collaboration has broadened the appreciation of the tactical and strategic thinking and behaviour of the participating stakeholders. The *Exploring the future* project has also revitalised discussions that had stranded in the problems of the present time.

The technique of backcasting gives a coherent idea of the diversity of obstacles and their relationship to one another. Identifying and naming the obstacles has proved to be an effective way to formulate the key issues for system innovation and to involve the stakeholders in the development of solutions. It is important to analyse the obstacles in depth to see if they really do stand in the way of progress. The nucleus of obstacles that finally remain are almost always easy to formulate, as are the eventual solutions. How to *actualise* the solutions is, however, far from simple and exploring options and avenues for solutions costs time.

Target visions stimulate creativity and lead to innovative ideas to realise these goals. But it also has become apparent that too radical steps will undermine the support. It has become clear that support for far-reaching innovations is made easier by making them tangible. The more concrete the transition point, for example one with a strong regional component, the easier it is to involve stakeholders. Another important factor is that of external developments. If there is a generally felt sense of urgency this influences the process in a positive way. The innovation projects can be carried in this way by autonomous trends that speed up the process of change. Factors that block change include stakeholders who play political games and the lengthy duration of governance (bureaucratic) processes.

11.5.2 Innovation projects

The innovation project *A taste of tomorrow* began at the end of 2003. In the spring of 2007 an evaluation was done on the basis of a network and stakeholder analysis and the intervention rationale of the project. The experience gained in past years has confirmed that the value of innovation projects is optimal when it fulfils three requirements: that it is firmly placed in the perspective of a target vision, that it relates directly to it's environment and includes the future users of the innovations. These three points deserve full attention when guiding and designing the projects.

First of all, it has to be clear which transition points and visions of the future are being addressed by the innovation project and how these relate to one another. The more concrete and tangible the methods used to address both the more successful the project. A number of explicit strategies are used to work on the transition point 'nil emission of pesticides' in *A taste of tomorrow* (see Box 11.2). This transition point has to be solved to realise the vision 'Urban agriculture' – an environment that has zero tolerance for pesticides. Not only that, but this vision of the future imposes crucial design criteria on the experimental work such as 'environmentally friendly' (energy and other aspects of sustainability, etc.), 'attractive' (choice of crop and diversity) and 'small scale' (which is also realisable via automation). In the pilot project for urban agriculture in Almere, a city not far from Amsterdam with a huge programme for urban growth in the coming decades, the vision itself is being realised. In this project the three aspects, culture, structure and technology (orgware, software and hardware) for this system innovation are dealt with.

A second critical factor for success in innovation projects is the degree to which the project group succeeds in creating or assembling enough critical mass, quality and originality from the scientific and business community to work on the transition points and new strategies (consisting out of methods and new technology). Often this requires surmounting institutional obstacles to do with finance, status and image. *A taste of tomorrow* has developed a diversity of stakeholder networks around specific themes such as organic farming, precision farming, biodiversity and urban agriculture. Having both organic and integrated systems under one roof has lead to inspiring cross pollination. The collaboration with regional institutions, schools, business partners and different research groups from Wageningen UR has lead to a number of new projects. These involve either fundamental research groups or commercial companies. Almost all important stakeholders operating on the boundary of the public domain and the rural area have been involved via the pilot urban agriculture project in Almere.

The third factor to bear in mind for a good innovation project is the ongoing, supportive relationship with the future users of the innovation. In the end it is they who have to take the subsequent steps towards implementation in practice. Support can be gained, fotr example, by generating as much spin off as possible that is usable in the field in the short term. Moreover it increases the involvement by making the larger process of change visible in interim steps. In the case of *A taste of tomorrow*, this insight has lead to major corrections to the experimental design by making allowances for short term questions that arise along the route towards the long term objective. Publishing brochures and guides for urban agriculture (see paragraph 11.5.3) has contributed greatly towards the building of a network involving local councils and is another example of the way by which future users can be encouraged to take interest in the project.

If we analyse the results in terms of output, outcome and impact it becomes apparent that the results often don't relate just to one of these categories, but contribute more or less at the same time to all of them. The direct results of the project (output); often also result in change of mentality and behaviour of participants (outcome) and/or will be used in practice (impact). The efforts to realise breakthroughs in systems (new methods and technology) often don't lead to tangible results in the short term. The contribution of bio-diversity in agriculture and polyculture to controlling plagues is still difficult to assess because these are complex ecological methods. A breakthrough appears to be most imminent in the precision farming techniques. Techniques are being developed and tested that will lead to robotised, localised and pesticide-free crop protection. From 2007 *A taste of tomorrow* is concentrating on this precision farming next to the use of bio-diversity. The search for other methods and technologies is in the meantime receiving attention in other research programmes or in new projects. These follow-up projects can be characterised as spin-off from *A taste of tomorrow.*

There has definitely been progress in the realisation of the vision of the future for urban agriculture. The pilot, *Agromere,* has realised a number of interim results which contribute to the growth of the network and increasing understanding of the opportunities and possibilities for urban agriculture. The Urban Agricultural Guide lists the initiatives in the Netherlands (Anonymous, 2007a) and the brochure '*Agriculture comes to town*' describes the added value of agriculture within Almere's city limits (Anonymous, 2006a). If all goes according to plan, within three years Almere will begin developing the first urban agricultural suburb, A*gromere.* A survey was done of the opinions and perceptions of 340 urban dwellers with regard to agriculture. Together with stakeholders in a follow-up workshop, future scenarios have been mapped out. Two of these scenarios, 'Boerenbrink' [*Farmers corner*] and 'Ecostad' [*Eco-city*] will be further developed into plans for specific city quarters (Anonymous, 2007b*). A taste of tomorrow* launched urban agriculture as a serious avenue for development in the Netherlands and it functions as one of the focal points in urban agriculture in the Netherlands. New initiatives have begun in both Lelystad and Amsterdam. The orientation talks with Amsterdam have lead to surveys of what the inhabitants want.

Finally, a critical factor for this type of project is the tangible (financial) and intangible support of commissioners. The high investment level and uncertain results are not an attractive proposition. Not only that, the question is how progress in these projects should be measured as the results are strongly dependent on progress in the process of change – and that is difficult to quantify. The Athena Institute has identified three elements that can be used as (supplementary) evaluation criteria and which need to be considered by projects: documenting learning experiences in developmental processes, showing the relationship between method and concrete results (for both positive and negative results) and identifying (more modest) successes along the route towards change (B. Regeer, personal communications).

11.5.3 Innovators' networks

In the autumn of 2007, the project *Value works* was evaluated, using the Most Significant Change method (Davies and Dart, 2005). This evaluation confirmed the value of collaboration between pioneering entrepreneurs as described in Chapter 6.4: the collaboration with innovators offered a unique chance to discuss on the course that the innovation should take with an eye to the future.

They discussed their target vision for multifunctional agriculture with representatives from government, agriculture, health care organisations, the environment and the recreational sector. The evaluation determined that: 'by bringing entrepreneurs together in a network, the strengths of the individual pioneers were combined enabling them to get close to government policy'. The contact between the network and policy makers has had a noticeable effect contributing to a change in thinking about multifunctional agriculture within the LNV Ministry. The network advised the Minister of LNV to create a national task force for multifunctional agriculture with representatives from all interest groups. This Task Force was inaugurated in December 2007.

The breakthroughs that are needed to realise the vision (transition points) are identified in a so-called breakthrough agenda and translated into concrete actions for the different players. Action lists are continually being prioritised and carried out. This is not only of importance for the current group, but also for those just beginning to explore the possibilities of multifunctional agriculture. The network presented a top ten of the obstacles formed by legislation and regulations to the LNV Ministry. The accompanying publication 'Ruimte in regels' [*Legal space*] offers a number of examples of creative solutions (Anonymous, 2006b). These were presented to the National organisation for local councils. The Ministry of LNV adopted creating space in legislation and regulations as a factor to bear in mind when making new policy.

Finally, when the potential of an innovation has been proved, it is a challenge to turn its niche position into an opportunity to be taken up by others in the field. The publication, 'Kansscanner' [*Opportunity scanner*] gives interested entrepreneurs advice on how to start a new activity, what skills they need to have, and what this means for their business management (Anonymous, 2007c). This was used in the project 'Plattelandsimpuls' [*Impulse for the countryside*] whereby 350 entrepreneurs, assisted by 'value workers', were guided in their quest for new product/market combinations in multifunctional agriculture.

Pioneering entrepreneurs are perfect ambassadors since they proved to be very convincing communicators of their vision and philosophy, showing powerful practical examples. Finally, we learned that collaboration with pioneers requires a flexible and dynamic approach with plenty of room for contributions by the participants.

11.6 Conclusion

Transition towards sustainable agriculture demands a coherent strategy, following two pathways; the first focussed on creating radical innovations and the second focussed on 'exploiting' the potential for innovation that pioneering entrepreneurs are already pursuing in practice. Interaction between the two pathways has obvious benefits: the strategic space for entrepreneurs and their partners in agribusiness is increased if both pathways are followed. Innovators could be interested in the new innovations arising out of innovation experiments and *vice versa*, the innovation experiments might benefit from the creativity of the innovating entrepreneurs. Factors contributing to the success or failure of this approach are outlined in this chapter. They have also been found to be key points in effective monitoring and evaluation.

Acknowledgements

The authors would like to thank John Grin of the University of Amsterdam for the inspirational role that he fulfilled while the Visions of the Future project was being implemented. He helped us on the way. Joske Bunders and Barbara Regeer, both working for the Athena Institute of the Free University of Amsterdam have made a valuable contribution during our discussions about monitoring and evaluation. The ideas which formed the basis for the two transition pathways would never have developed without the cooperation of the programme management of 'System innovations'. The authors would like to warmly thank their colleagues for their contribution during the many discussions, which have taken place right up until the present day.

References

Aarts, W., 1998. Een handreiking voor duurzame technologische ontwikkeling. SWOKA: The Hague, the Netherlands.

Anonymous, 2006a. Stad en landbouw een vruchtbare combinatie. Praktijkonderzoek Plant en Omgeving (PPO): Lelystad, the Netherlands. 16 pp.

Anonymous, 2006b. Ruimte voor regels. Praktijkonderzoek Plant en Omgeving (PPO), Lelystad, the Netherlands. 20 pp.

Anonymous, 2007a. Stadslandbouwgids, stadslandbouw onder de loep. Praktijkonderzoek Plant en Omgeving (PPO), Lelystad, the Netherlands. 20 pp. (see also www.waardenwerken.nl for other publications of the network of innovators [in Dutch]).

Anonymous, 2007b. Landbouw brengt vernieuwing in stadsuitbreiding. Syscope 15: 14-17. Available at: www.syscope.nl.

Anonymous, 2007c. Kansenscanner verbrede landbouw. Praktijkonderzoek Plant en Omgeving (PPO), Lelystad, the Netherlands. 20 pp.

Davies, R. and Dart, J., 2005. The 'most significant change' (MSC) technique: a guide to its use. 104 pp. Available at: http://www.mande.co.uk/docs/MSCGuide.htm.

De Wolf, P., De Buck, A. and Grin, J., 2006. Werken aan de toekomst. Beschrijving en evaluatie van het toekomstbeeldenproject in de plantaardige sectoren. Publikatie nr. 351. Praktijkonderzoek Plant en Omgeving (PPO): Wageningen, the Netherlands.

Geels, F., 2002. Understanding the dynamics of technological transitions: a co-evolutionary and socio-technical analysis. PhD thesis. University Twente: Enschede, the Netherlands.

Geels, F.W. and Kemp, R., 2000. Transities vanuit een socio-economisch perspectief. Rapport voor het ministerie van VROM. Universiteit Twente en MERIT, Universiteit Maastricht, the Netherlands.

Grin, J. and Van de Graaf, H., 1996. Technology assessment as learning. Science, Technology and Human Values 20: 72-99.

Jansen, J.L.A. and Vergragt, Ph., 1993. Naar duurzame ontwikkeling met technologie: uitdaging in programmatisch perspectief. Milieu 8(5): 179-183.

Kommers, M.A.W. and Hopster, G.K., 2004. Vernieuwing en verbreding. Praktijkonderzoek Plant en Omgeving (PPO): Lelystad, the Netherlands. 32 pp.

Poot, E., 2004. Innovatie voor intensivering. Praktijkonderzoek Plant en Omgeving (PPO): Naaldwijk, the Netherlands. 15 pp.

Rotmans, J., 2003. Transitiemanagement. Sleutel voor een duurzame samenleving. Koninklijke van Gorcum: Assen, the Netherlands.

Vogelezang, J., Wals, A., Van Mierlo, B. and Wijnands, F., 2009. Learning networks in Dutch agriculture. In: Poppe, K.J, Termeer, C. and Slingerland, M. (Eds.) Transitions towards sustainable agriculture and food chains in peri-urban areas. Wageningen Academic Publishers: Wageningen, the Netherlands, pp. Chapter 6 in this Volume.

VROM, 2001. Een wereld en een wil, werken aan duurzaamheid. Nationaal Milieubeleidsplan, Ministerie van Volkshuisvesting, Ruimtelijke ordening en Milieubeheer: The Hague, the Netherlands.

Chapter 12

Synthesising needs in system innovation through structured design: a methodical outline of the role of needs in reflexive interactive design (RIO)

Bram Bos and Peter Groot Koerkamp

Abstract

In order to make modern western animal production systems more sustainable, it is necessary to design systems that address multiple challenges at one time. It is not only profitability that is at stake, but also issues like the position and welfare of animals, the environmental impact, labour quality and quantity and biodiversity. Solving these issues in isolation of one another will likely conflict with, or even negatively influence the performance of systems in other ways. To prevent this from happening, a structural reorientation of the system at hand (a system innovation) is needed, which tackles the systematic linkages between desired and undesired effects. It is this linkage that makes it difficult, for instance, to improve animal welfare without increasing the detrimental environmental impact of that production system. The RIO approach (a Dutch acronym for *reflexive interactive design*) is a set of methods that is being applied by ASG WUR (Animal Sciences Group of Wageningen University and Research Centre) in order to interactively design and realise system innovations in complex and often controversial contexts like animal husbandry, in an effort to circumvent social and technical constraints for sustainable development. Determining the basic *needs* of live actors (animals, farmers, the general public and consumers) that are involved in a system is a key starting point for this approach. Needs may be different from (short-term) interests, since they are the expression of fundamental preconditions for a good life, as perceived by the actor, or engrained in the actor's biological make up. Short-term interests may be context dependent, and may change over time. RIO aims at the synthesis of the needs of these different actors, instead of weighing the pros and cons of the various interests. By taking needs as the central 'currency' for this design approach, actors – like animals and man – can be treated symmetrically. RIO connects the structured design approach of Van den Kroonenberg, in which needs are the departure point, with interpretive and constructivist approaches to technology assessment and development, in which more fundamental values, preferences and futuristic views of (human) actors take the place of short term interests. An important first step in this design approach for an (intended) production system or production chain is formulating the different actors' needs and desires and their translation into a Brief of Requirements (BoR). These requirements are quantified as far as possible, based on scientific and practical knowledge. The BoR lays the foundation for the design process itself. In this paper, the role of the concept 'needs' in the RIO approach is discussed.

We detail the way a BoR can be made for animals and how this entails interpretation and analysis. We furthermore show how animal needs can be brought on a par with human needs. In conclusion, we claim that the needs approach taken in RIO facilitates the design of production systems that address a multiplicity of challenges, without implicitly having to trade off one against the other.

Keywords: system innovation, reflexive interactive design, structured design, animal welfare, poultry

12.1 Introduction

Imagine a large egg farm where 30,000 relaxed hens are contentedly building their own nests, consistently laying good quality eggs every day, cackling happily with each other in groups, relishing dust baths, scratching around just like they would in the back yard. All this in a high-tech environment where a farmer can harvest the eggs with a minimum fuss and loss, where there is no problem with dust in the air, nor chicken manure everywhere, where maintaining and feeding the hens is almost completely automated. Not only that, this system doesn't cost the farmer very much more than the old way of working and the consumer is happy to pay that little bit extra for his eggs. Does this sound like a dream? This was the challenge for one research team from Wageningen UR, to see if this dream could be made a reality.

12.2 Problems in poultry farming

For a long time, post-war modernisation of Dutch agriculture and livestock production has been recognised world wide as a success story of massively increasing yields and diminishing costs. However, due to its narrow focus on volume and cost-efficiency, livestock production is increasingly confronted with a series of self-generated risks and unwanted side effects, like excess manure which pollutes soil and ground water, emissions of pollutants like ammonia and methane, controversial animal welfare and health issues, and downright crises caused by outbreaks of infectious diseases like Classical Swine Fever and Avian Influenza. Paraphrasing the sentry in Hamlet: 'There is something rotten in the agricultural state of Holland'.

The Dutch way of producing massive amounts of animal protein is an example of what Ulrich Beck (Beck, 1992 [1986]; Beck *et al.*, 1994) has called 'first modernisation'; the means of attaining progress that is characteristic for industrialisation and 20th Century mass production. Its success is accompanied by a range of self produced risks and side effects, to a degree that politics is more about the distribution of risks than the distribution of the wealth produced. However, Beck *et al.*, (Beck *et al.*, 1994, 2003) suggest that this result of modernisation is not inevitable: they propose 'reflexive modernisation' as the idea that progress can be maintained in a way that anticipates and prevents the occurrence of unwanted side effects by systematic reflection on the basic and hidden assumptions on which our modern production and consumption methods are built.

This 'master narrative' of reflexive modernisation was operationalised in a multidisciplinary project for the development of more sustainable husbandry systems for laying hens (called *Houden van hennen [Keeping/loving chickens])*. In this design project, sustainability was conceived as the challenge of addressing a multiplicity of needs *simultaneously*. These included animal needs; environmental requirements; positive contributions towards public acceptance; positive effects on the landscape; work satisfaction for the farmer; and a sound economical basis. Both the focus on the needs of different actors as well as a systematic approach that is oriented on the simultaneous integration or synthesis of these needs in a design, differentiates the project from regular technological R&D in first modernisation. Instead of solving issues one by one, and assuming the inevitability of trade offs between these issues (like for instance the trade off between animal welfare and work satisfaction or environmental concerns), the project looked for opportunities for synthesis in the design. The range of opportunities can be enlarged by identifying needs on a more fundamental level (thereby transcending supposed conflicts of interests in the short term), as well as looking for solutions that serve multiple functions at the same time (functional compatibility).

12.3 To know or not to know: that is the question

Although there are a lot of definitions of sustainability, one can safely argue that it is a normative ideal implying significant improvements to our current production and consumption methods in multiple dimensions. A common way to classify these dimensions is the trinity of 3P (People, Planet and Profit), meaning that social, ecological and economical aspects should be in harmony with, and not traded off against, each other. Projects like *Houden van hennen* aim at improving the sustainability of *current* systems by design. This implies at least two fundamental questions to work on that are more generally relevant to deliberate system innovation of (Grin *et al.*, 2004) or even 'transition' in (Rotmans *et al.*, 2005; Loorbach, 2007) societal systems. The first is how to improve on multiple dimensions of sustainability in design, the second is how to promote the realisation of suggested improvements within an *existing* socio-technological regime. These questions are closely related. On the one hand, simultaneous improvements in several dimensions is most likely to correspond to a redesign of structural features that have co-evolved and stabilised for decades within the socio-technological regime. On the other hand, the same regime will resist attempts to implement structural change in various ways. Thus, any attempt to redesign existing systems for sustainability will have the simultaneous tasks of both strategically dealing with, anticipating and transcending this resistance to change exerted by the existing socio-technological regime. The old fashioned idea that pure technological magic will do the job, no longer applies. It is therefore not enough to propose new technical arrangements (or 'innovations' as it is popularly called) and expect adoption, since these arrangements presuppose a parallel change in the surrounding structure.

The needs approach as adopted in *Houden van hennen* is a systematic way to address this issue. By taking needs as the central unit of analysis and design, it transcends (short term) interests that are essentially the product of the dominant regime. However, by interactive reflection on these needs, actors from within the regime are challenged to rethink some of the basic assumptions that structure their own behaviour and judgements of what is desirable and realisable. In this way, working from needs instead of interests facilitates the acceptability of new solutions, that would be rejected if they were judged solely from (perceived) interests. This is no guarantee for the complete dissolution of conflicts of interests, of course, but it is a good way to reduce them, by looking for possibilities for synthesis and by extending the time frame from short term to long term. In the next sections, we describe the way needs take central stage in a systematic design approach called RIO, that was adopted and partly developed in *Houden van hennen.*

12.4 Introduction to RIO

Reflexive interactive design (RIO) is an approach aimed at interactive design and realisation of system innovations in complex and value-laden contexts. The approach is based on various sources in innovation and political sciences (Grin *et al.*, 1997; Grin, 2005; Grin and Van Staveren, 2007; Schot, 1992; Rip and Kemp, 1998; Loorbach, 2007; Rotmans, 2003; Weaver *et al.*, 2000) and is characterised by the consistent attention paid to technical and social/institutional aspects of innovation.

RIO is *reflexive* for two reasons. The first one is that the approach is a practical realisation of *reflexive modernisation*, a term coined by Ulrich Beck (Beck, 1997), meaning that modern societies are increasingly forced to address the side effects and risks that originate from first modernity. This self-confrontation compels one to act differently. The second reason is that we arrange the design process in such a way that actors gain knowledge by reflection on needs and presuppositions and the analysis of dominant structures. In every-day life we work *within* those structures and on the basis of those presuppositions, mostly without problems. By making them explicit in transition design, the problem space and solution space for actors is increased (Voß and Kemp, 2005).

RIO is interactive because both the reflective as well as the design parts are done in interaction with the people concerned in the problem area or with the method chosen solve the problems. This interaction is necessary for at least two reasons: to prevent *substantive value bias* (Feenberg, 1999) – i.e. specific values implicitly becoming materialised into technological instruments – and to increase the chances of realising a structural reorientation: A technological innovation is not inherently successful and is not just a matter of bringing intelligent engineers and designers together. Innovation in general, but system innovation in particular, also demands a kind of social engineering: working on the structure in which a new technique or new practice is to prosper, designing strategic connections with and among allies, and positioning that innovation in society. A concrete design of a technical object or system (for example, an animal husbandry system) is not the final product in RIO, but a vehicle for structural changes that are likely to transcend spatial boundaries. Thus in RIO, *design* means more than just the work of an engineer, designer or architect.

12.5 Needs and animal welfare in RIO

For both methodical and material reasons, the needs of central actors in the system to be designed play a pivotal role in RIO. Methodically, needs are the starting point in the systematic design methodology by Van den Kroonenberg called *Methodisch ontwerpen* (Structured design: Siers, 2004; De Beer, 1997). Structured design (SD) is an attempt to make the design process of artefacts like buildings and machinery more rigorous and traceable. Van den Kroonenberg was dissatisfied with the rather intuitive way architects and engineers tend to translate functions and requirements into concrete designs and artefacts. SD emphasises the importance of a rigorous analysis of the actors' and users' needs, and their translation into an elaborate set of quantitative requirements, based on traceable sources. Thinking of solutions and design is postponed until this work is done. One of the benefits of this approach is that the requirements are formulated independently of the perceived solution space, which leads to a wider range of options. Another benefit is that the method stimulates more fundamental reflection on the needs of prospective actors.

A substantial reason for taking the needs of actors as a starting point lies in the origins and the current context for applying RIO: animal husbandry. One of the biggest current challenges is the amelioration of animal welfare. In order to design for animal welfare it has to be operationalised to specify eventual choices. The fact that animal welfare is an inherently controversial concept makes this requirement even more pressing.

Animal welfare however is often approached in negative terms: of the well-known five freedoms of Brambell (1965), only one has been formulated positively ('freedom to express normal behaviour'). Instead, (Bracke *et al.*, 1999) have proposed the use of the notion 'need' as a basis for (verifiable) criteria that can be used to assess animal husbandry systems in terms of their performance in animal welfare. Their basic assumption is, that if the needs of animals are fulfilled, their welfare is beyond doubt. This follows on from the proposition put forward by Bracke *et al.* (1999) that the only thing intrinsically relevant for animal welfare is their *emotional* state.

Since we cannot read this emotional situation directly, nor is the animal capable of explicitly and unambiguously communicating this with people, we have to rely on inferred signs of the emotional state of the animal, such as behaviour and physiological response. Bracke *et al.* (1999) assume that an animal's emotional state (for example, lust, hunger or fear) is the signalling or motivating part of an internal control mechanism, for which the output is a particular physiological response or certain behaviour. This output in turn produces an effect that changes the emotional state of the animal (gratification of lust, no hunger, no fear). Each emotional signal (or: motivation) which gives cause to a particular physiological response or to showing certain behaviour, we call the animal's *need*. Needs cannot be detected *directly* nor be measured: we attribute them to animals on the basis of what we observe.

Needs can relate to a certain desired goal (for example, eating and being satisfied, keeping warm), it may concern avoiding negative feelings (fleeing when threatened), but it can also include behaviour that in itself is meaningful for the animal (the so-called 'ethological needs', for example, scratching or dust bathing for chickens). If we can determine what the needs of animals are, we can arrange circumstances so that these needs are met or that the animals can meet their needs themselves. In this way the *bottom line* is: animal welfare is optimal when all needs are satisfied.

This specific interpretation of animal welfare by Bracke *et al.* (see for example Anonymous, 2001) is not definitive, however. There are a few suppositions and assumptions that are certainly open to discussion. For instance, the term 'need' itself remains a theoretical construction, which does not have an explanatory value in itself, nor does it refer to a clear apparent neuronal substrate. There have also been critics (Korte *et al.*, 2007) of the (widespread) presupposition, which is also used by Bracke *et al.* (1999), that an animal aims at homeostasis in all respects and that this would also be good for welfare. Korte *et al.* (2007) claim instead that *allostasis* might be necessary for animal welfare: a certain level of stress may be good for it. Another uncertainty involves our lack of insight into how animals react to different situations; why particular situations may or may not trigger certain behaviour. In a situation when an animal is faced with two or more adverse needs, we don't know why a choice is made for a certain response.

The concept 'need' is nevertheless very useful in designing and testing animal husbandry systems. Firstly, it is an adequate classification to gather all kinds of ethological and practical observations and to order them, and in this way get more coverage from objective scientific data (Bracke *et al.*, 1999). Secondly, it is an approach to animal welfare that principally starts with the desired goals for the animal (*'performance-based'* in animal welfare experts' terms), after which the design requirements ('design-based') are inferred. This corresponds to the systematic approach of structured design. An important third reason to make animal welfare operational in the RIO-approach in terms of needs is that in this way animals can be put into a production system under the same denominator as other interested (live) parties[19], both communicatively as well as normatively.

This approach was first applied in a previous RIO project on fattening pigs (the animal-directed chain design-project) which led to the *comfort class* approach and the *comfort class*-barn (Diergericht Ontwerpen, 2003; Welzwijn, 2006). Initially, the design process was oriented exclusively on the needs of the pigs, after which in subsequent projects other actors

[19] We reserve the term *need* expressly for living creatures. In our view, the economy nor the environment have 'needs', but these may result from the needs of one or more stakeholders as requirements.

and requirements were added. Subsequently, in a project for laying hen husbandry called *Houden van hennen* (2004) both the hens, the farmer and specific citizen/consumer groups were considered as actors with needs, that were treated equally as actors in and around the system to be designed.

12.6 Needs and human beings in RIO

Unlike animals, human actors can be asked directly about their needs. Needs, however, should not be confused with interests. Grin *et al.* (1997) indicate that in projects aiming at solutions for 'unstructured problems' (in which there is uncertainty about the facts as well as dissent over values) it is useful to investigate the human actors' (individuals but also organisations and institutions) *frames of meaning* (Schön, 1983). Frames of meaning consist of problem definitions and preferred solutions, appreciative systems (value systems) and overarching theories that help make sense of situations (Grin and Van de Graaf, 1996). By explicating these frames it may become clear that the actor's perceived current interests may be different from what he or she is essentially striving for in the long run. In a similar vein, the short term needs of human actors may differ from those in the long run, or those in the future.

It is important to look for these longer term needs in system innovative projects, since they presuppose other contexts and structures that fulfil future needs, without producing the sort of side effects we were used to in first modernisation. Explicating the actors' frame of meaning prevents short-term interests being automatically considered similar to long-term ones and that, in turn, enlarges the solution space.

By identifying the (human and institutional) actors' frames of meaning we actually identify the more fundamental (or future) needs of actors. By mentally suspending the current context, one can create room for reflection on aims and ambitions that surpass the present. A livestock farmer asked for his 'need' will probably mention cost price decrease. Reasoning from his current context – with a bulk market with bulk prices – this is absolutely legitimate and understandable. However, cost price decrease will often not be his main *need*. It is (one of the) solutions – for a need that can rather be described as: 'continuity of my farm', 'stable farm income' or 'passing on a viable farm to my successor'. Determining such needs plays a vital part in the 'theory of acting'

By formulating the requirements of human actors in *positive* terms, as is done in a Brief of Requirements (BoR), the design process and the dialogues surrounding it will be oriented on what actors want, instead of what they do not. This might seem trivial, but if we work in an unstructured problem area with a history of heavily contested solutions, this will turn out to be very meaningful. Furthermore, a (positively formulated) BoR can be used in later stages of a project as a yard stick for interactive evaluation of the (intermediate) results of a design process.

12.7 Structured design in RIO

The SD approach is inferred from Van den Kroonenberg, a Delft University Professor of Architecture, and has been described again by De Beer (1997) and Siers (2004). SD presupposes that good design (for a building or a technical installation) can be promoted by a thorough analysis of the needs that are to be met and the functions required to meet these needs. The basic idea is that you have to postpone formulating solutions (including the part creativity plays in them) until you know for sure what the needs are, what functions have to be fulfilled and what this requires. SD forces the designer or engineer to direct his reflection more precisely, before starting the actual design. The approach is an important tool to use so that you can reformulate the problem more accurately.

The latter is of great importance for two reasons. Firstly, people are naturally inclined to confuse (current) solutions with needs or requirements. You think that you need a car, but your real need is a (flexible) form of mobility, or more precisely, the possibility of moving yourself at any moment of the day. By thinking of specific solutions immediately, you limit the possible solutions. By abstracting the concrete solution and formulating the need behind it, more solutions may reveal themselves. The second reason is connected with it: if you want to design more complex systems (such as husbandry systems), many different interrelated functions have to be fulfilled. By assuming (existing) solutions for some of those functions, it will be difficult to fulfil one function without compromising another. Those functions which are good for the animal are then fulfilled at the expense of functions that are good for the farmer or the environment. Because SD goes back to the exact needs *behind* specific solutions, chances are that such trade-offs are not necessary. According to RIO, this is exactly what sustainable development is about: *searching for synthesis of needs* instead of repeating the mantra that sustainability is a trade off, or a 'balance', between People, Planet and Profit.

SD, therefore, is a good tool for designing sustainable production systems. The price is, however, that a lot of work is to be done first: the greater part of the job lies in the problem and system analysis: the definition of needs, functions and requirements. But eventually this will make it easier to connect the different needs of different actors.

The first stage of SD leads to a Brief of Requirements (BoR) for the building, apparatus or system to be designed. In the RIO-approach, the BoR is put together from sub-programmes for different actors in the (husbandry) system to be designed. In the project *Houden van hennen* (2005) they were the BoR for the Laying Hen, the BoR for the poultry farmer and the BoR for the citizen/consumer. This type of BoR is an (extensive) table in which, per need, the requirements for meeting these needs are formulated as precisely as possible. These requirements are quantified as much as possible and as many relevant references as possible are applied, to ensure that these requirements are both qualitatively and quantitatively based.

In principle, a BoR is specific to one actor, who is actually considered to function or to live in the system to be designed. Livestock farmers differ in their individual needs and preferences, but so do (production) animals – particularly if we talk about different breeds. In a broad design project, which doesn't deal with a specific situation, it is necessary to make explicit what kind of 'average' actor is assumed. Where possible, main aspects should be singled out which can be allocated a different weighting in specific cases. In *Houden van hennen*, for example, the BoR for the poultry farmer was divided into three sub-groups of needs: those of the *entrepreneur, animal care taker*, and *labourer*. In specific design projects special emphasis can be given to each of these needs, without making a completely new BoR.

A BoR is the starting point for the further design track in SD. The next step is determining the functions that are to be fulfilled on the basis of the needs that have been formulated in the BoR. These functions are then synthesised in a so-called morphological (function) diagram, which is used to define the structure (i.e. the order and the mutual relationships of the functions) of the design. Then an inventory is made of the possible solutions per function. The search may and should be as wide as possible. Creativity plays an important role when there is a lack of obvious solutions which fulfil the function as well as meet the requirements. On the basis of the structure chosen (the morphological diagram), the search is for combinations of solutions that are compatible with one another. At this stage synthesis of needs is possible by choosing those solutions that fulfil various functions simultaneously (and with that, possibly, also more needs). This is called *functional compatibility* (Simondon, [1958] 1989; Bos *et al.*, 2003).

The actual design stage only starts if there is a morphological diagram as well as a promising combination of solutions, which meet the (greater part of) requirements. A BoR is thus also used in the evaluation phase. In a quantitative assessment the draft designs are evaluated against the BoR: are the requirements met and to what extent is this so? In practice this means that in SD, 80% of the energy is not used for designing but for the preceding stages. The great advantage of this systematic approach is that there is a better guarantee that all needs in the design are covered and that fewer 'perverse connections' creep in: functions that imply improvements on the one hand, but have unintended and undesired side effects on the other. Moreover, the design process becomes more transparent for the outside world, as the least possible implicit assumptions are made (Bos, 2008). Next, making the needs and requirements more explicit in a BoR will encourage the repeated use of the work: a broad BoR is applicable to specific new contexts. Finally, the RIO approach encourages iteration between different phases: it is very possible that comparing intermediate design steps with a BoR will identify hitherto unnoticed conflicts, after which either the problem, or the BoR, or the design might be adapted.

12.8 Formulating a brief of requirements for structured design

Formulating a brief of requirements (BoR) within the framework of SD is thus a much more fundamental activity than simply making a shopping list on the basis of current – and already articulated – wishes. Formulating a BoR is a reflexive and interpretative activity in the case of both human and non-human actors. It is reflexive in the sense that the core of the work is the systematic reflection on assumptions as to articulated wishes, needs, requirements and solutions. It is interpretative, because signals of actors (for example, from interviews or behaviouristic studies) are translated into an understanding of the core preferences, needs and ideals (the 'theory of acting'). This reflexive and interpretative method (BoR) adds more detail to the systematic ordering of pre-existing knowledge. A brief of requirements has a specific structure, which is shown in Table 12.1, with one part taken from the BoR for the laying hen (Houden van hennen, 2005) as an example.

The key question a BoR must answer is: 'What needs does actor X have, what is required to fulfil these needs, and in what quantitative value or *range* can these requirements be expressed?'[20] Answering this question is not a linear but an *iterative* process: needs, requirements and quantifications are identified in correlation with one another. After all, we do not exactly know what the needs are in advance, and these needs may often be expressed indirectly in the literature or in discussions with actors in terms of solutions or requirements. It is the task of the creator of the BoR ('analyst') to clarify matters by consistently considering the nature of a scientific claim, ulfillmen experience from the field, a signal or a statement.

The analyst can draw on various sources, which, dependent on their nature, result in stronger or weaker statements as to needs and requirements. In a BoR for an animal the following sources are particularly useful:
1. (statements about) the current practice concerning animal husbandry (from inside as well as from outside the sector, including NGOs such as the Society for the protection of animals);
2. interviews with farmers and other people in the field;
3. data concerning animal behaviour in their wild counterparts;
4. practice manuals;
5. interviews with experts in the area of housing and welfare;
6. scientific articles in the area of housing and welfare;
7. ethological scientific literature (consumer demand studies and the like);
8. stress-physiological scientific literature;
9. ecological and zoological literature as to the natural habitat and way of living in wild counterparts.

[20] It is likely that during the design process it becomes apparent that not all needs *can* be met. Then a choice has to be made and accounted for within the BoR, or the latter should be adjusted. It is not wise, however, to assume certain impossibilities in advance, because these will preempt the design process that is to follow.

Sources 1 to 5 will mostly produce primary information on specific (and mostly commonly held) current solutions or requirements applied, or to be applied, in practice. Because this information originates from the field, it will not be clear in advance whether the requirements and solutions are actually related to the needs of the animal. They can also originate from the production objectives, from another of the livestock farmer's own needs (for example, labour circumstances or limitation of investment costs), from a societal demand with (possibly) anthropomorphic projection, from a combination of needs, or from obvious assumptions in the existing context. The purpose of a BoR is, however, to define the needs and requirements *per actor* as clearly as possible. A production objective *with* an animal should never be confused with the animal's need itself. In all of these cases the analyst will have to ask the following questions:

a. What is the nature of the claim (solution, requirement or need)?
b. If it is a solution, what need + requirement will it fulfil?
c. If it is a requirement, what need does it match and whose need is it in reality?
d. If it is a need: whose is it?

These sources are workable as a *starting point* for formulating a BoR, but in most cases the available information should be processed and analysed. This is the most important research activity. Whether the sources can also be used as a *basis* for a need or requirement in the BoR is another matter. Scientifically-based data are preferred, because they are less specific or non-specific for a certain situation and can, therefore, justify a stronger claim as to general validity than the first five sources. Moreover, this knowledge can be traced. Not all relevant practical knowledge and experience is scientifically based, however. In such cases one can draw on the first five sources. In the BoR the relative weighting of a source can be indicated in a separate column ('Type of source' in Table 12.1).

The sources 6-9 have indeed a stronger scientific basis, but these too will only find their way into the BoR after *interpretation*. In most cases, such articles will not express their claims in the format used in the BoR, since they develop out of experimental set ups with a different objective and because of the fact that the notion 'need' is not a standard notion in scientific ethological literature. An ethological article which indicates an animal's particular preference, will not, for example, always translate this preference into an animal's need, or this preference is connected to a complex set of requirements. An ecological or zoological article with statements about the lifestyle of wild counterparts cannot as such be considered relevant to the needs of domestic animals, because the latter may have acquired other preferences due to breeding and selection. Scientific literature is also embedded in a particular tradition (a scientific *paradigm*), which can be of great importance in the weight authors ascribe to certain needs or preferences. Lastly, there is a great deal of scientific literature that describes effects in animals which originate from specific housing systems: in such cases it should be carefully examined whether particular behaviour actually originates from an internal need of the animal or whether this behaviour is an instrument triggered by specific living conditions.

Table 12.1. Example of the structure of a Brief of Requirements (adapted from Houden van hennen, 2005). The column 'wish or necessity' indicates whether a requirement should be met in all circumstances (necessity), or could be met if possible (wish). In the case of actors that cannot speak

Code	Needs	Specific needs	Demand
LO	Suitable living environment for the laying hen		Sufficient space and facilities per hen to perform ethological needs
LO14	Presence of light and an optimal light quality to perform ethological needs	Optimal light spectrum, optimal light intensity and a minimal flickering frequency for the optimal functioning of the laying hen	Daylight spectra (inclusive UV)
LO15			Minimum frequency (invisible flickering for the hen)
LO16		Light with social recognition	Light spectrum and minimum light intensity needed for social recognition
LO18		Light during roosting	Dusk
LO18		Light during eating and drinking	Light environment
LO21	Day- and night rhythm	Light/dark cycles	The presence of a light/dark periodicity

(1) Poultry prefers fluoresced light + UV light to fluoresced light without UV (Moinard and Sherwin, 1999) and fluoresced light to light from light bulbs (Widowski *et al.*, 1992).
Chickens are (in contrast to humans) capable of seeing UV-A light (320< lambda <400 nm), they experience colours differently to humans. Hens exposed to light with UV, have lower (basal) levels of the stress hormone corticosteron.
Broilers prefer natural daylight to most other types of light (except warm white light) (See Kristensen *et al.*, 2002). Chickens are capable of seeing colours in daylight, but not in the dark. Nevertheless, in comparison with humans, they see better in the dark. Colours influence the activity of chickens, they are more sensitive to blue and red part of the light spectrum (see Lewis and Morris, 2000).

for themselves (like animals), the question of a requirement being a wish or necessity may allow some interpretative flexibility, and may be dependent on the ambition of the project.

Fixed or variable	Quantity	Source	Type source	Wish or necessity	Explanation
variable	280< lambda <780 nm daylight varies between 1,000-100,000 lux	1. (Prescottet *al.* Jarvis 2003). 2. (Maddocks et *al.*, 2001) 3. (Lewis and Morris, 2000)	Refereed articles	Necessity	(1)
variable	100 Hz	1. (Taylor et *al.*, 2002).	Refereed articles	Necessity	(2)
variable	Light with Uva spectrum (320 nm < lambda <400 nm), minimum 70 lux	1. (Moinard and Sherwin, 1999) 2. (Widowski et *al.*, 1992) 3. (Kristensen et *al.*, 2002) 4. (D'Eath and Keeling, 2003)	Refereed articles	Necessity	
variable	0.5-1.0 lux (see explanation)		Expert opinion	Necessity	(3)
variable	>60 lux (see explanation)	1. (Prescott and Wathes, 2002) 2. (Davis et *al.*, 1999)	Refereed articles	Necessity	(4)
variable	min. 8 hours continuous darkness	1. (Prescott et *al.*, 2003) 2. (Manser, 1996)	Expert opinion	Necessity	(5)

(2) Research has proven that it is unlikely that hens can detect the flickering of low frequency fluoresced lights. From 100 Hz onwards, chickens probably do not experience it as aversive. The level the hens still are capable of detecting is dependent on the light intensity and spectrum (Taylor et *al.*, 2002).
(3) The perch needs to be very well visible to jump to, for example using white colours.
(4) Eating at 200 lux is preferred to eating at 60 lux (Davis et *al.*, 1999).
(5) Alternating darkness and light periods (intermittent) results in aberrant sleeping behaviour (Blokhuis, 1983; Coenen et *al.*, 1988; Manser, 1996). Light periods of 22 hours and more, result in eye handicaps and blindness. Not more than 20 hours of light (minimum 14-16 hours of light is necessary for the egg laying).

This type of information gleaned from scientific literature has not become irrelevant, but it should be translated and interpreted if it is to play a part in a BoR.

Using only one source is not considered to be sufficient to make assertions about a need or requirement. It is better to base the assertion on different sources, preferably of a heterogeneous nature. A statement about a need or requirement becomes stronger if a basis can be found in practice or in scientific literature. That is why we recommend working according the principle of *triangulation* (Stake, 1995): finding at least three *independent* data sources or points which, together, can give power to a statement. It is justified to invest extra time, energy and money in substantiating the requirements, particularly those which, during the design process, are likely to be of great importance in the political, economic or practical sense.

12.9 Experiences and lessons learned

The focus in RIO on the needs of different actors, and their translation into a BoR is meant to open up the solution space for synthesising different needs in one design. Theoretically, this is done by either a redefinition of needs into more general or abstract terms or a search for functional compatibilities within the solutions that fulfil different needs. This synthesis is important if one wants to improve current systems in more dimensions of sustainability at the same time.

In *Houden van hennen*, this worked out well in several cases. For instance, most farmers would state they need *cost reduction* but cost reduction is a solution within a specific context, not an intrinsic need. If questioned further, the real need turned out not to be cost reduction, but more basic values like *continuity*, (*a reasonable income for) earning a living*, and *work satisfaction*. For many, cost reduction was the only conceivable option for survival in an increasingly competitive bulk market, which had been structured solely around price for decades. By questioning the self-evident assumption of cost reduction, other solutions become equally feasible, like a different distribution of the profits within the chain, or the creation of a separate market for eggs with added value. This solution was also the basis for further institutional actions after the designs were finished.

Something similar was found, when investigating the hen's need for nesting material. As a rule, most people would say that hens prefer straw to make a nest, and that good conditions for the hen's welfare include straw. Straw, however, is impractical for farmers, costly, and even damaging for the farmer's health because it generates dust. In reality, straw is more of a solution than a need. By taking a closer look, and by abstracting from the specific solution, this need can be better defined as: material that is transformable by the hen herself to make a suitable place to lay eggs.

Our third actor, the general public/consumer, is a special case. How can we say a consumer or a member of the public actually *needs* something in a system he will never actually be involved in? We can assume that most people hold certain opinions and preferences about the way farming should be done, and that they would like animals to be kept properly. Inclusion of this 'actor' from outside the sector itself was a strategic choice. One of the critical issues in animal husbandry in the Netherlands is how to market products that are produced differently at a higher price. One of the obstacles to doing so, is the established dichotomy within the sector between technical and economical issues – delegated to the production side – and societal and ethical considerations, which are delegated to the consumption side. Within this perspective, issues like animal welfare are seen as external requirements that have to be implemented on top of what is seen as technically or economically necessary. Any such external requirements represent additional demands that will increase costs if realised as an add-on to what is technically and economically required.

In order to prevent a replication of this dichotomy in the designs, we chose to extend the range of actors involved in the prospective new animal husbandry system by adding representatives from the general public/consumers. Their opinions and preferences regarding laying hen husbandry were investigated at the same level of elaboration, and were used at the same stages in the design process, as were the needs of the other two main actors: the farmer and the laying hen.

During the sessions we held with three groups from the general public, the central question that was addressed was how they envisioned their ideal way of keeping laying hens. However, we did not specify in which respect this ideal should be interpreted: ideal for the laying hen, ideal for the farmer, or ideal for themselves. This was done deliberately to prevent any bias towards a specific interpretation, for instance animal welfare.

An important result of these sessions was, that there was much more differentiation between the groups than the established opinion within the sector allowed for. It was generally thought that the general public (i.e. people not involved directly in agriculture) cherishes a romantic image of a few hens scraping happily around a small farm in a bucolic setting. However, we discovered a multiplicity of ideal images within our citizen panels. One of them was to some extent comparable with this romantic, traditional image, but it was only present in a subset of people who held traditional values in general. Industrial, dynamic and wildly natural images were present as well, and correlated to the different sets of values people held for their own lives.

The approach of continued questioning and delving into the consumers' reasoning as well as the emotional levels of judgement, also resulted in a better understanding of what people mean by their primary response to the question of what an ideal husbandry system for laying hens should be. For instance: one group stressed the importance of *nature*, or a natural environment for laying hens. If taken at face value, this could easily be interpreted as an environment that is close to the natural habitat of hens in the wild. Our in-depth questioning revealed that the reference to nature had rather to do with structural features like self-sufficiency and the pursuit of (ecological) balance, which can be realised in a heterogeneous mix of organic and technical elements.

Because of these more thorough assessments of the needs of these three actors, several requirements that seemed to contradict each other at first sight, could be reconciled. Not only that, some needs turned out to be more easily and cheaply achievable than had previously been thought. An important one is the need for the hen to express foraging behaviour. This requires space, but it doesn't require costly infrastructure. Within the dominant regime of laying hen husbandry, allowing hens more space is a costly affair, since it implies doubling the size of the same infrastructure. Because *Houden van hennen* was able to redesign from scratch, the space requirement could be met more cheaply, by adding a simple outside area (in the Plantation) or a terrace above the roost (in the Roundel).

12.10 Lessons learned: the theory

The design of system innovations, that are meant to improve multiple dimensions of sustainability, benefits both methodologically as well as normatively from a needs approach. Normatively, formulating the needs of prospective actors, implies the reflexive explication of values that will be embedded or neglected in technological instruments or institutional arrangements; it will stimulate the differentiation between short term interests and long term needs and will help to discriminate between the actual sources of proposed solutions. Together, this will increase the normative and political accountability of design projects that have an inherent political character along with their technical and scientific nature. Methodologically, the needs approach facilitates the symmetrical treatment of the needs of human and non-human actors in the proposed production and consumption systems. In animal husbandry, this will result in a clear differentiation between the origin of different requirements. This helps to prevent anthropomorphic projections on animal needs without necessarily discarding these projections as irrelevant. Next, by systematic reflection on and abstraction from proposed requirements and solutions to fundamental needs and values, the solution space for system innovative projects is enlarged. This increases the possibility of synthesising the ulfillment of different needs in a design, which might seem contradictory at first glance.

Formulating a BoR for the different prospective actors in a system design is an important step towards synthesising seemingly conflicting interests that are normally traded off against each other. Furthermore, formulating a BoR contributes to the growth of knowledge, in the sense that it brings scientific knowledge and practical experience from extremely diverse backgrounds together under one denominator. Statements as to knowledge and experience about solutions, requirements, behaviour, preferences and needs are translated, through interpretation and analysis, to a specific conceptual scheme, which assumes an existent basis category 'need' for human and non-human actors. Although a BoR does not make *new* statements about reality, it represents a systematic reordering of existing knowledge by active *interpretation* and *reordering* of existing claims about reality in a format that is very useful as an overall reference for design projects, as well as a criterion to judge the performance of existing systems.

The approach described here is relevant in design trajectories that involve multiple stakeholders. The focus on needs is a key element in the design methodology, but is in itself not sufficient for 'system innovation', because needs do not necessarily guide the structural layout of the set of functions in a system in a specific direction, whereas system innovation implies such a structural reorientation. To further develop RIO, work has to be done on the connection between a structural systems account on the one hand, and an actor/needs account on the other. Related to this is the question, which actors should be taken into account. From a sustainability point of view, the range of actors might be much larger than was the case in *Houden van hennen*, and might even comprise actors that have not yet been identified as such. Furthermore, the needs approach is limited to actors that are part of the system, or are served by the system, whereas it is conceivable that actors and entities that are detrimentally effected, rather than served by the system, should be given a more systematic role in the approach. In *Houden van hennen*, this was solved by the addition of extra (for instance environmental) requirements, but these requirements are not connected to a need.

Finally, in order to effectively reform existing structures, and contribute to system innovations, designs alone are not enough. The RIO approach should make us more systematically aware of how choices in the design process facilitate or inhibit implementation. A number of examples of failed system innovation projects exist which did not adequately anticipate resistance from the outside world [see for instance the case of Hercules; (Bos and Grin, in press)], or did not create enough external ownership of ideas.

Acknowledgements

This article was written for the project *Kracht van koeien* (BO-07-009-005), a policy supporting research project financed by the Ministry of Agriculture, Nature and Food Quality, and was subsidised by the KB7 programme Transitiekennis en –kunde. Special thanks to Sierk F. Spoelstra, Krijn Poppe and Karel H. De Greef for their critical comments on an earlier draft of this paper.

References

Anonymous, 2001. Scientists' assessment of the impact of housing and management on animal welfare. Journal of Applied Animal Welfare Science 4: 3-52.

Beck, U., 1992 [1986]. Risk society: towards a new modernity. Sage Publications Ltd.: London, UK.

Beck, U., 1997. The re-invention of politics. Rethinking modernity in the global social order. Polity Press: Cambridge, UK.

Beck, U., Bonss, W. and Lau, C., 2003. The theory of reflexive modernization. Problematic, hypotheses and research programme. Theory, Culture & Society 20 (2): 1-33.

Beck, U., Giddens, A. and Lash, S., 1994. Reflexive modernization. Polity Press: Cambridge, UK.

Blokhuis, H.J., 1983. The relevance of sleep in poultry. World's Poultry Science Journal 39: 33-37.

Bos, A.P., 2008. Instrumentalization theory and reflexive design in animal husbandry. Social Epistemology 22 (1): 29-50.

Bos, A.P., and Grin, J., 2008. 'Doing' reflexive modernization in pig husbandry: the hard work of changing the course of a river. Science, Technology & Human Values 33: 480-507.

Bos, A.P., Groot Koerkamp, P.W.G. and Groenestein, K., 2003. A novel design approach based on recursive control. Livestock Production Science 84: 157-170.

Bracke, M.B.M., Spruijt, B.M. and Metz, J.H.M., 1999. Overall animal welfare assessment reviewed. Part 1: Is it possible. Netherlands Journal of Agricultural Science 47: 279-291.

Bracke, M.B.M., Spruijt, B.M. and Metz, J.H.M., 1999. Overall animal welfare assessment reviewed. Part 3: Welfare assessment based on needs and supported by expert opinion. Netherlands Journal of Agricultural Science 47: 307-322.

Brambell, Committee. 1965. Report of the technical Committee to inquire into the Welfare of Animals kept under Intensive Livestock Husbandry Systems. Her Majesty's Stationary Office: London, UK.

Coenen, A.M.L., Wolters, E.M.T.J., Van Luijteleer, E.L.J.M. and Blokhuis, H.J., 1988. Effects of intermittent lighying on sleep and activity in the domestic hen. Applied Animal Behaviour Science 20: 309-318.

D'Eath, R.B. and Keeling, L.J., 2003. Social discrimination and aggression by laying hens in large groups from peck orders to social tolerance. Applied Animal Behaviour Science 84: 197-212.

De Beer, J.C.F., 1997. Methodisch ontwerpen. Academic Service: the Hague, the Netherlands.

Davis, N.J., Prescott, N.B., Savory, C.J. and Wathes, C.M., 1999. Preferences of growing fowls for different light intensities in relation to age, strain and behaviour. Animal Welfare 8: 193-203.

Diergericht Ontwerpen, Projectteam, 2003. Welzwijn in de toekomst, over varkenswensen voor varkensstallen. Wageningen UR: Wageningen/Lelystad, the Netherlands.

Feenberg, A., 1999. Questioning technology. Routledge: London, UK.

Grin, J., 2005. Reflexive modernization as a governance issue – or: designing and shaping Re-structuration. In: J.-P. Voß, D. Bauknecht and R. Kemp (Eds.) Reflexive governance for sustainable development. Edward Elgar: Cheltenham, UK.

Grin, J. and Van de Graaf, H., 1996. Technology assessment as learning. Science, Technology & Human Values 21 (1): 72-99.

Grin, J. and Van Staveren, A. 2007. Werken aan systeeminnovaties. Lessen uit de praktijk van InnovatieNetwerk. Assen: Van Gorcum.

Grin, J., Van de Graaf, H. and Hoppe, R., 1997. Technology assessment through interaction: a guide. Den Haag: SDU.

Grin, J., Felix, F., Bos, A.P. and Spoelstra, S.F., 2004. Practices for reflexive design: lessons from a Dutch programme on sustainable agriculture. International Journal on Foresight and Innovation Policy 1 (1/2):126-149.

Houden van Hennen, 2004. Laying hen husbandry – towards a happy hen life, proud farmers and a satisfied society. Wageningen UR: Wageningen – Lelystad, the Netherlands.

Houden van Hennen, 2005. Programme of demands – based on the needs of poultry farmer, laying hen and citizen. Wageningen UR: Wageningen, the Netherlands.

Korte, S.M., Olivier, B. and Koolhaas, J.M., 2007. A new animal welfare concept based on allostasis. Physiology & Behavior 92 (3): 422-428.

Kristensen, H.H., Prescott, N.B., Ladewig, J., Perry, G., Johnsen, P.F. and Wathes, C.M., 2002. Light quality preferences of broiler chickens. British Poultry Science 43 (5): S11-S12.

Lewis, P.D., and Morris, T.R., 2000. Poultry and coloured light. World's Poultry Science Journal 56: 189-207.

Loorbach, D., 2007. Transition management. New mode of governance for sustainable development. International Books: Utrecht, the Netherlands.

Maddocks, S.A., Cuthill, I.C. Goldsmith, A.R. and Sherwin, C.M., 2001. Behavioural and physiological effect of absence of ultraviolet wavelengths for domestic chicks. Animal Behaviour 62: 1013-1019.

Manser, C.E., 1996. Effects of lighting on the welfare of domestic poultry: a review. Animal Welfare 5:341-360.

Moinard, C., and Sherwin, C.M., 1999. Turkeys prefer fluorescent light with supplementary ultraviolet radiation. Applied Animal Behaviour Science 64: 261-267.

Prescott, N.B. and Wathes, C.M. 2002. Preference and motivation of laying hens to eat under different illuminances and the effect of illuminance on eating behaviour. British Poultry Science 43: 190-195.

Prescott, N.B., Wathes, C.M. and Jarvis, J.R., 2003. Light, vision and the welfare of poultry. Animal Welfare 12: 269-288.

Rip, A. and Kemp, R., 1998. Technological change. In: S. Rayner and E. L. Malone (Eds.) Human choice and climate change. Batelle Press: Washington D.C., USA.

Rotmans, J., 2003. Transitiemanagement: sleutel voor een duurzame samenleving. Van Gorcum: Assen, the Netherlands.

Rotmans, J., Loorbach, D. and Van der Brugge, R., 2005. Transitiemanagement en duurzame ontwikkeling. Co-evolutionaire sturing in het licht van complexiteit. Beleidswetenschap 19 (2): 3-23.

Schön, D.A., 1983. The reflective practitioner; how professionals think in action. Basic Books: New York, USA.

Schot, J.W., 1992. Constructive technology assessment and technology dynamics: the case of clean technologies. Science, Technology, & Human Values 17 (1): 36-56.

Siers, F.J., 2004. Methodisch ontwerpen volgens H.H. van den Kroonenberg. Wolters-Noordhoff: Amsterdam, the Netherlands.

Simondon, G., [1958] 1989. Du mode d'existence des objets technique. Aubier: Paris, France.

Stake, R.E., 1995. The art of case study research. SAGE: Thousands Oaks,

Taylor, N.R., Prescott, N.B., Jarvis, J.R. and Wathes, C.M., 2002. Can domestic fowl detect the flicker of fluorescent lights? British Poultry Science 43 (5): S13-S14.

Voß, J.-P. and Kemp, R., 2005. Reflexive governance: learning to cope with fundamental limitations in steering sustainable development. In: ESEE conference proceedings, Lisbon, Portugal.

Weaver, P., Jansen, L., Van Grootveld, G., Van Spiegel, E. and Vergragt, P., 2000. Sustainable technology development. Greenleaf Publishing: Sheffield, UK.

Welzwijn, 2006. Welzwijn wel doen. Vernieuwingsinitiatieven in de varkenshouderij. Wageningen UR: Wageningen/Lelystad, the Netherlands.

Widowski, T.M., Keeling, L.J. and Duncan, I.J.H., 1992. The preferences of hens for compact fluorescent over incandenscent lighting. Canadian Journal of Animal Science 72: 203-211.

Chapter 13

How to deal with competing claims in peri-urban design and development: the DEED framework in the Agromere project

Andries J. Visser, Jan Eelco Jansma, Herman Schoorlemmer and Maja Slingerland

Abstract

In the Netherlands the boundaries between cities and the countryside are extremely sharp. It is often not possible to go directly from one to the other because of (ring)roads or ditches separating both environments. The city of Almere (180,000 inhabitants) has to build 30-60,000 new houses in the next 15 years in areas that are now used for agriculture. But it is not only housing needs that have to be met, conservationists, water catchment authorities, and cultural groups all stake their claim on this area. In the Agromere project, involving all relevant stakeholders, we explored the possibility of developing new suburbs for Almere in which city and countryside are integrated in order to include most competing claims. We used the DEED framework that was developed to deal with competing claims of different stakeholders on natural resources. In the different phases of the project different methodologies were used; like scenario planning and stakeholder management. In this paper we describe the process, the results (process and design) and we reflect on the usefulness of the DEED framework and the role research played in this specific case.

Keywords: peri-urban design, DEED framework, Almere, scenario planning, stakeholder management, urban agriculture

13.1 Introduction

Nowadays more than 50% of the world population (3.3 billion people) lives in cities (Martine, 2007). In the Netherlands, this percentage is already higher than 75% (Brockerhof, 2000). In the Dutch urban environment the traditional functions of the countryside like food production, natural heritage and landscape are completely pushed aside resulting in an extremely sharp delineation between city and countryside. As a consequence, people become estranged from the realities of food production, nature and the basic values of rural live, like quietness, the natural interplay of light and darkness and the rhythm of the seasons (Slingerland *et al.*, 2003). As a result, the mental and physical distance between city (and city dweller) and the countryside is increasing.

This sharp delineation is partly the result of Dutch city planning processes (see Figure 13.1) New districts are designed on drawing tables by city planners and urban developers with a strong focus on their building task but with little eye for other perspectives like agriculture, conservation and so on. When areas are assigned for housing development all former functions like agriculture, conservation and recreation are removed. If necessary, this is done through dispossession. This approach has advantages for the developers: the building area can easily be redesigned for new houses, infrastructure, shops, starting from scratch without all of the legal problems associated with development within an existing framework.

Amongst other things, the separation of the urban and the rural worlds has its origins in their separation at the political level, where city development is the responsibility of the Ministry of housing and environment, whereas conservation and landscape development fall under the auspices of the Ministry of agriculture, nature and food quality (LNV) (Gordijn *et al.*, 2003). Another important factor in this sharp delineation is the lack of sufficient development land and the resulting large difference in price between land designated for housing and land for conservation or agricultural purposes (Luijt *et al.*, 2003). As a consequence, the weaker community functions like nature and landscape are pushed aside in favour of the stronger economic functions.

Today, there is a growing interest in re-establishing a connection between city life and country values represented by a green infrastructure and healthy food. For example in London the food strategy was launched (LDA, 2006) as a response to the fact that obesity and diet-related

Figure 13.1. Illustration of sharply delineated city-fringes in the Netherlands (source: Google Earth).

illnesses accounted for a huge number of premature deaths in London, with people on low incomes suffering disproportionately. Inspired by the London initiative, Amsterdam launched the Amsterdam food strategy. Recent findings show the health benefits of an accessible green environment around our cities. (Health Council of the Netherlands, 2005, Vreke *et al.,* 2006). These are examples of the growing trend in society which has to do with quality of life. This has lead to the question: is it possible to integrate agriculture into city development schemes? The research project 'Agromere' was designed to address this question, with the prospect of designing a more sustainable and desireable urban environment.

13.2 The integration of agriculture in city development in the Netherlands

In this chapter, we introduce the case of *Agromere,* an innovative city development design for the Dutch city of Almere. The objective of Agromere is to create a new suburb where agriculture is fully integrated into a city housing estate. Urban farming is already taking place in both developing and developed cities worldwide, including the Netherlands (Dekking *et al.,* 2007; Van Veenhuizen, 2006). In most cases urban farming is about local food production. In addition to food, urban agriculture can provide more services and activities. The farming sector is in a position to shape and manage the green fringe of the city. It can function as energy supplier, water buffer and processor of city waste. The city's need for health care facilities, care for the elderly, childcare and education services are already part of some agricultural enterprises in the Netherlands (Dekking *et al.,* 2007).

In order to integrate city and countryside successfully, new concepts are needed which are preferably supported by *all* relevant stakeholders. This requires a carefully planned and managed development process. For this purpose we used the DEED framework that was developed to deal with competing claims of different stakeholders on limited natural resources.

13.3 The DEED framework as workable approach

Creating innovative concepts with a broad support base for integrating city and countryside is a challenging task, since many stakeholders with different vested interests are involved. All these different stakeholders together make more demands on the land than there is land available to meet these demands. The main benefits that can be expected from these new concepts are found in the long term overall outcome, integrating progress in the three sustainability domains: people, planet and profit. For some stakeholders, short term, less ambitious benefits for their own specific interests may be more attractive to them, hampering their commitment to the long term goals.

In southern Africa competing claims on natural resources is an acute problem, with the poor being most vulnerable to adverse outcomes of such competition. The INREF programme 'Competing claims on natural resources' developed an interdisciplinary and interactive methodological approach for: (1) understanding competing claims and stakeholder strategies; (2) the identification of alternative resource use options; and (3) the scientific support for negotiation processes between stakeholders, with the aim of developing policy interventions that simultaneously improve livelihoods and the sustainable use of natural resources (Giller *et al.*, 2008). The focus in this programme is on the development and testing of a new interdisciplinary methodology that focuses on the role of science in supporting negotiation between and within different stakeholder groups at different levels. The programme developed the DEED framework, a repetitive cycle of investigation starting from Description and working it's way through Explanatory, Exploratory and Design phases (Figure 13.2). Each of the activities and methods used in these phases feeds into and benefits from negotiations between stakeholders, which forms the core of this approach. An important assumption behind the use of the DEED framework is that stakeholders have different worldviews and are driven by different values leading to different perceptions of their environment and of potential future visions for it. This assumes that there is not one objective reality but that there is rather a negotiated 'reality'. This plays a role in all phases. A second assumption is that local resource use not only depends on local stakeholders but also on interests of stakeholders that play at higher scale levels.

Although DEED is being developed for the situation in southern Africa, competing claims on natural resources are present in the Dutch urban environment as well. The Dutch city

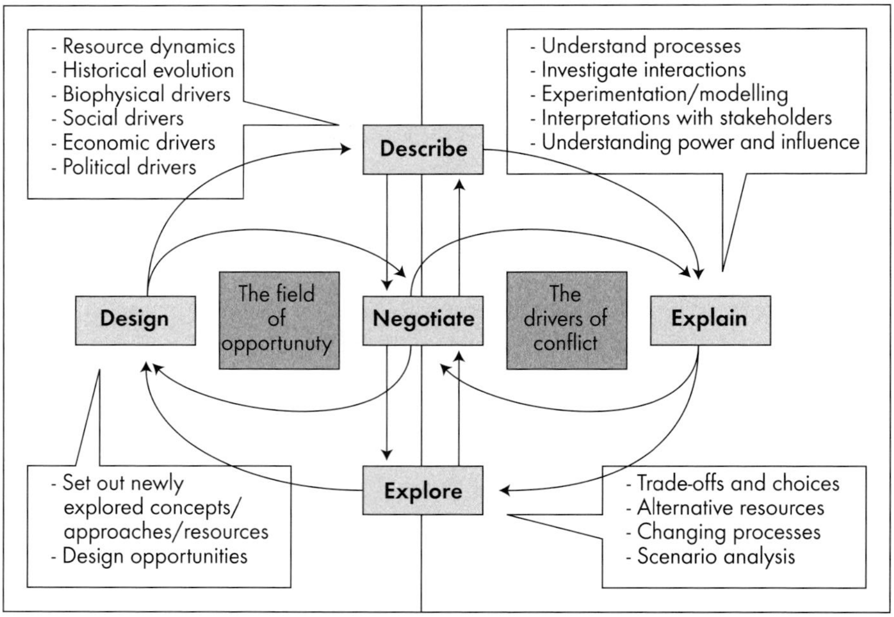

Figure 13.2. The DEED framework: methodological steps to be used in the analyses of competing claims in each locality.

of Almere has to expand because of the growing need for new houses in the West of the Netherlands and in the absence of other places to build. In 2030 with 400,000 inhabitants Almere will have become one of the major cities in Holland. In the area where the new houses are planned, there are several competing claims by a number of different stakeholders for use of the land. For the successful development of an innovative concept for urban farming in this specific area, it was essential that all relevant stakeholders from different scale levels participated in the design process and could fully contribute to the final result. In order to facilitate this, we used the DEED framework for our approach.

13.4 Agromere: towards a joint vision for urban agriculture

In the process towards a joint vision we used different methodologies and approaches in the subsequent phases of the DEED framework. How the different phases were traversed during the Agromere project is described below.

13.4.1 Descriptive phase

In this phase, the present situation in the northeast of the city of Almere (approx. 2,500 ha) was described as well as the historical context using (old) provincial development plans, CBS data and other relevant literature. An advantage was that the researchers could use the extensive database, library and archives of their research institute Applied Plant Research that has experience in the past of research in that area. The most important aspects were the following: The area described is a *polder landscape* that was reclaimed from the sea approximately 50 years ago. This land is one of the best agricultural production sites in the world. Agriculture, in the form of large-scale arable farming, is the main land use activity nowadays. The city of Almere wants to expand into this area with 40 to 60,000 houses as part of a national development plan, since land for new housing estates around the older large cities in the Netherlands is scarce. Besides the need for housing development there are currently other claims from different stakeholders for this scarce and, therefore, valuable commodity: claims for conservation plans, water catchment, preservation of cultural historic sites, industry and infrastructure. At this present time, the interests of the agricultural sector are not included in the developing plans, even though there are about 50 farmers active in the projected development area.

13.4.2 Explanatory phase

Agromere started in 2005 by creating a multi-stakeholder network in Almere. These stakeholders include local farmers, the city council of Almere and Zeewolde (nearby town), the province of Flevoland, nature conservation organisations and commercial city developers. All of these have different claims on that area north-east of Almere In this phase of the investigation, the major drivers behind these different claims on the different scale levels (local, regional, national and global) were analysed and described (Table 13.1). The

Table 13.1. Stakeholders and their claims.

Stakeholder	Interest/claim
City of Almere	Sustainable and attractive new building projects. Unique, green city. 60.000 new houses: national task
Municipality Zeewolde	Preservation of rural character
Nature organisations (3)	Nature development and management. Ecological connection zone (1000 ha)
Project developers (5)	Selling unique houses in new concepts: profit. Own 50% of the land
Department of economic affairs	Owns 50% of the land
Farmers	Continuity, are or will be bought out, look for new possibilities

role of the scientists involved in this case was to do a literature search and to conduct the interviews with the different stakeholders. A good insight in drivers behind the different claims is essential for the negotiation process. Claims may originate from any of the different scales which means that other stakeholders or persons (outside of the group) may be involved and need to be consulted in order to reach solutions (design phase). In fact the explanatory phase consisted of creating a greater understanding of the claims and their origins with the objective of increasing the room to manoeuvre in the next (exploratory and design) phases.

13.4.3 Exploratory phase

To explore the possibilities for the future of the land to the north-east of Almere, the approach using 'future scenarios' was used. The future scenario's approach is based on the assumption that it is important to construct systems or designs which are flexible enough to deal with future uncertainties, rather than to build on known certainties (Van der Heijden, 2005). All the stakeholders were interviewed to ascertain possible uncertainties in developing north-east Almere and the role of urban farming in this development. In a workshop, the different stakeholders explored together how agriculture and city development could be integrated under different future scenarios. Futuristic visuals were drawn for four possible communities of the future (Figure 13.3). It is important to emphasise that these images of the future are not the goal itself but rather used for inspiration. One of the important, positive, side-effects of working with future scenarios, is that the participants feel free to put aside their current vested interest and leave disagreements out of the discussion since these are often coupled to today's reality. Thinking of the future is all about planning for the next generation and results in a very high commitment of all participants. It promotes the search for common ground and provides room to manoeuvre for the design phase.

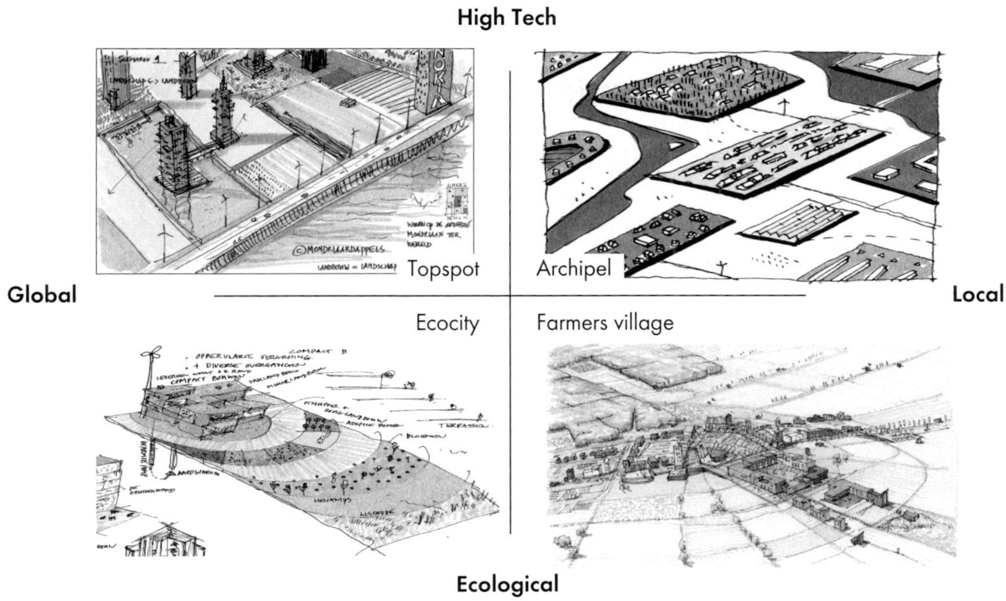

Figure 13.3. Artist's impression of the 4 scenarios.

After the workshop, for each of the four different future scenarios, scientists calculated its impact on the available resources (land, water) and the degree to which it accommodated the different claims (water, housing, nature, etc.) in the projected area. The results were discussed in a second workshop with all stakeholders during which the group verified whether the right assumptions were made. The result of this stakeholder consultancy was that all stakeholders together agreed that only two of the four future scenarios were realistic for the development of north-east Almere. They asked the researchers to integrate these two scenarios and to downscale the image to the level of a township.

Moreover the stakeholders concluded that it was important to operate on three different timescales: short term (1 year), mid-term (1 to 5 year) and long term (5 to 20 years). For the short term focus should be on land and real estate politics in the city. Two actions for the short-term were agreed with the stakeholder group:

1. The city of Almere has already got a city farm which uses future housing land for temporary agricultural production and at the same time provides a destination for excursions for all the primary schools in the city. The city council only issued permission to exploit the site for one year at a time and, as a result, the farm did not have any security for negotiating investment capital. In this way it was not possible to invest structurally. All stakeholders agreed that it was important to solve this problem first before thinking of urban agriculture on a large scale in Almere. The problem was raised at the city council (with support of citizens of Almere who were not directly involved in the Agromere case)

and now a structural solution has been found. Initiatives to achieve this came only partly from the group involved in the Agromere case. This clearly shows that not all action needs to be taken by the people directly involved. Often it is even more efficient to let other interested parties take action.

2. The city of Almere has a relatively a high percentage of green space, which is highly appreciated by its citizens. The costs of maintaining this are however very high and becoming a real financial burden for the city. The stakeholders concluded that involving farmers in the management might result in lower costs. A cycle-tour through the city was organised in which the city parks manager and scientists involved in Agromere made an inventory of suitable green areas in the city for this type of management. So far this has not resulted in any change of management.

Short term action was found to be essential to keep all stakeholders involved in the overall process.The focus for the mid-term was on the structural vision of the city for the development of the land to the northeast of Almere and, in the long term, on developing inspiring concepts for urban agriculture in order to be ready for when the housing development was due to start.

One important stakeholder, the future inhabitant of the housing development, was not present during the different stakeholder consultations. In order to get an idea of the needs and drives of possible future inhabitants, surveys were carried out in Almere among its residents. In the 2005 survey, 342 residents were interviewed by telephone to determine the possible demand for urban agriculture in Almere (Stobbelaar *et al.*, 2006). In 2007 an internet survey in Almere (n=562) explored the criteria for urban agriculture should it become part of an urban neighbourhood (Engelen, 2007). Some of the results of these surveys are shown in Figure 13.4.

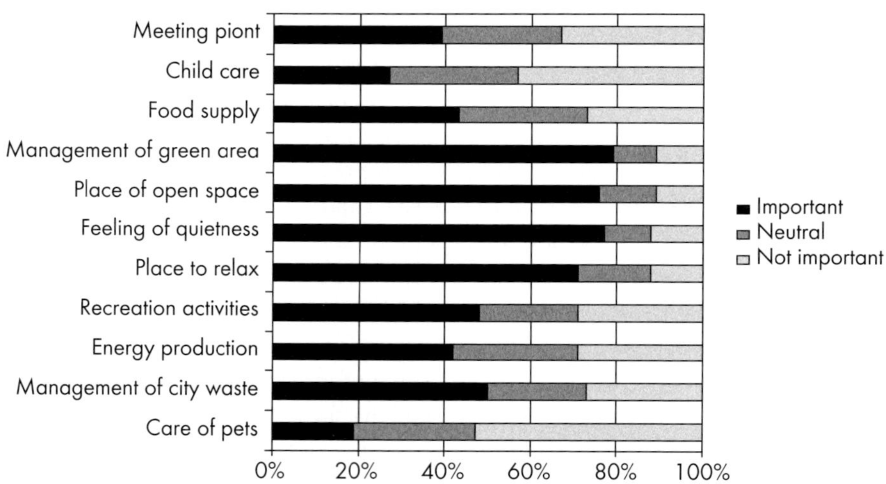

Figure 13.4. Importance of urban farming functions as perceived by urban dwellers (n=562) in Almere, the Netherlands (Engelen, 2007).

14.4.4 Design phase

In this fourth step we designed a virtual township based on the two, by the stakeholders selected concepts (from the future scenario approach) and the perceived desires of the potential inhabitants as they were determined by the two surveys we carried out in Almere. The final design of the township will not be presented here since it is still in development and in discussion with the stakeholders, but we will present some of the basic principles we used. The virtual township is 250 ha with an area of 75 ha for houses, leaving 175 ha for agricultural activities. The township will cater for approximately 5,000 inhabitants resulting in 30 houses per ha. We based the design on the needs of the prospective inhabitants of this township. These will be people who are attracted by the future scenario's 'Farmers village' and 'Ecocity' (see Figure 13.3). Based on research on lifestyles by Motiveaction (Lampert and Van der Leij, 2003) we divided the future inhabitants into 4 different groups: young people aged 25-35, families with children between the ages of 0 and 10, families with children between the ages of 10 and 20 and people over the age of 65. For all of these groups we made an inventory of the kinds of facilities (healthcare, education, daily products, etc.) that were deemed necessary. With these needs as starting point, we started to design a township in which agriculture provides, wherever possible, the different needs of the community. Important aspects of the design are the sustainability principles (PPP); nutrient cycles which are contained within the township, energy that is produced locally (greenhouse en biomass energy) and because of the local production of food, small transparent chains that will result in a small ecological footprint.

13.5 Lessons learned: the case

13.5.1 Multi-scale

The accent on an approach that addresses multiple scale levels of stakeholders was crucial as those at higher scale levels may have a large influence on the local process. For instance Almere city wanted to negotiate with the government for a new road/bridge connection to Amsterdam in exchange for the promise to build 60,000 houses. Hence any local initiatives to determine the way these houses should be accommodated around Almere could be interpreted by the city council as a threat to their negotiating power and hence be opposed. This issue needed to be resolved to provide negotiation space. Another example of a higher level involvement is that of the national development plan which stipulated the inclusion of an ecological zone in the local plans. This stipulation reduced the number of potential future scenarios but more importantly, the process of deciding on the location of the ecological zone ignored farmers' opinions and interests. This had a negative impact on the farmer's motivation to participate and undermined their hope for having any impact on the future. On the other hand, involving these same farmers from the outset in the Agromere project, gave them a forum to express their own ideas about the land on which their farms are situated. This resulted in their commitment to the project as 'finally they were taken seriously'.

13.5.2 Negotiation

Especially at the start of the multi-stakeholder process, negotiation with or between stakeholders has many aspects that are not directly related to the content or the task at hand (more claims than space) but rather related to people's place in organisations, relationships between and perceptions of stakeholders or even characteristics of the individual people involved.

The scientists have to negotiate their mandate as facilitators and the facilitator has to negotiate the participation of each of the stakeholders. The facilitator has to search out people within the stakeholder 'community' that are motivated, willing to be inspired, willing to learn and to act. If necessary, different people from the same organisation are invited/consulted in different stages of the process.

In the design phase, negotiation about the task takes on a much more dominant role because when it comes to implementation of the ideas, stakeholders start realising what the trade-offs will mean for them personally and especially those compromises that are to be expected in the near future.

13.5.3 Sequence and ordering of processes

The DEED framework rightfully puts the exploratory phase before the design phase. Going directly from explanation to design would have meant a lot of energy would have to be spent on negotiating many partial solutions between subsets of stakeholders that would have been difficult to successfully integrate afterwards. The exploratory phase was essential to create common understanding and common ground for a common future and this reduced the number of negotiations down to those that directly contributed to a comprehensive outcome.

In general the DEED framework assisted in sequencing activities, choosing appropriate methodologies in each phase and monitoring and managing the entire process.

Apart from learning about the process (the *how* question), was there also progress on the concepts (the *what* question)? Was it indeed possible to integrate agriculture into cities and to reconnect urban dwellers to the countryside? The future scenarios are examples of creative ways to shape this integration and, on paper, it seems to work. However no work has been done in practice with prospective citizens of the new township yet, therefore it is to early to draw conclusions about this integration of town and country and how it will in reality reconnect urban dwellers to the countryside. In September 2008 a workshop will be organised for the residents of Almere where they will be invited to design their own house and neighbourhood. The results of this workshop will only become available after publication of this book. This chapter reports on ongoing work, hence no final conclusions can yet be drawn on the 'what' question.

13.6 Lessons learned: is DEED a suitable theoretical framework?

13.6.1 The role of the scientist – facilitator, knowledge provider and knowledge broker

Although the DEED framework puts negotiation central it is not explicit about the role of the scientist in this negotiation process except that it assumes that he or she has some connection to local stakeholders. In the Agromere case it became clear that the scientists had acquired a mandate (in the eyes of all stakeholders) to play a role as facilitator. Yet the initial role of the scientist in this project was certainly one of stakeholder. The scientific interest in this project lay in exploring the potential for urban agriculture to solve problems that arise from the sharp delineation between city and countryside. Acquiring additional experience and knowledge in this field through dedicated participation in the process would assist the scientists in the long run to be better equipped to acquire and contribute to commercial projects that require this expertise. This increase in knowledge and ability will also be of benefit to the institution the scientist works for. When a scientist is involved in a multi-stakeholder process, he or she needs to be genuinely interested in the venture and, realising that a locally embedded outcome is essential to successful interventions in future, needs to have sufficient patience and skills to play the facilitator role and at least temporarily set aside his or her own agenda and research ambitions. The scientist has to wait and see if the results turn up anything personally or professionally advantageous and will need to refrain from trying to steer the process (too much). As a result the scientist-facilitator spends 80% of the time on stakeholder management and only 20% on research.

In the Agromere case the facilitator-scientist also provided access to other scientists who were co-opted as knowledge providers. We refer, for instance, to those scientists that were qualified to calculate the impact on land and water use for each of the four scenarios. These scientists do not need to be involved in the multi-stakeholder process at all as long as they provide the requested input. By involving these people, indirectly, the facilitator-scientist partly fulfilled his own agenda as a stakeholder.

Going from exploration to design and especially in the implementation phase the facilitator-scientist has two items on his agenda. He or she wants to:

- transfer the responsibility for the implementation process to other stakeholders: create ownership;
- become stakeholder himself for the execution of some of the future work.

This change of role is not easy or automatic and it remains to be seen if this is possible or whether the facilitator-scientist has to maintain his role and/or gradually move into the role of knowledge broker assisting the stakeholders in accessing relevant scientists that can fulfil the role of knowledge providers.

In fact one should ask another more fundamental question regarding the DEED approach: 'who can best take the responsibility for the process, who can take or be given the role of facilitator?' In fact it is not always convenient to be a scientist-facilitator as it may conflict with the role as scientist-knowledge provider. However the Agromere case showed that the scientists could fulfil both roles as long as it is made explicit, when each role is assumed. In a meeting the scientist-facilitator can step into the role of knowledge provider for a few minutes then step back into the facilitator role for the rest of the meeting. Another option is to pay a professional facilitator to tackle an occasional, specific task in the multi-stakeholder management process (e.g. chairing a session on future scenarios) or, on an itinerant basis, pay a scientist to do specific research (e.g. doing interviews with Almere's residents). Most other stakeholders would have difficulty gaining sufficient trust from the others to be seen and accepted as an impartial facilitator for the 'common good'. If a scientist cannot be the facilitator, who else can do the job?

The assumption underpinning the DEED framework is that the scientists connect to the negotiation process by providing scientific insights to *all* stakeholders for the common good. As well as that, they investigate those questions that are of relevance to specific stakeholders who, thereby, become more or less the scientist's clients. In this case, once they have chosen a client to work for, scientists will have problems being impartial. In fact the role of a scientist as a facilitator is not a specific goal of the DEED approach. The idea is rather that scientists provide relevant tools for all phases of the project and ensure that relevant research topics/ questions are addressed using those tools. In the southern African programme, for which the DEED approach was conceived, the role of the scientist was specifically meant to be one of support for the weaker and poorer local parties to better articulate their claims and to empower them to have their claims either honoured or properly compensated for.

13.6.2 Tools to be used

Stakeholders differ in their ways of expressing themselves, level of knowledge, level of training, etc. To overcome these differences and to prevent some stakeholders having more say and influence than others, three strategies were of particular importance:

1. use of visual scenarios or images;
2. alternate bilateral meetings and multi-stakeholder meetings;
3. alternate informative and decision-making meetings.

The Agromere case clearly showed that employing professional designers to make attractive visuals of the four futures scenarios that the stakeholders put forward, played a significant role in inspiring and motivating all participants. The participants had a visible product which could be presented with pride to anybody. The importance and efforts of the multi-stakeholder process which could easily de dismissed by outsiders, suddenly had a presentable and very attractive result. The visuals gave a tangibility to these future visions, and increased the understanding of these future visions for the stakeholders themselves, as well as proving to be an effective means of communication to outsiders. The advantages and drawbacks of all the scenarios became visible and this lead to two explicit proposals for improved scenarios which better reflected their strongly desired features. One can imagine that in remote areas of southern Africa pictures or visuals could equally play a major role in communication and in increasing understanding of the negotiated outcomes of the exploratory sessions. The Agromere experience strongly promotes use of images as tools in the explorative and design phase of the DEED framework.

In Agromere's case it was quickly discovered that bilateral meetings between facilitator and individual stakeholders were needed to increase understanding of the stakeholders' real interests and to create the trust that the stakeholders interests would be taken seriously in the larger process. These meetings were also needed to convince stakeholders that they would benefit from participating in the larger process or at least risk losing something if they didn't participate. Other reasons to meet bilaterally were to explore specific expertise, former experiences and the current relationships between the stakeholders.

The Agromere project also made clear that information sharing should not be mixed with decision-making. Sometimes stakeholder representatives need time to discuss the implications of the presented information with their constituency. Sometimes information is needed to be further processed, such as making the visuals of the result of the explorative scenario work, before any decisions could be made. In fact it was desirable first to explore all possible and desirable future visions without being forced to decide on any one of them, which would inevitably lead to all kinds of constraints. Decision-making in itself should be very clearly marked by a transparent decision process, followed by attaching responsibilities to the implementation of the decisions. There is nothing more de-motivating than making decisions that lead nowhere, because nobody takes responsibility for enacting them.

13.6.3 Negotiation and learning

For the DEED framework the alternating bilateral and multi-stakeholder meetings and the separation of information sharing from decision-making should be made explicit for the negotiation concept. In fact within the DEED framework diagram negotiation is presented as a central box without explicit content or features. The framework needs to be strengthened by explaining and perhaps setting guidelines for the negotiation process itself. Just as there are methods in the four phases, there are methods and tools that specifically address negotiation and these should be made explicit.

Another fundamental question is whether learning is possible and if so by whom? Will those stakeholders that pursued the DEED framework approach together in the Agromere project be able to do so again with other stakeholders dealing with other problems? Will the scientist-facilitator be able to guide a similar process in another situation? Should negotiation and learning be together in the middle of the diagram instead of negotiation alone?

References

Brockerhoff, M.P., 2000. An urbanising world. Population Bulletin 55 (3).

Dekking, A., Jansma, J.E. and Visser, A.J., 2007. Urban agriculture guide; urban agriculture in the Netherlands under the magnifying glass. Wageningen University, Applied Plant Research: Lelystad, the Netherlands. 20 pp.

Engelen, Chr., 2007. Integratie van stadslandbouw en wonen [Integration of urban agriculture and living; Summary in English]. Student thesis RSO 80436. Wageningen University: Wageningen, the Netherlands. 105 pp.

Giller, K.E., Leeuwis, C., Andersson, J.A., Andriesse, W., Brouwer, A., Frost, P., Hebinck, P., Heitkönig, I., Van Ittersum, M.K., Koning, N., Ruben, R., Slingerland, M., Udo, H., Veldkamp, T., Van de Vijver, C., Van Wijk, M.T. and Windmeijer, P., 2008. Competing claims on natural resources: what role for science? Ecology and Society 13(2): 34. Available at: http://www.ecologyandsociety.org/vol13/iss2/art34/

Health Council of the Netherlands, 2005. Nature and health: the influence of nature on social, psychological and physical well-being. The Hague, the Netherlands.

Gordijn, H., Derksen, W., Groen, J., Lára Pálsdóttir, H.L., Piek, M., Pieterse, N. and Snellen, D. 2003. De ongekende ruimte verkend. NAi Uitgevers: Ruimtelijk Planbureau, Rotterdam, the Netherlands.

Lampert, M. and Van der Leij, B., 2003. Sociale structuurschets: leefstijlen in Almere. Dienst Maatschappelijke Ontwikkeling: Almere, the Netherlands. 19 pp.

London Development Agency (LDA), 2006. Healthy and sustainable food for London. The mayor's food strategy. May 2006. Available at: http://www.londonfoodstrategy.org.uk/upload/pdf/LDA_Food_strategy.pdf.

Luijt, J., Kuhlman, J.W and Pilkes, J., 2003. Agrarische grondprijzen onder stedelijke druk. Stedelijke optiewaarde en agrarische gebruikswaarde afhankelijk van ligging. Landbouw Economisch Instituut: The Hague, the Netherlands.

Martine, G., 2007. State of world population 2007. United Nations Population Fund (UNPF): New York, USA. 99 pp.

Slingerland, M.A., Klijn, J.A., Jongman, R.H.G. and Van der Schans, J.W. (Eds.), 2003. The unifying power of sustainable development. Towards balanced choices between people, planet and profit in agricultural production chains and rural land use: the role of science. WUR-report Sustainable Development. Wageningen University, Wageningen, the Netherlands, pp. 19-24.

Stobbelaar, D.J., Warnaar, M., Jansma, J.E. and Rossing, W.A.H., 2006. Urban oriented agriculture: the case of Almere (the Netherlands). In: H. Langeveld and N. Roling (Eds.) Changing European farming systems for a better future. New visions for rural areas. Wageningen Academic Publishers: Wageningen, the Netherlands, pp. 159-163.

Van der Heijden, K., 2005. Scenario's: the art of strategic conversation. 2nd Edition. John Wiley: Chicester, UK. 365 pp.

Van Veenhuizen, R., 2006. Cities farming for future. In: Van Veenhuizen, R. (Ed). Cities farming for future, urban agriculture for green and productive cities. RUAF Foundation, IDRC and IIRP, ETC-Urban agriculture: Leusden, the Netherlands, pp. 2-17.

Vreke, J., Donders, J.L., Langers, F., Salverda, I.E. and Veeneklaas, F.R., 2006. Potenties van groen! De invloed van groen in en om de stad op overgewicht bij kinderen en op het binden van huishoudens met midden- en hoge inkomens aan de stad. Alterra rapport 1356. Wageningen UR: Wageningen, the Netherlands.

Chapter 14

Governmental strategies and sustainable transitions: monitoring systems for the prevention of animal disease

Catherine J.A.M. Termeer and Geert van der Peet

Abstract

One of the most important issues facing the intensive livestock sector is that of animal disease epidemics which not only cost a large number of animal lives, but also cause social unrest. For more than ten years now, farmers, agricultural organisations, commercial companies in the food chain, researchers and government have been searching for new modes of cooperation to realise a monitoring system as one step in the prevention of animal disease epidemics. In this chapter we look at the development of a monitoring system as an example of the transition to sustainable intensive livestock farming. We will analyse this development with a framework in which a distinction is made between change that is episodic, discontinuous and intentional and change that is continuous, evolving and incremental. In the case of episodic change, actor(system)s are thought to be unable to adapt their underlying structures to the new demands for sustainable development. Drastic outside interventions are needed. In the case of continuous change actor(system)s are viewed as emergent and self-organising. The role of government is more that of giving power to changes by participating, certifying and sense making. Moreover, when governmental actors opt for central steering or control they can even undermine some of the best innovations and some of the most adaptive processes.

Keywords: animal disease prevention, continuous change, episodic change, monitoring systems, governance

14.1 Introduction

In 1997 swine fever broke out in the Netherlands and in the 14 months that followed, the sector faced a complete export ban. Nearly 650,000 pigs from 429 infected farms were destroyed, more than a million were destroyed as a preventive measure, and a huge 8 million suckling pigs had to be destroyed because the nursery farms were bursting at the seams as a result of the complete transport ban for livestock. It was not just swine fever that made 1997 a dramatic year for animal farmers. In that same year the first case of mad cows disease (BSE variant of Creutzfeldt-Jakob's disease) was diagnosed in a cow in the Netherlands. As a result, the Ministry of Agriculture, Nature and Food Quality (Landbouw Natuur en Voedselkwaliteit

crafted by Van de Ven and Poole (1995) who deduce four basic understandings of change: life cycle theories, teleological theories, dialectical theories and evolutionary theories. These theories are classified along three dimensions: change units, change sequences and generative mechanisms. Weick and Quinn suggest to add the tempo of change to these typologies, characterised as rate, rhythm and pattern of activities. They do that by elaborating on the distinction between episodic and continuous change. In their article they compare both points of view by addressing the main properties of a comprehensive change theory. In Table 14.1 we have translated and elaborated on these characteristics in the context of transitions towards sustainability. In this way we have broadened the scope from organisations to networks and policy domains, we have related the process to sustainability and we have paid special attention to governmental actors in the role of change agents. We have also added some examples of intervention strategies based on the social-cognitive schools of change management (Termeer, 2007).

14.3.1 Episodic change and transitions

The label 'episodic change' is used for short periods of fundamental change in values, structures and activities, which then become a basis for a new period of equilibrium (Weick and Quinn, 1999; Boonstra, 2004). During this period, also called a transition period, an old system (structures, procedures, beliefs, activities) is replaced by a new system. Episodic change is closely associated with planned intentional change, in all its varieties.

The basic premise underlying episodic change is that organisations and even people are inert. Inertia is defined as the inability of an organisation to change as rapidly as its environment (Weick and Quin, 1999). In the case of sustainability, organisations or networks or policy communities are not able to adapt their underlying structures to address the new demands of sustainability. Indeed, organisations tend to perceive new demands as threats which unintentionally mobilise defensive routines resulting in the stagnation of learning processes (Argyris, 1990). Inertia is reinforced by organisations or networks that combine characteristics like tightly-coupled interdependencies; powerful beliefs embedded in (sub) cultures; imitation as major learning strategy; and constraints on actions by institutionalised norms (Weick and Quinn, 2004).

The perceived inertia provides the main justification for interventions. After all, planned change would not be necessary if people had succeeded in creating continuous adaptations. Extrapolating from this diagnosis; the best option to start a change traject is to follow Lewin's prescription for phasing change: *unfreeze-transition-freeze* (Weick and Quinn, 1999).

Because the existing equilibrium is the main impediment to change, it must be disrupted in order to enter the transition phase. However many scholars stress the difficulty of unfreezing patterns (Van Eeten, 1999; Schon and Rein, 1994). Explanations for this are found in the defensive routines of people (Argyris, 1990) or the recalcitrance of institutions (Selznick, 1957).

Table 14.1. Transitions as episodic and continuous change (inspired by Weick and Quinn).

	Transition as episodic change	Transition as continuous change
Conceptualisation	Change is infrequent, discontinuous and intentional. Dramatic change in which an old system is replaced by a new one. Transition to a new equilibrium	Change is constant, evolving and cumulative. Change is never finished. Transition is not part of the vocabulary.
Motor of change	External threats, changed environmental demands.	Logic of attraction to continuous updates, endless modifications in social practices and numerous small adaptations
Values for justifying interventions	As people, organisations and networks are inert and fail to adapt to a changing environment outside interventions are needed. The emphasis is on adaptation.	As people, organisations and networks are self-organising, interventions make changes more visible and reduce blockages to adaptation and learning. The emphasis is on adaptability.
Intervention theory	Focus is on inertia. Necessary change is created by deliberate interventions. Equilibrium breaking is needed. Unfreeze, transition, refreeze.	Focus is on change. Interventions are a redirection of what is already under way. Freeze, rebalance, unfreeze.
Role of governmental actors	Prime mover who creates change.	Participant. Sense-maker who redirects change in the direction of sustainability.
Strategies of governmental actors	Loosening up fixations; restructuring organisations; changing meaning systems; implementing new procedures; setting new standards; building social and political commitment.	Encouraging reflection; making conflicts productive; developing new language; alertness to exclusion; coupling and de-coupling.

For this purpose the social-cognitive schools of thought provide some interesting insights. They introduce the concept of social and cognitive fixations (Van Dongen *et al.*, 1996; Termeer and Kessener, 2007). In social fixations, reflection by and about the participants is not possible, nor is reflection possible on the mutual rules of interaction. With cognitive fixations, the contents are fixed and there aren't any openings for other interpretations/meanings. When fixations occur, people are no longer able to reflect and to change their behaviour within the existing context. Trying harder does not suffice. It is only possible to break through fixations by allowing a confrontation to take place in the part of the interaction that still allows for some diversity (Termeer and Koppenjan, 1997). This is the principle of context variation (Voogt, 1991). In the case of cognitive fixation, the intervention is aimed at involving new actors or creating new game rules. Whereas with social fixations a good intervention strategy is to facilitate the contribution of new ideas, context variation is counter-intuitive for many people because many interventions are aimed precisely at emphasising the things that are locked in.

When interventions succeed and fixations have been thawed, new openness arises. In the old forms of planned change this openness was immediately filled with new structures, new procedures and new values. During the transition phase the old program was replaced by a new one, and with it the risk of resistance and new fixations. New forms of intentional change try to avoid these temptations and cherish the openness in order to encourage and restart processes of learning (Boonstra, 2004).

The transition period ends with the re-freezing phase. The focus is on re-establishing stability, building and maintaining social and political commitment and preventing people relapsing into previous routines.

In episodic change, change agents are the prime movers who create change. Within organisations these roles are played by top managers assisted by specialised change managers. In the case of sustainable transitions, governmental actors can be viewed as change agents. However, this point is not unequivocal. Firstly, the role of government in loosening up inertia needs some consideration. When governmental organisations are not involved they can take on the intervention role. However, as we will see in the example of the monitoring system, governmental actors can also form part of the inert system. No matter where they come from it's impossible for participants to break through blockages themselves. After all, only Baron von Münchhausen has ever managed to pull himself out of quicksand by his own hair... At best, actors involved can recognise fixations and invite an 'outsider' in to play the role of intervener (Termeer, 2007). Secondly, recent political research shows the disadvantages of government as a central steering unit. If a government organisation tries to impose a reality with a great show of power, offering resistance is possibly the most rational reaction to be expected by the people affected. Under the motto, 'from government to governance' a shift is taking place from hierarchical and well-institutionalised forms of government towards less formalised forms of governance in which networks have grown in importance (Rhodes, 1997; Pierre, 2000; Blatter, 2003; Klijn, 2005; Arts and Van Tatenhove, 2005). This narrative stimulated the formation of, and proposals for alternative governing strategies, such as network management, interactive governance, collaboration, deliberative policy making and partnerships among other things (Kickert *et al.*, 1997; Edelenbos, 2005; Mandell and Steelman, 2003; Hajer and Wagenaar, 2003; Fischer, 2003; Diamond and Liddle, 2005).

14.3.2 Continuous change and transitions

A description of continuous change is that of an ongoing process of adjustments to, or experiments with everyday contingencies, exceptions, opportunities or unintended consequences. It's grounded in continuing updates of social practices. Each shift in practice creates the conditions for further breakdowns and innovations. Change is emergent, meaning that it is the development of new patterns of organisation without *a priori* intentions (Weick and Quinn, 1999). There is no beginning or end point. The fact that the changes are micro does not mean that they are trivial. Small continuous adjustments can cumulate, amplify and create second or even third order change. Organisations can produce continuous change by improvisation, translation and learning (Weick and Quinn, 1999). Even large bureaucratic organisations, with structures too rigid to adapt to fast-paced change, have people somewhere who are already adjusting to the new environments. In contrast to episodic change, continuous change provides a micro perspective on organising. The concept of transitions is not part of the vocabulary. At best, in a retrospective way, people can make sense of a period as a transition.

Extrapolating from this conceptualisation of change, diversity is seen as source of dynamism and innovation (Termeer, 1993; Van Dongen *et al.*, 1996). Confrontation with different realities, different people or different forms of interaction can lead to ideas arising about new meanings and new options for behaviour. Encountering variation is the engine for change. Contradictions, a good design, surprises, constructive conflicts, 'strange people', unexpected acts, new models, crises, unfamiliar interactional settings or research outcomes are what fuels and allows for learning. It brings the kind of richness that does justice to the complexity of socio-technical systems. Referring to Ashby's law of requisite variety (only variety can beat variety) it takes a complex sensory system to register a complex environment (Weick, 1995). Changing is not only about talking differently, it is also about thinking and acting in a highly flexible way to be able to grasp the variables in an ongoing flow of events. Continuous change approaches change from a voluntaristic perspective, meaning that its underlying assumption is that people have freedom of choice to organise their environment. People change to new positions, not because they are forced to do so, but because they are attracted to them (Cooperrider, 1999).

In the context of continuous change two justifications for intervention can be identified. Firstly, there is the redirection of changes that are already underway (Weick and Quinn, 1999). In our case it means recognising changes that promote sustainability and making them more salient. Secondly, intervention can be aimed at preventing exclusion of people, meanings and game rules (Van Dongen *et al.*, 1996). Phrased in positive terms this is about creating conditions for effective learning processes that enable people to connect with their surroundings in a variety of ways (Van Dijk, 1998).

The norm of preventing exclusion has been subject to a number of misconceptions. Seeing variation as the engine for change does not mean that everything must always remain open.

At regular intervals, meanings will have to be stabilised. This is necessary to come to decisions and actions. So it is that stability can be defined as a temporary agreement that people come to at a particular moment in that local situation (Wierdsma, 2004). Conflicts might arise at the point where a phase difference occurs between people about varying or updating and stabilising. In this sense there is a built-in different of pace between policy makers and entrepreneurs: the policy makers slow down updating so they can fix norms in legislation, and the entrepreneurs accelerate updating so they can continuously adapt their production process to changed circumstances.

In the face of continuous change the system is already unfrozen. Further efforts of unfreezing could disrupt what is essentially a complex adaptive system that is already working. Weick and Quinn suggest that the most plausible change sequence might be freeze, rebalance and unfreeze. For managers or change agents, accustomed to planned change, a different mind set is needed.

To freeze continuous change means making sustainable adaptations visible and revealing patterns in the sequence of events. An important first step is to make open-minded observations in order to obtain an accurate and rounded picture of the way people in different contexts are working on the problems surrounding sustainable agriculture. It is essential to have as broad a view as possible of the diversity of initiatives, meanings, connections, etc. It is not simple, especially in a society that has scarcely accepted contemporary networks. Topics like transition towards sustainable agriculture are hard to comprehend and are ambiguous. Situations are usually not free of obligations for the people concerned. They have substantial economic and political/administrative interests and they often will not take the time or make the mental space for careful open-minded observation. Many examples exist of policy makers who, after having visited some projects, report that there is nothing new out there.

To rebalance is to reinterpret issues as opportunities for sustainable development and make them more salient. An interesting strategy is the use of the logic of attraction. As the name implies people change to a new position because they are attracted to it or inspired by it (Cooperider and Whitney, 1999). There are various strategies for redirecting and stimulating learning and development, such as encouraging reflection, making conflicts productive, seizing dynamics or creating a communal language (Termeer and Koppenjan, 1997; Termeer, 2001). Some other strategies are the use of appreciative inquiry, the alertness to exclusion and coupling and de-coupling. Sometimes interventions are needed when people look at developments with the same frame as the one they used to produce a problem and they are, therefore, unable to see its opportunities.

Finally it is also important to unfreeze and resume improvisation, translation and learning. Freezing and rebalancing have had their function but, in order to prevent 'copy and pasting', processes of ongoing adaptations must be continued. For example a recombination of societal services and agriculture is not the solution for every farmer and in all local situations. The challenge is to resume learning that might result in new combinations.

Following the insights of continuous change we can conclude that transitions cannot be managed. However, because governmental actors feel responsible for, and are involved in different ways in interactions that have to do with sustainable agriculture, it is not even desirable for members of government or public servants of the ministry of LNV to place themselves outside the transition process. If one extrapolates from this idea, government parties that used to regard themselves as central actors will shift their emphasis from steering to participating in networks, chains and activities. Participating is a way of making different realities possible and being involved in that process (Hosking, 2002). It is not aimed at increasing the possibility of accepting decisions nor at improving the quality of the consensus. Aside from being a participant, an important role for governmental actors is the role of interpreter. It is about recognising sustainable developments and making them more visible and salient. In this way governmental actors can use their resources to authorise new meanings and label sustainable patterns in experiments and local adaptations. This is how a shift occurs from initiating transitions to certifying them.

Participation involves obligation. After all, interactions between people are often organised around resources. To be able to participate, the government will have to deploy its own resources, not with the intention of limiting space but with the intention of creating space. This implies an intervention using its own means or resources. In principle the government can use all its policy instruments, relationships or knowledge to achieve this goal. This is on condition that the deployment of these means contributes to an interpretation that is as complete as possible, or more precisely, to the prevention of the exclusion of meanings. In this framework it doesn't matter which means are used as long as the following basic conditions are satisfied (Campell as cited in Weick, 2000): it should animate people and get them moving; it should generate experiments that reveal opportunities; provide direction; encourage continual updating through improved situational awareness and closer attention to what's actually happening; it should facilitate respectful interaction in which trust, trustworthiness, and self-respect can flourish.

14.4 The case of the monitoring system revisited

14.4.1 Chronological description of the case

As outlined in the introduction, the motivation for setting up a monitoring system was born out of the different animal disease crises that have occurred since the middle of the 1990s. The outbreak of Swine Fever prompted the government and the agricultural business sector to set up a communal programme in which a monitoring system and animal health care measures would be directed at prevention rather than eradication. Several attempts to set up a monitoring system failed, and there was an outbreak of an unknown disease in dairy cattle in 1999. At this point, parliamentarians called the Minister of LNV to order by asking questions on the matter. In the Waalkens/Ter Veer motion the proposal was made to set up a central registration site for veterinarian treatments – including the prescription of medicines for farm animals – under the auspices of the Food and Goods Authority (VWA). The object of the motion was to improve government insight into trends and developments in the animal husbandry sector.

On 18 November 2002, the Minister reports to parliament that he supports the ideas behind the motion. He indicates that he will execute the intention of the motion by means of an approach which will gather all the necessary information for a well-founded risk analysis. This comprises of a plan to get specially trained veterinarians to visit farms periodically and he will formulate regulations to ensure that it happens. To that end, the VWA [The Dutch Food and Consumer Product Safety Authority] will develop proposals for a system to be set up by both the government and the business sector to collate the data collected during the periodic visits to the farms. The object of the exercise is for the VWA to analyse the data and, on the basis of that data, publish the trends that are relevant to agricultural policy-making. Not only that, they will assess the relevance of other data that might possibly indicate a danger to the rest of the chain, and decide whether it should be saved in the database. Furthermore, that it is his intention to set up a registration system for symptoms that could possibly form a threat to food safety or animal health.

On 7 March 2003 the Minister explains to parliament how he plans to execute the Waalkens/Ter Veer motion and what the latest developments are. He reports that both the cattle and the pig sectors are working on gaining insight into trends and developments to improve the safety of their products and secure sales. The sectors are doing that with their own quality control systems: KKM for cattle and IKB for pigs. Not only that, the government and the business sector are both working to set up a farm monitoring system. Given this development, the Minister stakes his reputation on a development route that includes farm visits by qualified veterinarians to identify risk profiles which have been set by the government and the sector together. The VWA will then distil the relevant trends and developments out of the centrally registered data.

On 12 December 2003, the Minister reports on progress to parliament and submits a report by the LNV Expertise Centre which outlines the various costs and income from possible future activities surrounding veterinary checks and farm monitoring. He also reports that the setting up of a registration point has been delayed by the Avian Pest crisis. He promises parliament that he will have more information in mid-2004.

This report results in strong reactions from the business sector which leads to extensive talks with this sector with support from Wageningen UR. The business sector is concerned because it is already working hard to assemble data, that this process will be hindered if LNV comes with proposals for collecting data for its own prospective data system with its own analysis of trends and developments. The sector's distrust of the government is great, as is shown by the fear that the government will close down farms for an unnecessarily long time on the hint of a suspicion [of an outbreak]. Furthermore, they declare that the proposed policy to set up the monitoring system conflicts with the policy programme promulgated in September 2003 with the title, 2004-2007 *Vitaal en samen* [Vibrant and together].

After extensive discussions within LNV, this results in 2004 in a letter to parliament on 5 November 2004 with new criteria and a modified approach. The central ideas from *Vitaal en samen* are the key: the business sector is itself primarily responsible for maintaining animal health and food safety, and thus for the collection of data; the bureaucratic burden should be reduced and there should be a reduction in regulations. To this end, according to the advice from the Administrative Burden Reduction commission, there will be a stock-take of all information streams that already exist in the chain, and a check to see if this information is adequate for the government to efficiently execute its responsibilities. In consultation with the business sector, the motion will take further shape under the auspices of the LNV Expertise Centre and Wageningen UR. The business sector will set up a private information registration system taking into account the existing information streams within the government and the business sectors. With this system, the sector fulfils the government's information requirements. Not only that, but the government and the business sector make specific arrangements to guarantee the reliability of the information and the government will ensure that this is observed. Emerging diseases will receive special attention as requested in the motion 'slijters'. In this case, it is important that clear arrangements are made beforehand about the role of the government and that of the business sector. The government has the primary responsibility in incidents when acute action is necessary as in contagious animal diseases or risks to public health, when the business sector cannot be expected to be able to fend off the danger by itself, as in the case of some zoonoses or contaminations (residues). According to proposed EU regulations based on a law on hygiene, animal farmers have to keep a record of information that is relevant for food safety (and animal health). Should the prospective private information gathering system prove inadequate, LNV will consider applying new instruments such as periodic farm visits.

The LNV Expertise Centre and Wageningen UR cooperative project will provide the description of the stakeholders via the methodology of *interactive technology assessment* to arrive at a private monitoring system with roles for both the sector (specifically the primary sector, veterinarians and animal health services [GD]) and the government with regard to the accumulation, management and analysis of the data. Possible bottle-necks are made subject to discussion to enhance trust: by animal farmers for the government and the animal health authorities, by veterinarians and animal health authorities and to establish distribution of the costs. This quickly results in a functioning data analysis system in the dairy farming sector that is placed under the auspices of the GD. In the pig farming sector, they are still experimenting with a similar system whereby veterinarians independently collect data, without making use of the GD data analysis system.

On 23 October 2006 the Minister informs the House how he has executed the Waalkens/Ter Veer motion. Conforming to the modified approach as explained in the letter of 5 November 2004, the motion is addressed by means of a private information registration system with a monitoring programme that is run by the GD. In the beef cattle sector the data is coupled to relevant databanks (Rendac, NRS, I & R, the milk inspection service and GD). In this way, they have made it possible to monitor trends and developments in animal health in the long term. It also means that surprise developments and dubious trends can be addressed [as soon as they are noticed]. With respect to the role division between the business sector and the government, the initial responsibility for action lies with the business sector. The involvement of government is guaranteed by mutually establishing whether further investigation is needed on the basis of perceived trends and developments. To this end there will be a quarterly meeting about trends and developments between LNV and VWA with GD and the sector. As well as that, a registration centre for emerging diseases and possibly dangerous symptoms forms an early warning system for animal health. The responsibility for addressing these signals lies primarily with the sector itself. The government's responsibility is to ensure that this system exists and that it functions as it should. This registration centre is allied to the voluntary registration centre run by GD that is part of the existing monitoring system. On top of that, the GD experts meet weekly to discuss acute problems and incoming information and observations. On a monthly basis, the VWA receives information that has been anonimised and thus remains well informed about the developments that have to do with animal health.

14.4.2 Perspective of episodic change

From the point of view of episodic change, the implementation of a monitoring system has failed. Ten years after the political decision to set up a monitoring system no real results have been reached yet. Members of Parliament's attempts to accelerate the process have failed. Due to lack of support, the ministry's proposals were postponed. The only result is a new compromise to assign responsibilities which should lead to the development and control of a monitoring system. In episodic change, inertia or the relapse into old patterns are the main cause of failed transitions. We mention below possible causes of the endless discussion on the new monitoring system.

First we will look at the relapse into the old neo-corporatist type of agricultural policy. Up until the end of the twentieth century, agricultural policy was formed in a closed policy community, the so-called iron triangle comprising the Ministry of Agriculture, farmers' organisations and agricultural specialists in parliament. The triangle was based on a strong consensus on the agriculture modernisation paradigm. Outsiders, who were not specialised in agriculture, were not interested in or not allowed to have a say in these matters. However, with the growing concern for societal values, the iron triangle itself, as a solid, neo-corporatist arrangement for policy-making became the subject of criticism and was judged to be an important reason for problems in the agricultural sector (Frouws, 1998). In spite of many attempts to break through this arrangement and open up the policy process for previous outsiders, the monitoring case shows that the iron triangle is still very dominant.

A second explanation for inertia is found in the focus on combating. In our case it is remarkable that different stakeholders, including the government, underwrite the importance of animal disease prevention, yet spend most of their time and energy on optimising the strategy for the efficient eradication of animal diseases. Despite the government's policy to implement prevention strategies, combating strategies still dominate the stage.

A third explanation for inertia can be found in what we call institutionalised forms of distrust. One example in our case is embodied in the distrust farmers have of the government. They are afraid that even if there is the slightest suspicion of the outbreak of a disease, farms will be isolated for an unnecessarily long time. So they do not have any incentive to report suspicious circumstances associated with animal diseases. Transferring the responsibility for setting up a monitoring system from the government to the business sector did not change this attitude at first.

The process of excluding variety is another cause. Excluding variety is not characteristic of the agriculture network. Frissen even stresses that exclusion is more or less institutionalised in public governance. Many of our steering and control instruments are directed at combating variety (Frissen, 2003). The intention of any efforts is to freeze that which moves, by wanting to check and control it. As the history of the monitoring systems shows us, people tend to exclude diversity and, therefore, run the risk of excluding possible change. Although the farmers' organisations and the meat industry were already setting up their own monitoring systems, the government decided to take over. By focussing on their own version of a common monitoring system, the government excluded possible partners and undermined their motivation in such a way that they provoked a mood of resistance and distrust.

Inertia can also be caused by deadlocks or fixations. One dominant stagnation point in our case involves the question of ownership and sharing of costs and benefits. This results in a power play in which all actors concentrate on defending their own borders, limiting their responsibilities and enhancing their own power. Farmers distrust both the new role of government and the new role in which the meat industry takes the initiative. The *Gezondheidsdienst Dieren* (GD) [Institute for Animal Health] claims an important role for itself due to its experience and present work on monitoring systems. Veterinarians who are trying to assume some of the GD's tasks, resist its claims. Farmers, in turn, do not accept the GD's monopoly position for fear of escalating costs.

During the policy trajectory some interventions had been undertaken to break through this inertia and to unfreeze the fixations. One of them was the introduction of the method *Interactive Technology Assessment* (ITA) to facilitate an interaction between stakeholders that would lead to an innovation that fulfilled the stakeholders' requirements (Grin and Van de Graaf, 1996). In this case, they drew up a rough definition of the problem with a possible direction for the solution. Subsequently, they used interactive construction methods to realise a tenable design in which technical, social, economic and structural aspects were taken into account. This meant not imposing a blue print, but achieving the design by an iterative succession of interviews and analyses and by continually adjusting the possible direction for finding the solution. This continued until agreement on the solution was reached. It is a feature of the ITA that solutions are not achieved by negotiation but by searching for the common goals underlying apparently contradictory interests. This means that existing conflicts between stakeholders are avoided so that they no longer form a predetermined barrier to the realisation of the transition. Reflecting back on the ITA it can be concluded that they had to deal with many more deadlocks and fixations than expected.

14.4.3 Perspective of continuous change

In order to discuss the case from a perspective of continuous change we have to reformulate the transition goals. It is not about the realisation of a sustainable system including monitoring but it is about the process taken to arrive at a way of organising the agricultural sector so that, in a continually changing world, it will result in sustainable development that also takes account of consumers, the general public and other important stakeholders. From this point of view we talk about a transition in which animal disease prevention is an integral part of the thinking, acting and interacting of all actors involved. The development of a monitoring system is a small but significant step in this process. Small changes can create conditions for further breakdowns and innovations. Talking about a monitoring system brings the abstract discourse about prevention more down to earth and can motivate people to experiment with prevention strategies.

Following the storyline of continuous change you get a picture of drastic changes in a sector that still survives, despite 'urgent problems for more than 30 years'; of livestock farmers who have very different ways of giving meaning to developments and adapting their businesses to them; of new social parties that have become involved in animal farming and are looking for links between animal farming and what they believe is important. None of the actors involved doubt the urgency of prevention. Getting 'prevention' in farmers' and politicians 'mind is no longer an issue. Ten years ago sharing responsibility with private actors was very controversial as government was used to doing it on their own as they did with combating policy. Nowadays nobody questions shared public-private responsibility. From this point of view many 'small' changes can be observed.

The various calamities in animal husbandry like Swine Fever, Foot and Mouth Disease, Avian Disease, which lead to the destruction of many animals, has caused enormous economic and social damage. An increasing public interest in the sector has been created with a critical judgment over the way animals are kept. The animal husbandry sector has reacted, conscious that something needs to happen. A realisation has grown within the sector that the gap that has evolved between them and the general public has to be bridged. The concern that the sector will not survive another calamitous outbreak of animal disease has lead to measures that reduce the risk of another outbreak. This was displayed in 2000 when a group of pig farmers entered into a dialogue with the community. The pig farmers and animal rights groups got together to work on a license to produce (Backus and Van der Schans, 2000). Instead of shutting themselves off from the community, the pig farmers now investigate their options together with community organisations for accommodating their desires in a competitive globalising market. Small steps have been made which result in innovation agendas for the different sectors. With these innovation agendas the sectors themselves have assumed their responsibility and work on a sustainable, competitive form of animal farming.

In spite of setbacks, the government has become conscious of the fact that it is no longer able to solve the complex social problems that the animal husbandry sector faces. The government has adjusted its policy with a view to its new role. The sector is responsible for its own future. Government plays a facilitating role as the sector attempts to achieve its goals. One explicit government action has been the initiating of dialogue about the future of animal husbandry by holding meetings in diverse regions in the autumn of 2004. This has resulted in a final congress where government and the business sector have agreed on plans for the future of animal husbandry. Government acted as facilitator and the business sector made commitments aimed to effect self-determination and transparency. The Ministry has staked its wherewithal on facilitating in innovation agendas as is shown with its facilitating of research and the opening up of innovation regulations for pioneering animal farmers.

As a result of government withdrawal, its own organisations, such as the former GD have been privatised. This means that the new GD has to operate in the market and has to earn its way in the market as well. This is an institutional transition that has lead to [new] blockages that need to be cleared. The GD is still seen as branch of government and has to win over feelings of mistrust; the GD has to charge market prices and this is not understood nor accepted either by governmental organisations or the sector; the GD is seen as a monopolist that is judged by some parties as not having to conform to market forces; a shrinking number of animal farmers en veterinarian practices who see the GD as a competitor.

The increasing competition in the world market has necessitated mergers between abattoirs. A consolidated situation has arisen whereby one abattoir has 70% of the national pork market in it hands. A chain leader has emerged who has taken responsibility for pork production. The size of the abattoir implies a huge investment capability which requires that the risks should be minimalised as much as possible. That has resulted in a far-reaching fulfillment of the responsibility for transparency, improved food quality, set-up of tracking and tracing, preventive measures against calamities such as contagious animal diseases.

14.5 Concluding reflections

14.5.1 Lessons learned: the case

In the transition to sustainable agriculture in general and specifically the transition to an integrated system of animal disease prevention, we concentrate on the role or roles of actors from the Ministry of LNV and other relevant authorities. In recent years, successive ministers had a clear opinion about what they did *not* want to be, namely an authoritarian director: 'In the knowledge that the future is not determined from The Hague and that different parties in society are prepared to develop initiatives, the role of the government is not that of leader'[21]. In defining their roles in a negative way (what *not* to be) the ministers were complying with current political, scientific and societal insights. In letters and speeches from the ministry of LNV new roles are described in terms such as facilitate, inspire, help, support, urge, offer space, eliminate restrictions and connect. We can conclude from this that the development of new roles, and with them the identity of the ministry, is still under construction.

It is, however, precisely this construction process that complicates the governmental role in the transition process. The farmers are confused about the government's new role and they also distrust the new role taken on by the meat industry and the possible repercussions for their own business. The *Gezondheidsdienst Dieren* (GD) [Animal Health Service] is in transition from being a governmental organisation to a market oriented organisation. Their experience and current work on monitoring systems plays an important role for the future system logic. Farmers and, particularly, veterinarians distrust the GD because they see it as having a monopoly position and as being too expensive. This results in farmers not accepting the GD in their new market oriented role and veterinarians trying to take over the tasks of the GD.

14.5.2 Lessons learned: the theory

In this chapter we have described a framework to analyse transitions from the perspective of episodic and emergent change. Frozen inert systems, too rigid to adapt have to be unfrozen to be improved. The remaining question is how do both perspectives relate to one another. On this point, Weick and Quinn argue that most managers are much too focussed on inertia. If managers, or in our case governmental actors, take more notice of emergent change and its effects they can be more selective about new policy and regulations. To take notice also means to become more aware of personal assumptions about inertia. 'The wise leader sees emergent change where others see only inertia and pretexts for planned change... the problem is that when top management opts for planned change it often discards some of its best innovators, some of its best innovations and some of its most adaptive processes' (Weick, 2000: 238).

[21] The Cabinet's vision for the future of intensive livestock farming 19.12.2003

What we have revealed in our case is that it are not only governmental actors who have their assumptions on change and preferences for change strategies. The other actors in the policy network also develop their own diagnosis and preferred governmental strategies. When the former minister of LNV began working on this problem by placing the responsibility for animal friendly and environmentally friendly livestock farming back into the hands of the business sector and that of the consumer he met a lot of resistance. The green political parties argued that the Minister was doing too little to force changes in livestock farming; they believed that the community wanted politicians to act on its behalf. They expected the Minister to stick his neck out more. They missed a steering role from the government and they felt that things were happening too slowly.

Within the role of participant and interpreter, government actors will often achieve relatively small gains. These are not the huge heroic deeds that allow politicians to score extra points in the run for office. Emergent change consists of ongoing accommodations, adaptations and alternations that produce fundamental change. New realities, relationships or games are not mastered overnight. This requires time and patience. People need to be able to experiment in their own situation, to see how things work when done differently and share these experiences with colleagues. This is in contrast to the inherent need for (dramatic) results which drive public managers and politicians. The challenge is to make sense of these small gains in what Yanow named 'a spirit of passionate humility' (Yanow, 2003: 246).

References

Argyris, C., 1990. Overcoming organizational defences: facilitating organizational learning. Allyn and Bacon: Boston, USA.

Arts, B. and Van Tatenhove, J., 2005. Policy and power: a conceptual framework betweem the 'old' and 'new' policy idioms. Policy Sciences 37: 339-356.

Backus, G.B.C. and Van der Schans, J.W., 2000. Varkenshouders in dialoog met de samenleving. Report. LEI: The Hague, the Netherlands.

Blatter, J., 2003. Beyond hierarchies and networks: institutional logics and change in transboundary spaces. Governance, an International Journal of Policy, Administration and Institutions 16(4): p. 503-526.

Boonstra, J.J. (Ed.), 2004. Dynamics of organizational change and learning. John Wiley & Sons Ltd.: West Sussex, UK.

Cooperrider, D.L. and Whitney, D., 1999. A positive revolution in change: appreciative inquiry. Wiley: New York, USA.

Diamond, J. and Liddle, J., 2005. What are we learning from the partnership experience. Public Policy and Administration 20(3): 1-3.

Edelenbos, J., 2005. Institutional implications of interactive governance: insights from Dutch practice. Governance, an International Journal of Policy, Administration and Institutions 18(1): 111-134.

Fischer, F., 2003. Reframing public policy: discursive politics and deliberative practices. Oxford University Press: Oxford, UK.

Frissen, P.H.A., 2003. Een jaar later, een pleidooi voor ontregeling. Bestuurskunde 7: 314-322.

Frouws, J., 1998. The contested redefinition of the countryside, an analysis of rural discourses in the Netherlands. Sociologia Ruralis 38 (1): 54-68.

Grin, J. and Van de Graaf, H., 1996. Technology assessment as learning. Science, Technology and Human Values 20: 72-99.

Hajer, M.A. and Wagenaar, H. (Eds.), 2003. Deliberative policy analysis. Understanding governance in the network society. Cambridge University Press: Cambridge, UK.

Hosking, D.M., 2002. Constructing change: a social constructionist approach to change work. Katholieke Universiteit Brabant: Tilburg, the Netherlands.

Kickert, W.J.M., Klijn, E.H. and Koppenjan, J.F.M. (Eds.), 1997. Managing complex networks strategies for the public sector. Sage: London, UK.

Klijn, E.H., 2005. Networks and interorganizational management. In: The Oxford handbook of public management. Oxford University Press: Oxford, UK, pp. 257-281.

Korsten, A. and Hoppe, R., 2006. Van beleidswetenschap naar kennissamenleving: voortgang, vooruitgang en achteruitgang in de beleidswetenschap. Beleidswetenschap 20(4): 34-72.

Lindblom, C., 1959, The science of muddling through. Public Administration Review 19 (2): 79-88.

Mandell, M.P. and Steelman, T.A., 2003. Understanding what can be accomplished through interorganizational innovations. Public Management Review 5(2): 197-224.

Pierre, J. (Ed.), 2000. Debating governance. Oxford University Press: Oxford, UK.

Rhodes, R.A.W., 1997. Understanding governance: policy networks, governance, reflexivity and accountability. Open University Press: Milton Keynes, UK.

Rotmans, J., Loorbach, D. and Van der Brugge, R., 2005. Transitiemangement en duurzame ontwikkeling: coevolutionaire sturing in het licht van complexiteit. Beleidswetenschap 19(2): 3-23.

Selznick, P., 1957. Leadership in administration, a sociological interpretation. Harper and Row, New York, USA.

Termeer, C.J.A.M., 2001. Hoe je een boer aan het lachen krijgt. In: T. Abma en R. In 't Veld (Eds.) Handboek beleidswetenschap. Boom uitgevers: Amsterdam, the Netherlands, pp. 366-375.

Termeer, C.J.A.M., 1993. *Dynamiek en inertie rondom mestbeleid; een studie naar veranderingsprocessen in het varkenshouderijnetwerk.* Dissertatie. Vuga: The Hague, the Netherlands.

Termeer, C.J.A.M. and Koppenjan, J.F.M., 1997. Managing perceptions in networks. In: W.J.M Kickert, E.H. Klijn and J.F.M. Koppenjan (Eds.) Managing complex networks, strategies for the public sector. Sage: London, UK.

Termeer, C.J.A.M., 2007. Vital Differences. On public Leadership and societal innovation. Inaugural Speech, Wageningen University: Wageningen, the Netherlands.

Termeer, C.J.A.M. and Kessener, B., 2007. Revitalizing stagnated policy processes. Journal of Applied Behavioral Science 433: 232-256.

True, J.L., Jones, B.D. and Baumgartner, F.R., 2007. Punctuated equilibrium theory: explaining stability and change in policymaking. In: P.A. Sabatier (Ed.) Theories of the policy process. Westview Press: Colorado, USA, pp. 155-187.

Van de Ven, A.H. and Poole, M.S., 1995. Explaining change and development in organizations. Academy of Management Review 20(3): 510-540.

Van Dijk., N., 1998. Organiseren zonder grenzen, een grenzeloos bezit. In: Pasman, H., Snijders, G.M.F. and Van Andel, W.J.M. (Eds.) Overnemen een hele onderneming. Kluwer: Deventer, the Netherlands, pp. 389-413.

Van Dongen, H.J., De Laat, W.A.M. and Maas, A.J.J.A., 1996. Een kwestie van verschil. Eburon: Delft, the Netherlands.

Voogt, A.A., 1991. Managing of social cognitive configurations in a multiple context. In: R.J. In 't Veld, L. Schaap, C.J.A.M. Termeer and M.J.W. Van Twist (Eds.) Autopoiesis and configuration theory: new approaches to societal steering. Kluwer: Dordrecht, the Netherlands.

Weick, K.E., 1995. Sensemaking in organizations. Sage: London.

Weick, K., 2000. Emergent change as a universal in organizations. In: M. Beer and N. Nohria (Eds.) Breaking the code of change. Harvard University Press: Harvard, USA, pp. 223-243.

Weick, K.E. and Quinn, R., 1999. Organziational change and development, episodic and continious changing. Annual Review of Psycholoy 50: 361-386.

Wierdsma, A., 2004. Beyond implementation: co-creation in change and development. In: J.J. Boonstra (Ed.) Dynamics of organizational change and learning. John Wiley & Sons Ltd.: West Sussex, UK.

Yanow, D., 2003. Accessing local knowledge. In: M. Hajer and H. Wagenaar (Eds.) Deliberative policy analysis. Understanding governance in the network society. Cambridge University Press: Cambridge, UK.

Chapter 15

Institutional innovation and stakeholder engagement: linking transition management in the North with development in the global South

Jim Woodhill

Abstract

Transition management in the North and 'development' in the South[22] potentially have much to learn from each other. Yet they remain relatively disconnected fields of theory and practice. This chapter points to learning options by reflecting on the evolution of institutional innovation and stakeholder engagement within the development sector. The chapter draws attention to emerging ideas within international development, many rooted in vast experience with participatory methodologies. The perspective on institutional innovation and multi-stakeholder processes presented is influenced by current thinking on complexity, evolutionary processes in human systems and advances in the understanding of human cognition. Such a perspective challenges the extent to which transition can be technologically driven, expert-led or 'rationally' planned.

Keywords: stakeholder engagement, institutional innovation, learning dynamics

15.1 Introduction

For nearly half a century, the development sector has been struggling with the challenges of 'transition' in the agricultural systems and rural areas of countries with developing economies. The results have been mixed (Andriesse *et al.*, 2007; Easterly, 2006). The unfortunate reality is that, despite the world having sufficient food at present, vast rural populations remain severely impoverished and food insecure (World Bank, 2008). Further, the current use of natural resources for agriculture is more often than not unsustainable and the mass out-migration from rural areas to mega-cities creates many serious problems. These failures go hand in hand with the 'economic miracles' of many Asian countries that have unquestionably lifted many out of poverty. The Asian successes contrast markedly with economic stagnation in Africa and varied success in Latin America. However, in all regions and countries widening inequality and ecologically unsustainable resource use (World Bank, 2008) hang over us like a dark cloud.

[22] I will use the term South as shorthand for countries with developing and transitional economies who have received development assistance from multi and bilateral donors.

The past two decades of agricultural and rural development have included much work on participatory approaches aimed at empowering people to design and manage their own development. This has occurred against a changing backdrop of development policy and strategy variously pushed by donor countries, multi-lateral agencies and the governments of countries in the South.

Catalysing and managing 'transitions processes' can be seen as being at core of what agricultural and rural development in the South tries to do. Yet, 'transition' is not in the development lexicon. The worlds of transition management in the North and development in the South appear to operate largely in ignorance of each other. For a long time the knowledge push has been from North to South. This chapter offers a reading of the development journey in the South, particularly experiences of participatory approaches for stakeholder engagement, which are rich in lessons for transition management in the North. It will focus on the evolution towards institutional innovation in the development sector and the implications of this for multi-stakeholder processes. The chapter offers a conceptual perspective on institutional innovation and then outlines a practical framework for guiding the design and facilitation of multi-stakeholder processes. In this way we aim to link transition approaches in the North and development in the South. The key message is that these two largely disconnected fields of work and discourse have much to learn from each other. To illustrate this, the chapter will reflect on the challenge of institutional innovation and the potential of multi-stakeholder processes. To ground the discussion, the example of linking poor small-scale producers to modern markets will illustrate key points.

15.2 A perspective on the history of rural development

The history of support for agricultural and rural development in the South can be understood in terms of three main eras: a technological era; a local participatory era; and now an institutional era. This shift of focus is important to understand in relation to the current challenges facing development and in building the case for 'institutional innovation'. It also potentially offers insights for transition management in terms of focusing our efforts.

15.2.1 The technological era

This era characterised the earlier development efforts of the sixties, seventies and early eighties. It was driven by a concern over the capacity of the world to feed the rapidly growing population in the South; a view that more productive and efficient agriculture was a precursor to industrial development; and, a relatively unquestioned belief in the potential science and technology to overcome poverty. The focus was on improving crop varieties, the use of chemical fertilisers and pesticides, infrastructure and agricultural extension. This was largely driven top down with an assumption that the solutions to problems lay with the expertise of the scientists. A significant impetus was given by the establishing of the Consultative Group on International Agricultural Research (CGIAR) in 1971 and its global network of crop based international research centres (CGIAR, 2009). This paradigm gave rise to the Green Revolution in Asia and without doubt also to a massive increased potential for food production.

15.2.2 The local participatory era

Despite the overall increased potential for food production, by the mid eighties it was becoming clear that a disciplinary and technologically based approach also had severe limitations. What scientists may have thought was a good idea often proved impractical or unacceptable to farmers and local people. Further it was becoming clear that at a local level much more was required than simply technological solutions (Chambers *et al.*, 1989). These limitations gradually gave rise to the paradigm of participatory development as it is understood today. It began with the idea of Rapid Rural Appraisal (RRA). This involved a multi-disciplinary team of experts spending a week rapidly assessing the problems and needs of local people as a basis for designing hopefully more holistic and effective intervention projects. Within a few years, this inspired what became known as Participatory Rural Appraisal (PRA) where the focus was on local people doing their own analysis and planning and at least in theory driving their own development process (Chambers, 1994).

PRA, and what is now known as Participatory Learning and Action, led to a massive development of participatory methods and tools. These highly visual techniques were designed to support often illiterate communities learn about their changing contexts, engage in collective planning processes, manage implementation and monitor results. In the late eighties and early nineties, a whole paradigm of participatory development evolved, which included a clear agenda about strengthening the voice of the poor, women and marginalised groups in development. There was a shift from seeing development as a technological process to one of local empowerment. Associated with this was a widespread training agenda on participatory approaches (Pretty *et al.*, 1995). This filtered through many parts of the development sector, leading to widespread use of facilitated stakeholder learning and engagement processes on all themes and at all scales. Participatory approaches in the development sector also drew inspiration from the participatory methodologies from the United States, such as the 'Technology of Participation' (Spencer, 1989), the work on experiential learning of David Kolb (Kolb, 1984) and, for some, the Soft Systems Methodology of Checkland (Checkland, 1981).

15.2.3 The institutional era

From the mid nineties on, the limitation of local level participation and empowerment started to emerge. The critiques ranged from simplistic assumptions about community harmony (Guijt and Kaul Shah, 1998) to outright abuses (Cooke and Kothari, 2001). From another perspective, more fundamentally, the constraints lay in the disconnection between the focus of participation aficionados on development and poverty reduction at a local level and the higher scale institutional factors over which local people often had little influence. These include land tenure, inequities in global trade of agricultural commodities, impacts of economic structural adjustment programmes on local service delivery, the rigidities of aid systems or the consequences of corruption.

The institutional era has also been associated with a major change in discourse by aid agencies, NGOs and academics. The need to focus on higher level 'policy' issues led many NGOs to shift their attention from just direct service delivery and support to a more political agenda often framed in terms of '(human) rights based approaches' (Cornwall and Nymu-Musembi, 2005). Aid agencies moved away from a focus on local level projects to give more attention to 'good governance', sector wide approaches', decentralisation and donor harmonisation and aid effectiveness.

The last decade has also seen a major focus on 'market driven' rural development. This reflects a view that it is ultimately markets and entrepreneurial activity that will lift people out of poverty. The big questions then become how to create institutional arrangements that can 'make markets work for the poor' (Ferrand *et al.*, 2004) or, for that matter, 'work for the environment'.

Take, for example, the issues of linking small-scale producers to modern markets (Vermeulen *et al.*, 2008). At the consumer end, societal norms and values around concern for the poor in the South are a major influence on agribusiness behaviour. Private and public food safety standards influence the entire chain. At the producer end, farmers' capacity to organise and work collectively to achieve economies of scale and meet standards is critical. Foreign investment regulations influence the penetration of supermarkets in emerging economies and taxation arrangements have a big influence of the incentives for small scale producer to engage with the formal economy.

Such institutional innovation along an entire value chain requires constant and effective communication and coordination between different actors and much joint learning. It cannot be imposed by government or generated only by well intentioned efforts of NGOs working at a local level. Even if large businesses do have a sense of corporate social responsibility, many institutional changes are needed to overcome token window dressing. The work of the Re-governing Markets Programme (2008) showed how few actors along the chain engage with each other to understand each others' problems and seek win-win solutions.

At first glance such work to improve the institutional arrangements of a value chain might seem like a clear problem with clear actions that need to be taken to improve the situation. If only it were so simple! Despite much effort to work on this institutional level change is slow and difficult. The next section discusses in detail how some in the development sector are starting to think differently about the notion of institutional innovation. One of the major failings of development interventions has to been tackling such challenges in an engineering orientated way with a belief that social (and economic) change can somehow be planned and executed in a linear fashion (Beinhocker, 2005; Easterly, 2006).

15.3 Institutional innovation

Easterly (2006), in his devastating critique of the aid system, argues that there is a need to reorient from a planning mindset to a searching mindset. This seemingly simple observation points to a world of radical change in how 'development' is to be understood, the role of government and in how different stakeholders collaborate together. Much of this hinges on developing a capacity for 'institutional innovation' (the definition of which I tackle in the next sub-section).

In response to the questionable track record of development effectiveness, several areas of theory and practice are opening up that may enable a radical re-understanding of the challenges and strategic options within the current institutional era. These 'new kids on the development block' also have potential to inform transition management in the North.

This section discusses three essential themes central to institutional innovation: understanding institutions; recognising complexity; and accounting for power. This leads to the idea of collective distributed cognition as a principle for institutional innovation.

15.3.1 Understanding institutions

The starting point for institutional innovation is clarity about what institutions actually are. Despite the importance of the concept, it is often understood differently in theory (Hodgson, 2006) and in practice with a persistent confusion between the concepts of 'institution' and 'organisation'. For that reason it is important to create a shared language that can then enable deeper but practical dialogue about institutions.

Based on the institutional economics work of North (North, 1990, 2005), institutions can be understood as the 'rules of the game' that make ordered social life possible. Language, currency, marriage, property rights, taxation, education systems and laws are all examples of institutions. By definition, institutions are the more stable and permanent aspects of human systems. Many institutions have evolved without much conscious design, and they interrelate with each other in a complex network. The rules of language make it possible for laws to be established, and these laws are then upheld by courts and policing systems. People obey laws because of a whole system of societal beliefs, values and norms. Our lives are embedded in this highly complex web of social institutions, and we take many of them for granted, often not questioning their origin or the underlying assumptions and beliefs on which they are based.

In working with stakeholder groups around change processes, such as value chain development, it has proved necessary to be more explicit about different institutional dimensions. The framework given in Figure 15.1 has proved helpful in prompting a deeper analysis of important institutional factors. For pragmatic reasons it deliberately takes a broad perspective on what institutions are. This means including organisations and regular patterns of behaviour alongside the notion of institutions as 'rules'.

Formal and informal institutions are equally important, and often reinforce each other. Institutional analysis often focuses too much on formal rules, such as policies and laws. This framework shows the importance of asking questions about a wider set of factors that interact to shape the incentives for actors to behave in particular ways.

One key issue for linking farmers in the South to markets through value chains is food safety. This can be analysed using the framework as follows (Vermeulen *et al.*, 2008; Woodhill, 2008). Consumer beliefs ('meaning') – perhaps about the health risks of genetically modified organisms– and buying behaviour ('action') have a significant role in shaping business strategy and government policy making ('control'). A framework for scientific understanding and research ('meaning') underpins food quality and safety regulation and procedures. Organisationally, government agencies are responsible for food safety issues, and many different businesses interact along the value chain ('association'). Government food safety agencies are mandated to develop policies and establish rules and regulations, while the agrifood industry independently develops its own policies, standards and rules to meet consumer demands and legal requirements ('control'). These arrangements lead to the institutionalisation of supporting actions, such as regular monitoring of imports by a food safety authority or agribusiness introducing bar coding and tracing services ('action'). Some behaviours ('action') by different actors, including corruption, may disregard the formal rules and be driven by informal customs and rules ('control').

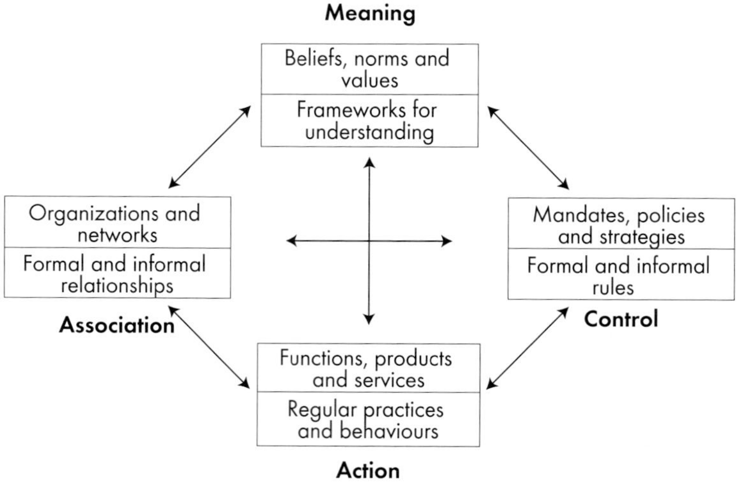

Figure 15.1. A framework for exploring the complexity of institutions.

Essentially any social change involves changing the incentives for how people and organisations behave which in turn means changing institutions. However, what a deeper understanding of institutions quickly makes clear is that incentives for behaviour come from a complex and highly interconnected web of institutional factors (North, 2005). Informal institutions, over which government, for example, has little influence, may be just as influential as formalised institutions such as laws. This complexity has major implications for policy making and planned change, and is a key reason why narrowly conceived policy interventions often fail or have all sorts of unintended and perverse impacts.

15.3.2 Recognising complexity

In the development sector the struggles that have occurred around trying to support institutional change through linear models of planning and intervention has led to a growing interest in the potential role for complexity thinking (Mowles *et al.*, 2008; Ramalingam *et al.*, 2008; Wageningen International, 2008).

Certainly ambitions for guided or directed institutional innovation need to be firmly grounded in an understanding of the complexity and uncertainty of human systems. Human societies, with their economic, social and political systems, like ecological systems, are complex and adaptive. Unlike ecological systems, in human systems the actors are also conscious and purposeful (reflexive) in how they act within the system. This adds yet another dimension of complex dynamics. Complexity thinking has enormous implications for how to consider institutional innovation and what may or may not be possible in terms of purposeful goal driven change.

The essence of complex adaptive systems is that order emerges 'bottom up' through the independent yet coordinated action of many individuals. Through a complex network of feedback mechanisms the system is constantly evolving in response to changes in both the internal and external conditions. Small inputs into the system can have very large (non-linear) impacts on the system's overall behaviour as can small changes in starting conditions. Complex adaptive systems also exhibit patterns of behaviour linked to what are called attractors. An attractor is what a system settles towards in a state of dynamic equilibrium that can be seen as a particular (and often complex) pattern of behaviour (Beinhocker, 2005; Ramalingam *et al.*, 2008).

There is still much debate about the transference of complexity theory from the physical sciences to human systems (Paley, 2007). Nevertheless there are some very clear implications for 'development' and 'transition management'. First, human systems are not predictable. No amount of up-front scientific analysis is going to elucidate cause and effect relationships that can be used to comprehensively plan social (including economic) change. This might seem like stating the obvious. However, the attention given to evidence based policy, good and best practices, results based management and even transition 'management' all reflect an underlying linear science mindset based on gathering data from experience about cause and effect relations in order to design an intervention that will have a predictive outcome. For a deeper understanding of the link between different levels of complexity and cause and effect relationships the work of Snowden and Kurtz on the Cynefin Framework is very insightful (Kurtz *et al.*, 2003; Snowden and Boone, 2007).

15.3.3 Accounting for power

A 'second generation' participatory development associated with the institutional era of development has brought much more attention to power dynamics (Cornwall and Pratt, 2002; Guijt and Kaul Shah, 1998). In essence this means recognising that institutional innovation is a 'political project'. It involves disrupting established personal, economic and decisional power dynamics.

There are many different frameworks and perspectives on power analysis that have emerged over recent times in the development sector, particularly from rights-based organisations which are increasingly focused on understanding power relations as a way to strategise. One such example is the power cube developed by Gaventa (2005) that looks at invisible, hidden and visible forms of power and how these play out in open, closed and claimed spaces for decision making.

There is no room here for a longer explanation of power and how it relates to processes of change. However, the role of power dynamics in complex adaptive systems is clearly an emerging area for exploration. This is a domain where complexity science as it has developed in the natural sciences has to be further developed for relevance in the social sciences.

15.3.4 Distributed collective cognition

If development, institutional change or transitions are to be goal directed and intentional, what are the mechanisms for such guided social change? Emerging from the points raised in the previous sections society seems confronted with a considerable dilemma. On one hand it is clear that there are serious risks for society – be it growing inequality and endemic poverty, climate change or ecosystem collapse – which demand a goal orientated and purposeful change to mitigate negative consequences. On the other hand history, complexity theory and political science raise serious questions about the degree to which social, economic and political change can be controlled and directed. Is there some middle ground? Some way of enabling human systems to evolve in more desirable ways?

It seems that the only real option for tackling this dilemma is to enhance societies' learning capacities (Woodhill, 2002; Woodhill and Röling, 1998) in ways that enable a greater responsiveness and resilience to emerging risks. Such learning capacities, it is argued, lie in creating mechanisms for distributed collective cognition (Oswick *et al.*, 1999; Röling and

Jiggins, 2001). In essence this means tackling 'transition' processes bottom up by distributing understanding, improving feedback linkages and enhancing capacities for adapting to change in a dispersed and non-hierarchical, yet coordinated, manner. This aligns with the basic ideas of how complex adaptive systems function and with ideas of participatory and discursive, rather than representative, democracy (Beck, 1997; Dryzek, 1997; Giddens, 1994; Held, 1996).

It is against this conceptual background that the chapter now introduces a more practical framework related to the facilitation of multi-stakeholder processes.

15.4 A framework for stakeholder engagement

The ideals of more participatory forms of governance, ownership and development partnerships have logically led to the establishment of a very wide variety of stakeholder processes, dialogues, learning alliances and roundtable forums. Local level processes of participatory development that evolved during the 80s and 90s have now been complemented by participatory processes that often work across multiple scales and sectors (Hemmati, 2002).

Despite good practical, theoretical and ethical justifications for such processes of stakeholder engagement their success and impact is far from assured. Very often such processes do not realise their potential. There is often a severe lack of capacity to design, lead, facilitate and support such processes in ways that can lead real learning and change. There are high risks of such processes simply reinforcing existing power relationships at the expense of poor and marginalised groups. There are also huge questions about how these processes fit within a wider understanding of governance and formal processes of government and policy making.

However, no matter how difficult it may be to foster effective processes of stakeholder engagement and dialogue it seems that there are few desirable alternatives. It is clear that markets alone will not drive equitable development, protect the environment or ensure social injustice. It is equally clear that governments alone do not have the understanding, capacity or effective power to solve complex problems in a top-down way. It seems critical then to invest in new forms of society-wide learning, participatory governance and stakeholder engagement that arguably give the best chance of working towards sustainability and social justice (Leeuwis and Pyburn, 2002; Röling and Wagemakers, 1998; Wals, 2007). The risk of such processes failing is that those with power and resources may resort to authoritarian means of trying to protect their narrow interests as crises unfold.

From this wider perspective on stakeholder processes, over the last six years or so, Wageningen International's Capacity Development and Institutional Change Programme (formerly the International Agriculture Centre) has developed a framework to guide the design and facilitation of multi-stakeholder processes MSP (Wageningen International, 2009). This framework has been built on the heritage of participatory development, reflection on numerous stakeholder processes, direct experience of processes facilitation and the learning that has accumulated from regularly conducting capacity development programmes for those at the front line of interactive process management.

The intention of the MSP framework is to guide facilitators, process managers and leaders of stakeholder groups in the task of designing and supporting a process that is unique to the demands of a specific situation. It offers the theoretical ideas, principles, practical tools and generic process elements that experience has shown optimise the chances for effective and productive stakeholder engagement.

As illustrated in Figure 15.2, the framework is built up of three main elements, (1) the underlying theoretical assumptions, (2) the dynamics of change and (3) the core process. Each of these elements will be briefly explained. While the practical tools can be found elsewhere, the core building blocks are explained here.

15.4.1 Theoretical assumptions

Any multi-stakeholder process is by definition an interdisciplinary undertaking. It will touch on many of the philosophical and pragmatic questions that challenge the social sciences. Whether conscious of it or not all stakeholders entering such process bring to it a whole baggage of theoretical assumptions about how change happens and how they can influence it. Any conscious design of a stakeholder process is also based on a set of theoretical assumptions. Further, much of what goes on as stakeholders learn together is a questioning, challenging

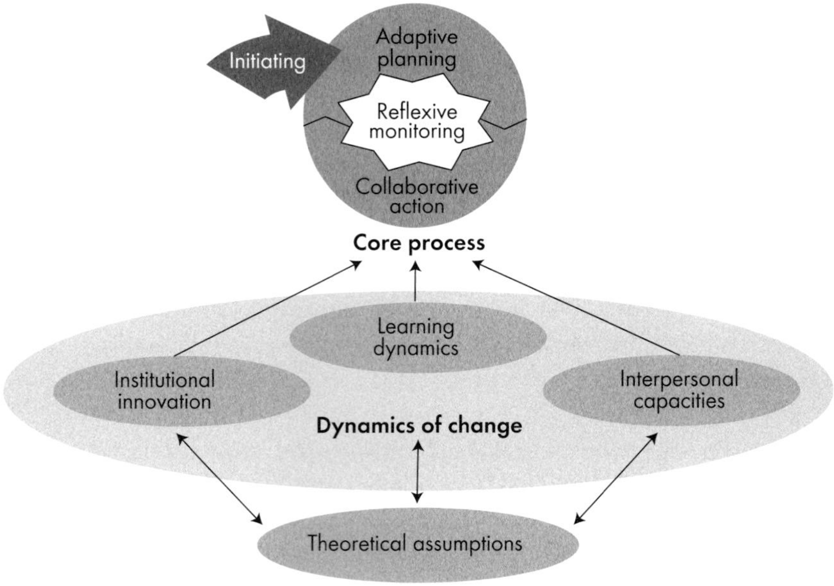

Figure 15.2. A framework for facilitating multi-stakeholder processes.

and reconfiguring of assumptions. Consequently any competent and responsible process facilitator needs to get to grips with theoretical assumptions in three ways. Firstly, they need to be aware of their own assumptions. Secondly, they need to have some understanding of the assumptions others may be bring to the situation and be able to help make these explicit. Thirdly, they need some grasp of the theories that underpin the intervention strategies and tactics they may employ or advocate. For example theories about the validity of scientific versus other forms of knowledge, governance, complexity, human motivation and rationality, or conflict management, just to mention a few.

15.4.2 Dynamics of change

The framework is based the idea that facilitated change occurs as a dynamic between individual capacities, institutional innovation and the processes of learning (which may be highly political) in which stakeholders engage. The idea of social change always being a tension between individual freedom and initiative and the constraints of social structures and institutions aligns closely with Giddens structuration theory (Giddens, 1984). The element of learning dynamics focuses on the approaches and methodologies that can be used to engage stakeholders in developing their own competencies for collective action to change institutional arrangements. As will be further elaborated, it should be clear that learning as it is understood here is not seen as politically neutral or free of power dynamics – quite the opposite.

These three elements of the dynamics of change are intended to convey a systemic approach to facilitated change processes that move beyond three classic failures. One is the failure when social change is seen as essentially an individualistic affair with the assumption that if the mindsets and behaviours of individuals change society will change. As important as this may be, such an individualist approach will often not change deeply embedded institutional blockages. Two, at the other extreme is the failure of policy. Here, the assumption is that change can be driven top down by the legislative power of government. However, without the understanding and support of individuals, significant change is often difficult. Policy processes often totally misunderstand the power and influence of informal institutions over which they have more marginal influence and the complex adaptive nature of human systems. Three, is the failure of simplistic methodological approaches and standardised facilitation techniques that engage naively with the political, power, psychological and cognitive dimensions of change.

Embedded within the framework for stakeholder engagement are 'tools for thought'. These are simple models that capture core theoretical ideas in ways that can be used by stakeholders in learning and dialogue processes to enrich analysis and learning. These tools for thought cover themes such as power, institutional analysis, conflict, complexity, communication, leadership and cognition. Their use aims to bridge theory and practice when engaging with stakeholders on the dynamics of change. They are a way of making the conceptual world of the social sciences more accessible and useful in everyday interactions.

Institutional innovation

The concept of institutional innovation has already be introduced and explained. In terms of the framework three points are worth making. First, institutional innovation is usually at the heart of what any stakeholder processes tries to tackle, although it will often not be framed in this way. Second, knowing how to design and facilitate a stakeholder process, and being effective in doing so, depends to a large degree on how well the institutional context has been understood. Third, institutional factors, themselves, are often a critical constraint for effective multi-stakeholder processes. In other words some institutional change may be necessary to make possible the stakeholder processes necessary to bring about the wider institutional innovation being sought. This means stakeholder processes are inevitably dynamic, iterative and adaptive.

Learning dynamics

The element of learning dynamics deals with how individuals, groups, communities, organisations and societies learn to be adaptive and resilient in the face of the opportunities and risks of a changing environment. This element of the framework looks at how a wide range of participatory methodologies and tools can be used to create processes appropriate to a unique stakeholder situation. Underpinning this is the experiential learning theory of Kolb (1984) and the notion of single, double and triple loop learning (Argyris and Schön, 1978; Bawden and Packham, 1993; Schön, 1983). The focus of this element is not just on single learning events or workshops but rather on how a whole set of activities and engagements link together over an extended period of time in a dynamic change process. This recognises that to be effective, building the understanding, trust and capacities for multi-stakeholder processes requires more than just 'multi-stakeholder workshops'. One on one meetings, interactions between the constituents of single stakeholder groups, communication with the wider constituencies, expert working groups as well as multi-stakeholder interaction are all important. Figure 15.3, while still an oversimplification, illustrates this more sophisticated notion of stakeholder engagement.

It is important to realise that learning dynamics cannot be separated from power and political dynamics. Empowerment of a particular stakeholder group may often be a precondition for any effective multi-stakeholder engagement. Such empowerment may relate to capacities for engagement or to the group's power in a wider political context. This means that some multi-stakeholder processes at times may have a clearly partisan objective. However, this partisan approach is embarked upon from the perspective of looking for ways to open up space for working constructively with other stakeholder groups rather than engaging primarily in adversarial politics or negotiated settlement.

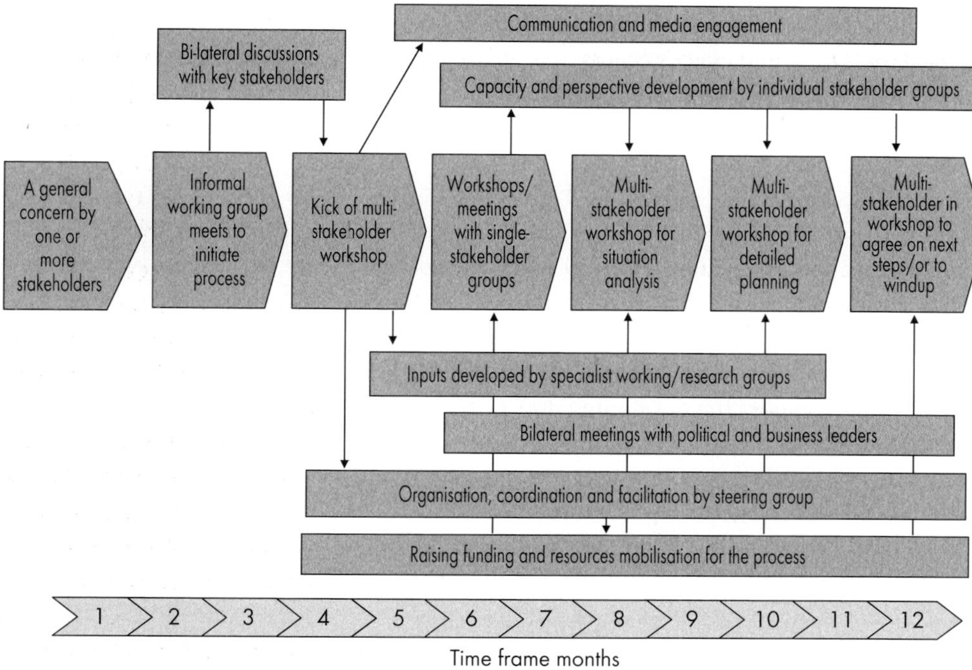

Figure 15.3. Illustration of the diverse activities of a multi-stakeholder process.

Interpersonal capacities

Ultimately, effective multi-stakeholder processes do depend on the capacities of individuals; on their capacities for communication, conceptual analysis, self reflection, leadership and facilitation. It is naïve to believe that methodologies or good facilitators alone will lead to effective multi-stakeholder processes if these core human capacities for collaboration and collective action are weak. Unfortunately they are often very weak. Entire education systems from primary school through to university education for the professions have focused largely on knowledge acquisition and technical skills at the expense of these capacities for effective human interaction.

In multi-stakeholder processes facilitators, innovation brokers and free-network actors all have critical roles to play and their competencies are one critical factor in a successful process. However, the leadership styles and outlook of those leading or representing various stakeholder interests will also be critical. The more self-aware all stakeholders are and the better their understanding of effective communication strategies, the more chance there is for creating shared understanding and trust.

This element of the framework brings together a diverse range of theoretical perspectives and practical tools for enhancing interpersonal and team dynamics, self development, leadership, facilitation, communication and conflict management.

15.4.3 Core processes

The MSP framework also includes a generic process model. This offers a set of process elements grouped according to four iterative phases, see Figure 15.4. These elements have been derived from experience. This experience has also shown that process failure is often due to, at least in part, inadequate consideration of these elements. The point here is not to offer any sort of blue-print for a multi-stakeholder process, rather it is to provide ideas and principles that support the development of processes that are adapted to (and adaptive of) the unique needs of a unique situation.

The four phases of the generic process model are: initiation; adaptive planning; collaborative action; and, reflexive monitoring[23]. These phases deliberatively mirror the action learning/ research cycle of planning, acting and reflecting. While in theory it may make logical sense to start with the initiating phase, in reality things often do not happen like this. It may be, for example, that the starting point is a small successful activity, or that a failed planning process leads back to the deeper questions about initiating.

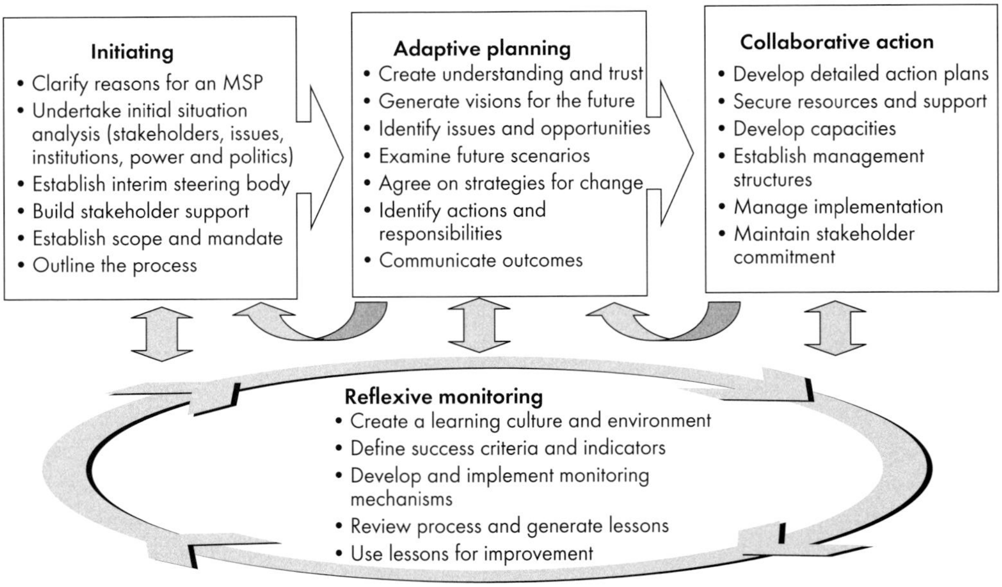

Figure 15.4. Key elements of for designing and facilitating and multi-stakeholder process.

[23] For a valuable perspective on stakeholder processes action research and the role of science see Giller *et al.* (2008).

Initiating

The initiating phase in particularly critical. This is when the feasibility of a process needs to be checked, legitimacy established, the politics of a situation understood, key stakeholders engaged and the initial scope, mandate and process agreed on. Mistakes and misjudgements at this early stage can spell disaster or at least create a lot of unnecessary difficulty. From reviewing many such processes with participants it is clear that often some very basic mistakes are made at this stage.

Adaptive planning

Most, if not all, processes require a phase whereby an understanding of the different issues and needs of the various stakeholder groups is created; where the wider environment is examined, the future vision and ambitions explored and current problems assessed. This is the basis on which strategies and actions can be agreed upon. However, this is not conceived of as a linear planning processes. It is fully recognised that taking action will lead to new insights and the revision of assumptions, goals and strategy. Hence the term adaptive planning. While beyond the scope this article, the framework explicitly deals with the question of what it means to plan adaptively in the context of high levels of complexity and uncertainty.

Collaborative action

A big risk for multi-stakeholder processes is that they stay at the planning and visioning phase and do not actually lead to action and change. A very different dynamic, set of resources and even skills is needed for the action or implementation phase of a stakeholder process. This is where different groups may have to commit themselves to serious investments, time input and difficult change. Maintaining the interest and commitment and ensuring the right incentives requires careful attention.

Reflexive monitoring

Very few stakeholder processes effectively embed monitoring into the process. We use the term reflexive monitoring here to refer to a type of monitoring that enables the actors to learn about their process as it unfolds and to adapt it. It is important to monitor not just the anticipated outcome of the process, but also the expectations and quality of the process itself. Engaging stakeholders in a discussion about what for them would constitute a quality processes and then setting up systems for monitoring and regularly reviewing this can be a very powerful tool for improving the processes.

15.5 Conclusion

Transition management and development are largely about changing institutional arrangements of the past. Given the problems of today, this does not make sense. Finding ways to be more effective in goal-directed institutional innovation is essential if societies are to be resilient in face of the risks posed, for example, by unsustainable agricultural systems, social inequality or climate change. However, major challenges remain in approaching institutional innovation in ways that (1) align with what is now understood about the complexity of human systems, (2) take on board the new insights into human cognition and (3) are not naïve about the influence of power. These three dimensions taken together suggest that new paradigms of goal-directed 'transition' are needed. Despite such new understanding, language and implicit assumptions about how change happens often reflect a paradigm that places technology, experts and 'rational' analysis as the central driving forces. Moving beyond this old paradigm opens up huge questions about the degree of control humans have over their destiny. What seems clear is that future change strategies will need be 'systemic' in nature – not trying to pre-plan and control change but rather creating conditions that encourage co-evolution in particular directions. In linking the domains of science, politics and public policy there remains a great deal to be understood about how to do this in ways that can enable effective and resilient responses to emerging risks.

This is a challenge that arguably demands new capacities for dialogue and collective learning that can bring a new dynamic between government, citizens, business, civil society and researchers. The risk of failing in this challenge is a scenario where humans have unleashed a cascade of social and environment change that is indeed beyond their own capacities of resilience. Associated with such failure is the likelihood of authoritarian responses to crises, which, from a democratic perspective, are highly undesirable. Not only that, complexity theory would suggest that it probably will not work either to avoid the crises.

It is certainly no panacea, however, it seems clear that working to improve the theory, methodologies and human capacities that underpin various forms of multi-stakeholder engagement is a good bet for enhancing the resilience of human systems. As illustrated by the rich content on transition management in this book, and the small snapshot into development in the South offered by this chapter, there is potentially much to be gained by a closer learning link between these two fields. Wageningen UR is at the nexus of both and, therefore, in a great position to widen the bridges.

References

Andriesse, W., Giller, K., Jiggins, J., Löffler, H., Oosterveer, P. and Woodhill, J., 2007. The role of agriculture in achieving MDG1 – A review of the leading reports. Wageningen International: Wageningen, the Netherlands.

Argyris, C. and Schön, DA., 1978. Organizational learning: a theory of action perspective. Addison-Wesley: Massachusetts, USA.

Bawden, R.J. and Packham, R.G., 1993. Systemic praxis in the education of the agricultural systems practitioner. Systems Practice 6: 7-19.

Beck, U., 1997. The reinvention of politics: rethinking modernity in the global social order. Polity: Cambridge, UK.

Beinhocker, E.D., 2005. The origin of wealth: evolution, complexity and the radical remaking of economics. Random House: London, UK.

Consultative Group on International Agricultural Research (CGIAR), 2009. Home page. Available at: www.cgiar.org.

Chambers, R., 1994. The origins and practice of participatory rural appraisal. World Development 22: 953-969.

Chambers, R., Pacey, A. and Thrupp, L.A. (Eds.), 1989. Farmer first. Intermediate Technology Publications: London, UK.

Checkland, P., 1981. Systems thinking, systems practice. John Wiley & Sons: Chichester, UK.

Cooke, B. and Kothari, U., 2001. Participation: the new tyranny? Zed Books: London, UK.

Cornwall, A. and Nymu-Musembi, C., 2005. Why rights, why now? Reflections on the rise of rights in international development discourse. IDS Bulletin 36: 9-18.

Cornwall, A. and Pratt, G., 2002. Pathways to participation: critical reflections on PRA. Briefing paper, Institute for Development Studies, Brighton, UK.

Dryzek, J.S., 1997. The politics of the earth: environmental discourses. Oxford University Press: Oxford, UK.

Easterly, W., 2006. The white man's burden. Why the West's efforts to aid the rest have done so much ill and so little good. Penguin Books: London, UK.

Ferrand, D., Gibson, A. and Scott, H., 2004. Making markets work for the poor – an objective approach for governments and development agencies. ComMark Trust: Woodmead, South Africa.

Gaventa, J., 2005. Reflections on the uses of the 'power cube' approach for analyzing the spaces, places and dynamics of civil society participation and engagement. MBN Secretariat: The Hague, the Netherlands.

Giddens, A., 1984. The constitution of society. Polity: Cambridge, UK.

Giddens, A., 1994. Beyond left and right; the future of radical politics. Polity: Cambridge, UK.

Giller, K.E., Leeuwis, C., Andersson, J.A., Andriesse, W., Brouwer, A., Frost, P., Hebinck, P., Heitkönig, I., Van Ittersum, M.K., Koning, N., Ruben, R., Slingerland, M., Udo, H., Veldkamp, T., Van de Vijver, C., Van Wijk, M.T. and Windmeijer, P., 2008. Competing claims on natural resources: what role for science? Ecology and Society 13(2): 34.

Guijt, I. and Kaul Shah, M., 1998. The myth of community: gender issues in participatory development. Intermediate Technology Publications: London, UK.

Held, D., 1996. Models of democracy. Polity: Cambridge, UK.

Hemmati, M., 2002. Multi-stakeholder processes for governance and sustainability: Beyond deadlock and conflict. Earthscan: London, UK.

Hodgson, G.M., 2006. What are institutions? Journal of Economic Issues 40: 1-25.

Kolb, DA., 1984. Experiential learning: experience as the source of learning and development. Prentice-Hall: Englewood Cliffs NJ, USA.

Kurtz, C.F. and Snowden, D.J., 2003. The new dynamics of strategy: sense-making in a complex and complicated world. IBM Systems Journal 42: 462-483.

Leeuwis, C. and Pyburn, R. (Eds.), 2002. Wheelbarrows full of frogs: social learning for rural resource management. Koninklijke Van Gorcum: Assen, the Netherlands.

Mowles, C., Stacey, R. and Griffin, D., 2008. What contribution can insights from the complexity sciences make to the theory and practice of development management? Journal of International Development 20: 804-820.

North, D., 1990. Institutions, institutional change and economic performance. Cambridge University Press: Cambridge, UK.

North, D., 2005. Understanding the process of economic change. Princeton University Press: Princeton, NJ, USA.

Oswick, C., Anthony, P., Grant, D., Keenoy, T. and Mangham, I., 1999. Collective cognition or social construction: a dialogic analysis of organisational learning. Third International Conference on Organisational Learning. Lancaster, UK.

Paley, J., 2007. Complex adaptive systems and nursing. Nursing Inquiry 14: 233-242.

Pretty, J.N., Guijt, I., Thompson, J. and Scoones, I., 1995. Participatory learning action – A trainer's guide for participatory learning and action. International Institute for Environment and Development: London, UK.

Ramalingam, B., Jones, H., Reba, T. and Young, J., 2008. Exploring the science of complexity: ideas and implications for development and humanitarian efforts. ODI Working Paper 285.

Regoverning Markets Programme, 2008. Small-scale poducers in modern agrifood markets. Available at: www.regoverningmarkets.org.

Röling, N. and Jiggins, J., 2001. Agents in adaptive collaborative management: the logic of collective cognition. In: L.E. Buck, C.G. Geisler, J.W. Schelhas and E. Wollenberg (Eds.) Biological diversity: balancing interests through adaptive collaborative management. CRC Press: Boca Raton, FL, USA, pp. 145-170.

Röling, N. and Wagemakers, M.A.E. (Eds.), 1998. Facilitating sustainable agriculture: participatory learning and adaptive management in times of environmental uncertainty. Cambridge University Press: Cambridge, UK.

Schön, D.A., 1983. The reflective practitioner. Basic Books, New York, USA.

Snowden, D.J. and Boone, M.E., 2007. A leader's framework for decision making. Harvard Business Review 85: 68-76.

Spencer, LJ., 1989. Winning through participation. Kendall/Hunt: Dubuque, USA.

Vermeulen, S., Woodhill, J., Proctor, F. and Delnoye, R. 2008. Chain-wide learning for inclusive agrifood market development – A guide to multi-stakeholder processes for linking small-scale producers to modern markets: International Institute for Environment and Development, and Wageningen University and Research Centre: Wageningen, the Netherlands.

Wageningen International, 2008. Navigating complexity – Report on innovation dialogue. Wageningen International – Programme for Capacity Development and Institutional Change: Wageningen, the Netherlands.

Wageningen International, 2009. Building your capacity to facilitate multi-stakeholder processes and social learning. Available at: http://portals.wdi.wur.nl/msp/

Wals, A., (Ed.), 2007. Social learning towards a sustainable world. Wageningen Academic Publishers: Wageningen, the Netherlands.

Woodhill, J., 2002. Sustainability, social learning and the democratic imperative. Lessons from the Australian landcare movement. In: C. Leeuwis and R. Pyburn (Eds.) Wheelbarrows full of frogs. Social learning in rural resource management. Royal van Gorcum: Assen, the Netherlands, pp. 317-331.

Woodhill, J., 2008. Shaping behaviour – How institutions evolved. The Broker 10: 4-8.

Woodhill, J. and Röling, N.G., 1998. The second wing of the eagle: how soft science can help us to learn our way to more sustainbale futures. In: N.G. Röling and M.A.E. Wagemakers (Eds.) Facilitating sustainable agriculture: participatory learning and adaptive management in times of environmental uncertainty. Cambridge University Press: Cambridge, UK, pp. 46-71.

World Bank, 2008. World development report 2008 – Agriculture for development. The World Bank: Washington D.C., USA.

Chapter 16

Transition: contradictory but interacting processes of change in Dutch agriculture

Jan Douwe van der Ploeg

Abstract

Transition describes the shift from one form of equilibrium to another. This puts transition forces at the heart of social struggle. In this chapter we point out the fallacy that transitional processes are singular, linear and easily isolated. It is our contention that in Dutch agriculture today, three processes are present at the same time. They interact with each other and develop in different directions. These three processes can be globally labelled as the industrialisation of agriculture, the repeasantisation and, thirdly, the deactivation of the sector. We argue that these processes, although apparently controllable and under control, appear to have ramifications that are not understood by either scientists, politicians or the sector itself. We highlight the importance of the paradigm differences between novelties and innovation, as this helps to explain the differences between the processes of repeasantisation and industrialisation as they are illustrated by the InnovatieNetwork and the experience of a large territorial cooperative, the Northern Friesian Woodlands (NFW). We look at how the S-curve model for describing transition processes is used to explain the innovation process in the state-run programme InnovatieNetwork, and identify the dichotomy of interest which is implicit in an a government 'controlled' innovation process. This leads to observations about the efficacy of 'state' innovation programmes as opposed to those that are bottom-up initiatives (as in the NFW case), and the usefulness of the S-curve to illustrate these two phenomena.

Keywords: transition, repeasantisation, industrialisation, deactivation, S-curve model

16.1 Introduction

A beautiful, varied and relatively unpolluted landscape adorns the very northern part of the Netherlands. In order to ensure its continued viability and attractiveness, a group of 6 independent organisations, all dealing with conservation, landscape and agriculture have consolidated their efforts in a combined organisation called Noardlike Fryske Wâlden. (The local Friesian words for Northern Friesian Woodlands). This organisation represents the public and private owners of 50,000 hectares of agrarian and conservation land and involves 850 local farmers. As a bottom-up initiative, it has as its goal the maintenance of the natural beauty of the area as well as supporting local initiatives to develop sustainable means of utilising and preserving their natural resources: land, water, flora and fauna. By combining their resources, these organisations play a major role in the development of the region and the preservation of its unique character. The NFW is a flourishing example of the repeasantisation trend within which the development and combination of many different 'novelties' play a major role. The many novelties tend to flow together in an irreversible transition towards sustainability.

16.2 Transition as a complex phenomenon

Although it is rarely made explicit, existing theories about transition all seem to be based on the assumption that a transitional process is fundamentally a unique, singular, well defined and well demarcated phenomenon. In other words, at any particular point in time and space there is just one transitional process in progress (or not). Besides that, it is assumed that the overall direction of the process is known. Thus, in the 1990s the transition of the former Soviet Union was described as moving *towards* a market economy, just as we can speak in retrospect about the transition of marine transport as moving from a system based on sailing ships *towards* one based on steam-driven ships.

This is not true for the transition processes in the Dutch agricultural sector today. One of its primary and highly problematic characteristics is that it is the object of at least *three* transitional processes[24] that are *simultaneously* present. These processes have been labelled repeasantisation, industrialisation and deactivation. All three interact in highly complex ways: they are mutually competitive and are even likely to exclude each other. At the same time they are developing in differentiated ways and directions. The more one particular process impacts upon food production and the landscape, the more the other transition processes are activated. This has caused a range of new contradictions to emerge from the complex dynamics formed by the simultaneous presence and mutual interaction of these diverging processes of transition. As a consequence, the resulting outcome cannot be predicted (let alone in a linear way). Nobody knows exactly where the 'winners podium' is located, nor how to explicate the terms that define 'victory' and distinguish it from 'defeat'.

Scientific knowledge and the associated expert systems play an important role in the transitional processes that are currently reshaping considerable parts of agriculture. However, the ramifications of their intervention can be quite different from the one that is assumed by the scientists and experts themselves[25]. Whilst scientists, experts and politicians believe that a process of transition might be triggered by, and will unfold according to (and as) a 'master

[24] This chapter focuses on agriculture and food production. However, I believe that many aspects that it discusses also have a more general application. The feature of mutual competition between different transitional trajectories can also be found in the current 'energy transition' and 'transport transition'.

[25] This, of course, has been a constant feature of modern agricultural sciences. A telling example of this is that productivity growth in agriculture is nearly always represented as a (more or less direct) reflection of previous agricultural research that was transmitted to the agricultural sector through a range of innovations. Consequently, the need for future increased productivity in farming is also translated (in a unilateral way) into an agenda for agricultural research and the consequent need for extra funding.

plan' – the real role and impact of such a plan can be quite different from what is assumed. Worse still, the real role and impact is barely understood and there doesn't seem to be any interest in understanding them.

Figure 16.1 shows the three transitional processes that are currently reshaping agriculture. These processes can be observed everywhere; however, their relative weight and especially the way they interrelate with and come to contradict each other will differ from place to place (just as it will differ along the time dimension)[26].

The industrialisation of agriculture is a multidimensional and multi-level process (as are the other transitional processes). Its most visible aspect is probably represented by the current processes of farm enterprise expansion (magnitude) and the associated increase in scale[27]. It implies a standardisation of the agricultural labour process and an associated artificialisation of nature (as contained in both the objects of labour and context). Industrialisation basically disconnects farming and food production from natural systems: ecological capital is replaced by industrial and financial capital, embodied in inputs, technologies, loans and their interrelations. It transforms farming into an artificial system, the agro-industry, the banking circuit and the large expert-systems. It is a process driven by many internal (e.g. 'entrepreneurship') and external drivers. Further industrialisation implies a widespread

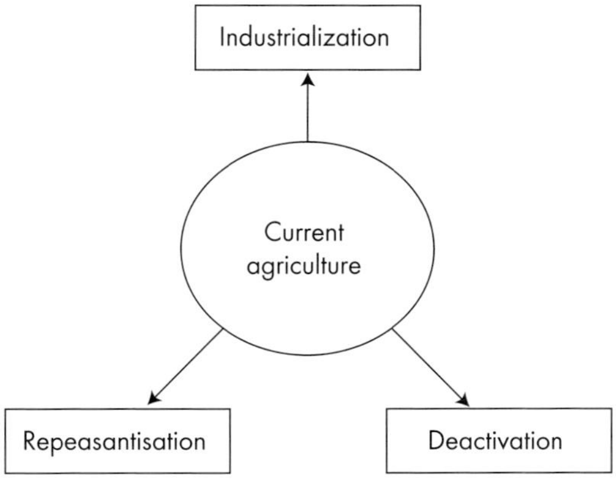

Figure 16.1. Interconnected and competing transitional processes.

[26] I am fully aware that representations, such as the one in figure 1, are far from neutral. In a way we have internalised 'geometrics' and notions such as 'progress', the upper side (and especially the upper right angle) represents the 'future' and 'development', whereas the 'lower left' angle represents the opposite: 'the past', the 'old', the 'inefficient' side of the equation, etc. Hence, we might turn the representation within in figure 1, in such a way that 'repeasantisation' moves to the upper right' angle and comes to represent the future.

[27] Following in the footsteps of Hayami and Ruttan (1985) I distinguish here between the magnitude of a farm and its scale. Scale explicitly refers to the relation *between* the total number of labour objects and the labour force. Magnitude as such refers to the total number of labour objects, often expressed as the volume of production.

disappearance of crafts and artisanal products, whilst simultaneously affecting landscapes, biodiversity and the quality of food. Ironically, it also introduces highly unfavourable tendencies in the main markets: sharp price fluctuations increasingly undermine the predictability required for long term investments, while margins are reduced to a minimum.

While industrialisation represents a progressive distantiation of farming from nature and the use of ecological capital, repeasantisation represents a *return* to using ecological capital: in which farming is increasingly 're-oriented' on nature. This allows for, and often occurs as, the active construction of new degrees of autonomy vis-à-vis the main markets for the factors of production and for non-factor inputs. The same also applies to relations with consumers with new degrees of autonomy being sought (or the relevance of remaining autonomy is revaluated) through, for example, on-farm processing, direct marketing and engagement in new networks that (re-)link town and countryside. Thus dependency on agro-industry and banks is reduced, whilst the relevance of 'cultural capital' is increased. One characteristic of this pattern is 'self-regulation' (based on available social capital)[28]. At higher levels of aggregation the process of repeasantisation represents a remarkable reversal in main trends: employment is stabilised while levels of Value Added are increased (Van der Ploeg, 2008); and sustainability is also frequently enhanced (Marsden, 2003).

Deactivation represents a reduction of agricultural activities and often an associated increase in other economic activities intended to develop the countryside as a 'space of consumption' (leisure industries, nature reserves, areas for high quality rural dwellings and probably, in the near future, large scale production of bio-energy). This reduction is often agriculture's retreat from one area matched by renewed or new farming activities in other areas (wherever located), that offer more favourable conditions (of whatever kind) whereby the activities often become more intensified and larger scale.

It might be argued that deactivation, repeasantisation and industrialisation *all* represent a specific project for transforming agriculture (and associated activities). And, although they are mutually exclusive (both as project and as practice), they are *simultaneously* present in agriculture today. In this way, we are not facing just one process of transition, but throughout Europe (and to a degree elsewhere as well) we are confronting three, mutually contrasting but interactive transitional processes. This makes the reality of transition far more complex than is normally assumed in the 'textbooks' and makes 'transition' quite *different* from the models assumed in literature.

The multilayered, complex and highly variable encounters between these different transitional processes imply, in the first place, that transition definitely cannot be conceptualised as a

[28] I use the term social capital here in the widest possible sense. It entails shared prospects and values and it flows into and reproduces, a range of social networks. It also embraces collective agency and knowledge – often local knowledge obtained in the process of transition itself (as is very clear in the cases of organic farming and the newly emerging territorial co-operatives).

singular and ultimately well-delineated shift. In retrospect it might easily be represented as such, but as a current phenomenon, it cannot be reduced to a single and progressively unfolding S-curve shaped 'shift' from one stage to another (as in Rotmans, 2003: 12; see also Dewulf *et al.*, 2009, this volume: Chapter 2)[29].

It is clear that transition involves social struggle: it is an ongoing encounter, involving competition and mutual transformation of different, mutually contrasting transitional processes. In this respect transition is *identical* to social struggle; the latter is not an additional aspect of transition (affecting it positively or hindering it) but rather is at the very core of it. The different and mutually competing transitional processes express, and meet, different sets of interests; they are based on particular and continuously changing coalitions; they present different trajectories that meet contrasting aspirations and which contain particular and mutually contrasting patterns for the spatial, temporal and social distribution of benefits and costs. They are, in short, *competing modes for ordering the future*. The notion of competition (i.e. between contrasting transition trajectories) is central here. It expresses the idea that transition is, and always was, a tough struggle to obtain a dominant position. Within such struggles there will always be claims that only one of the many *modes for ordering the future* is superior (or necessary, or unavoidable). These claims are most often grounded in economic and/or technological determinism and are often a powerful weapon in the struggle, but they are nothing more than a weapon (albeit, paraphrasing James Scott, 'a weapon of the strong')[30].

Transition not only involves social struggle, it is also *driven* by this socio-political struggle. In this respect it is important to note that the different coalitions and the vested interests no longer follow the once classical lines: for example labour versus capital and farmers versus the state. Transitional processes such as the ones summarised in Figure 16.1 are all partly rooted in different segments of a wider array of state organisations. Consequently, the socio-political struggle is partly a struggle within the state itself. The same might be said about the agricultural population which is divided into different segments. Each segment is aligned with a particular transitional trajectory which is, in turn, associated with specific trends within civil society as a whole. Particular coalitions emerge which periodically co-operate with each other and at other times, engage in sharp clashes. The balance between the different coalitions (and the relative weight of each of the competing transitional projects and processes) is very hard to predict. Dramatic and unexpected events (like the outbreak of FMD and its eradication) can drastically reshape such balances (see Van der Ziel, 2008).

[29] This especially applies when one intends to play a constructive role in ongoing transition processes.

[30] During processes of transition such claims are often presented as the 'truth', as a scientifically grounded necessity that excludes any alternative. The irony is, of course, that in transition processes the economic trends and technological pathways are often deeply shaken and reshuffled, giving rise to *new* trends and pathways.

16.3 Repeasantisation as a transitional process

There are many different 'flows' that, taken together, constitute the richly chequered and worldwide process of repeasantisation that we are witnessing today. I will limit myself here, for the sake of brevity, to one form. That is the 'redefinition of farm enterprise boundaries' (I use here an expression coined by Ventura and Milone, 2004)[31] as currently occurring especially, although far from exclusively, in Europe. This process is also known and categorised as farmer-driven rural development (Van der Ploeg *et al.*, 2002) and/or the shift towards multifunctionality (Renting et al, 2008). These terms are all somewhat ambiguous, but ambiguity is, as one might argue, a feature intrinsic to every process of transition.

There are several, mutually supporting reasons for classifying the changes that we are witnessing as 'repeasantisation'. Firstly, grass-root driven (or 'endogenous') processes of rural development are occurring as an actively constructed response to the difficult, if not hostile, environment in which farmers have to operate. It is a struggle for autonomy, progress and prosperity in a context that increasingly imposes dependency, deprivation and the lack of any prospects. The more the transitional process of industrialisation proceeds, the more repeasantisation will be triggered. Secondly, within and during this struggle, a resource base is wrought (and networks are constructed) that embodies and represents higher levels of autonomy. One main way in which autonomy is enlarged is through the creation of a new balance between commodity and non-commodity relations. The latter is beginning to prevail over the former, which not only represents a rupture with the 'entrepreneurial script' of farm development, but also represents a *positive* move in terms of repeasantisation. Thirdly, this same rebalancing often occurs as a result of the regrounding of farming upon nature (Verhoeven *et al.*, 2003). Fourthly, the same process often uses novel ways to increase the 'technical efficiency' of production, by means of the typical 'peasant' quality of craftmanship. Finally, it is very telling that through the *combination* of these elements, total Value Added (VA) at the level of the enterprise is maintained, if not significantly enlarged, *without any major scale increase.* The 'take over' of other units, a central feature of entrepreneurial farming, is not needed in this renewed 'peasant' model of farm development.

This richly chequered and multi-facetted process of 'repeasantisation' clearly represents a transitional process. It not only remodels the units in which primary production is located, it also radically transforms the networks and mechanisms that link these units with agro-

[31] These authors refer to the processes of deepening (meant to augment Value Added (VA) per unit of end product), broadening (aiming at the creation of additional VA through the integration of non-agrarian activities into the farm) and regrounding (novel forms of cost reduction that diminish the flows of external inputs, while improving the use of internally available resources). In and through these processes the farm enterprise is both extended and refigured and the internal and the external 'techno-economic relations' are reshuffled. Understanding the newly emerging economic rationalities and dynamics, and the emergent shift to 'economies of scope', is best gained through the perspective of neo-institutional economics.

industry, consumers, expert-systems, farmers' unions and state organisations[32]. In some instances the process even directly impacts upon the context as a whole, thus redefining major features of e.g. particular markets (this is theoretically discussed in Porter, 1985; an empirical illustration of the effects on the labour market is given by Cabello Norder, 2004).

At higher levels of aggregation (e.g. the regional level) processes of repeasantisation have a significant impact in terms of the number of farms, total employment, total generated VA and multiplier effects (Van der Ploeg *et al.*, 2002; Heyman *et al.*, 2002; Van der Ploeg, 2008). They also have a significant impact on the quality of life in rural areas (Milone and Ventura, 2008). An interesting feature, that indirectly confirms that we are dealing here with a far-reaching and radical transitional process, is that, at the theoretical level, a paradigm shift is needed to fully understand the magnitude, relevance, impact and mechanics of this process. This is reflected in the polemics in the Netherlands about the reach and relevance of the processes of deepening and broadening and is also clearly reflected in ongoing debates within the realm of rural sociology, for instance that between Goodman (2004) and Van der Ploeg and Renting (2004).

One of the main mechanisms of repeasantisation processes is novelty production. Novelties are, in a way, *deviations from reigning rule sets:* they may be deliberately created or simply be the unexpected outcome of the messiness of life. Novelties can be new practices, new artefacts or changed definitions of a particular task or situation. A key element is that they entail the promise that things can be done better. Materially, novelties produce a more or less visible (but mostly nearly invisible) rupture with existing routines and rules. Thus, novelties are 'change agents in disguise', or 'undercover agents' that help to spur and consolidate repeasantisation processes. They are the vehicles that help to construct and to extend autonomy[33], to 'shift boundaries' and/or to increase 'technical efficiency'. These features are often encapsulated in what at first sight seem to be 'monstrosities', such as 'improved manure' (Van der Ploeg *et al.*, 2006) or an 'electric fence alongside the canal' (Swagemakers, 2002).

In transition studies, and especially the literature on strategic niche management, novelty production has rightly received a lot of attention. Here I will draw attention to one particular feature. The potential strength and relevance of novelties is not intrinsic but resides, instead, *in the actively constructed inter linkages between different novelties.* I will refer to such inter-linkages by using the notion of 'the web'.

Figure 16.2 shows a web of interrelated novelties. It shows how an initial novelty – improved manure – has been translated into a wide range of interrelated novelties (for a detailed

[32] One important feature of this process is the emergence of a wide variety of new or alternative food supply chains and networks – (see Renting *et al.*, 2003; Wiskerke and Roep, 2007; Morgan and Sonnino, 2008).

[33] There are 'paradigmatic' differences between innovations and novelties. These are especially important since they are highly illustrative about the more general differences between industrialisation and repeasantisation. For an extended discussion see Wiskerke and Van der Ploeg (2004: chapter 1) and Van der Ploeg (2008: chapter 6).

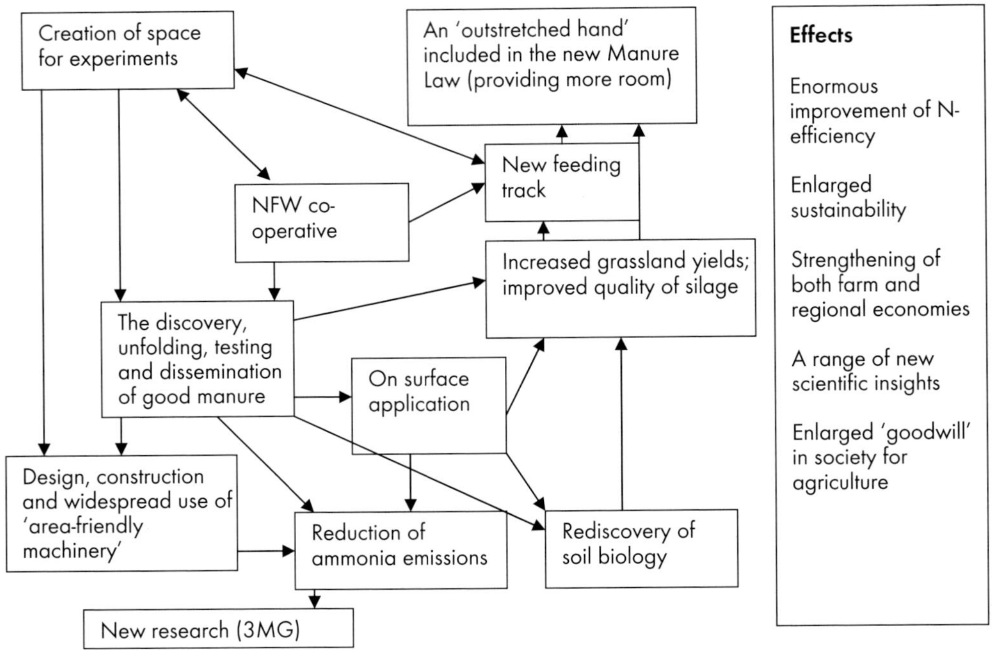

Figure 16.2. A web of interrelated novelties.

description see Van der Ploeg, 2008: chapter 7). This web might be considered as an unfolding (and not immediately visible) programme that is multi-layered: it involves and reshapes the practice of farming, it constitutes the core of the activities in the territorial co-operative (NFW) in which this web was germinated and came to blossom. It subsequently translated into pioneering scientific research (Sonneveld, 2004; Reijs, 2007) and into some major modifications of Dutch agrarian policy.

A decisive feature in this construction is that it was not driven from one single 'locus of control'. It is, instead, grass-root driven, spontaneous and, to a degree, guided by an unfolding 'narrative' that links the many initiatives and experiments into a self-propelling process. Because it is not planned, it allows for unexpected outcomes and, wherever possible, these outcomes are intelligently woven together into a seamless web – after which the emergent web gives rise to new novelties.

Equally important is that these changes are located within the immediacies of time and space; that they depart from the messiness of everyday life and prove themselves to be capable of emerging from within the many difficulties of everyday life (as e.g. the stranglehold of regulatory schemes imposed on farming and the economic squeeze on agriculture) and to go steadily beyond them. Finally, and closely interrelated with these features, is the fact that every novelty is by definition small, even seemingly irrelevant (and thus probably 'innocent').

However, when tied together into a larger and expanding web (i.e. a 'system innovation')[34], they start to produce important changes and reversals, as was the case with the Northern Friesian Woodlands territorial co-operative (NFW) from which the example in Figure 16.2 is derived.

Figure 16.3 briefly indicates the dimensions involved in the 'intelligent interweaving' of novelties into widening webs (these dimensions are discussed in Roep *et al.*, 2003 and, in an Italian context, in Ventura and Milone, 2005). The point I want to stress here is that the key feature of a 'niche' is not to be found in its relative 'exclusivity'. It is the 'inter-linkages' (or 'dimensions') that link it to the context that are decisive. These dimensions (i.e. the activities located and unfolding along them) re-pattern the interrelations between the 'niche' and its context. Simultaneously, they drive the expansion of the web forward. This has to do with the fact that there is no strictly defined plan that specifies a constellation located somewhere in the future. It is the 'ordering principles'(such as effective reformism, heterogeneous knowledge management, integration, etc.) instead that carry forward an unfolding and highly variable (if not 'mouldable') narrative (that every now and then is synthesised into a working plan).

This feature at the level of the 'niche', and the three features that characterise the construction of a web of novelties, are all highly remarkable when we compare them with other, now dominant approaches to transition and niche management, which are discussed in the next section.

16.4 'Uncaptured' transitions compared to 'controlled' transitions

Transition theories that are currently *en vogue* centre on transition as a shift from one well-known equilibrium to the next, equally well known but superior equilibrium. Consequently,

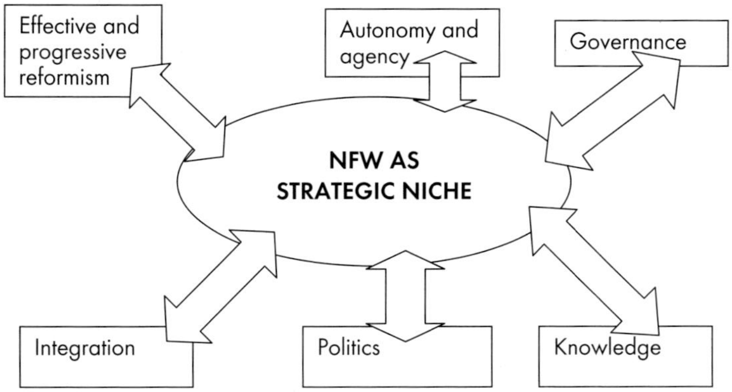

Figure 16.3. The dimensions of strategic niche management.

[34] Or, combining the conceptualisation of Arie Rip with one of the dimensions entailed in Figure 16.3, 'a configuration that increasingly performs better'.

transition is often graphically depicted as an S-curve. A current example is the transition from today's agricultural systems towards a new, sustainable agriculture.

In the Netherlands several programmes have been evolving during the last decade that explicitly aim at experimenting with, and trying out, so-called 'system innovations' and, subsequently, aim to specify strategies for successfully implementing such 'system innovations' or 'transitions' (examples of such programmes are Transforum, Innovation Network for Green Space and Agrocluster). I will discuss one particular programme, especially since its mechanics (and the associated discussions) have been carefully documented by Van Staveren (2007). I will especially discuss such state-induced 'exercises' in 'transition management' because I think they are far less innocent and harmless than they seem to be when considered in isolation[35].

The S-curved shift from 'current constellations' towards a new, more promising 'future' plays a central role in these programmes. This is substantiated in three ways[36]. Firstly, the lower left side of the S-curve is described, in the 'transition vocabulary' developed in these programmes, as 'corroded society' i.e. 'a society stuck in its routines'(Van Staveren, 2007: 26). It is characterised by 'old routines and traditional perspectives' (*ibid.*: 31)[37]. Consequently, it is full of 'persistent problems that cannot be resolved within the reigning frameworks (...)'. Building in one way or another on this lower left side has been made taboo: 'Starting from practice is most probably not innovative at all' (*ibid.*: 29), whilst whatever 'step-by-step strategy for change represents a major conceptual error' (*ibid.*: 87)[38]. Secondly, the upper right side (the new equilibrium) is created as 'a point at the horizon that can guide the path

[35] I make a special reference here to similar exercises done in the 1950s. State services like the 'socio-economic extension', the 'socio-cultural extension' were created as a result of a range of interrelated studies that were meant to spur the implementation of spatial reorganisation plans that, at that time, were heavily contested by Dutch farmers. These preparatory studies from the 1950s don't stand up to serious scrutiny. Nonetheless they gave rise to the abovementioned institutions for 'counter insurgency'. See Frouws and Van der Ploeg (1973) for an extensive discussion. In line with this historical precedent I believe that the 'research' done by organisations such as Transforum and Innovation Network may very well give birth to a new range of state interventions in society that are essentially at odds with democracy, subsidiarity, well constructed knowledge and, more generally, with the essentials of civil society.

[36] In what follows I quote Van Staveren's study. It is not a critique on the work of Grin and Van Staveren, but rather an attempt to use van Staveren's description to examine the vocabulary and the underlying approach adopted by the Innovation Netwerk programme. I have some personal experience with another programme, i.e. Transforum. It has been as disappointing as the case described by Van Staveren. However, in order to avoid any personal bias I will refrain here from any reference to the mechanics of Transforum.

[37] Without wanting to discuss whether such a characterisation is scientifically valid (or even defendable), I want to point here to a somewhat worrying resemblance with former 'modernisation theories' that accompanied the big modernisation process of agriculture between the midst of the 1950s and the 1980s. The point of departure was consistently 'rooted' in the 'traditional farmer'. Later on such theories (especially the 'diffusion of innovation tradition') were completely rejected. Nonetheless, the same epitaph is now being applied to a complete society.

[38] Here again, in this case there is a remarkable historical resemblance to the so-called Schultz-thesis that assumed that 'traditional' farming had an inherent 'technical ceiling' that allowed for no further (endogenous) progress whatsoever, implying that only far reaching and *external* interventions could carry change.

of the trajectory' (*ibid.*: 28). In other words: 'there is to be constructed a guiding image at the horizon' (here the upper right of the S-curve clearly is identical to this 'horizon') (*ibid.*: 70). In order to be able to construct such an image 'space is to be created within the future'. Current society, with its opportunities and limitations, should be intellectually 'killed off', so that it can be reborn. A 'new idea' is needed, which represents 'a rupture in reigning trends and tendencies' (*ibid.*: 65); it literally 'puts current thinking upside-down' (*ibid.*: 52). This new idea is to carry a 'strong brand name' and is, therefore, assumed to have a 'mental owner' (*ibid.*: 34) as well (I will come back to this intriguing element). The construction of the needed images is 'not to follow initiatives that are already present' (*ibid.*: 66)[39]. It is rather the other way around, it is 'like a voyage to the moon' – a telling metaphor that stresses the rejection of the current terrestrial problems and solutions. Thirdly, there is the combining element; the link between societal reality on the one hand and the 'pioneering idea' on the other. This link is defined as 'approach route' – as if we were dealing here with an airplane approaching a runway for landing. Here, on this 'approach route' (that is to be carefully prepared and controlled) the 'new idea' is to operate as a 'crow bar'(*ibid.*: 33). It is to provoke a 'carrier wave' (*ibid.*: 63).

Obviously, the shift from one equilibrium towards the next one could be a chaotic if not potentially dangerous process. Take, for example, the sort of shift that would occur in a change of equilibrium involving rivers embedded in a well developed system of dikes and other defence works in an area of dry land. This shift, i.e. the bursting of the dikes, starts slowly: water seeps out from under the dikes, weakening them (albeit scarcely visible). Then there is an acceleration, a first hole (and no little boy to put his finger in it as the American narrative would have us believe), that will be quickly followed by a complete burst. At this stage the 'shift' is already completely uncontrollable. Nothing can be done anymore. Only after the new equilibrium has been settled (inundated land), only then can strategies for recovery be designed and implemented.

Potentially, the shifts we are discussing (and that are symbolised with the S-curve) are highly disruptive. Current societies are riddled with such potential shifts. They come from many sides, and are rooted in many, often highly contrasting sets of interests and prospects. These need to be managed and this is done through specific socio-technical regimes that specify what is legitimate. However, these same regimes can block the shifts that are advocated as necessary by particular lobbies (of whatever kind) or which emerge as 'objectively necessary' (given a specific rationality).

In short: there is a current view that 'shifts' should be controlled. 'Transitions' are reduced, in the approach developed within these state programmes, to:

[39] Insiders will recognise this position as that of the Van der Zwan commission that was supposed to support innovation processes in Dutch agriculture. All existing innovative processes were considered to be not 'innovative enough', among other things because they were already present. Any notion of innovation as *an ongoing and probably self-strengthening process* was critically missing here. As a consequence, the Van der Zwan programme faded into obscurity.

1. those 'shifts' that are considered to be acceptable by currently existing power centres;
2. those which are implemented in a tightly controlled way;
3. those which reshuffle large segments of the social and natural worlds in order to be aligned with dominant interests; and
4. those which are organised in such a way that frictions and resistances are assumed to be minimal.

Programmes such as TransForum and InnovatieNetwerk are the pilots which test the methodologies for 'controlled transitions' and translate this into 'transition management'. It is as much about preventing the bursting of the dikes as it is about channelling the rising water (the increasing social and natural pressures) in ways acceptable to dominant interests[40].

The following elements are important issues in controlled transition processes. Firstly, there is the issue of 'ownership' (often referred to in terms of 'mental' or 'conceptual ownership'). In the InnovatieNetwork, this 'ownership' is clearly specified as belonging to the state programme. It can only be shifted to others, once they have accepted and internalised it. Even then, 'control remains in the hands of the director [of the programme]') (Van Staveren, 2007: 29). The 'director' clearly functions here as a metaphor for the powerful but benign State. Van Staveren goes on to say that being a director and managing transition is a 'craft' (*ibid.*: 86). It cannot be shifted towards others. It requires 'specific competencies'. As a matter of fact these competencies cover a wide range; they are synonymous with the social roles of the 'visionary, the artist, scientist, liaison officer, strategist, designer, system leader [*sic*], director, co-creator, independent facilitator and entrepreneur' (*ibid.*: 127, where these roles represent the many hats that should be worn by the director). In short: the director (i.e. the state) is 'everything': it absorbs and appropriates all skills available in wider society in order to 'boost' transition ('boosting', 'supercharging', 'pushing' are the frequently used words to describe the *process* of transition). And as if this all isn't enough to secure control (now symbolically and later materially), there is the 'institutionalised partnership' (referred to in Dutch language as 'maatschap') (see e.g. *ibid.*: 67) that is to support and to govern the assumed 'transition': it is composed of '35 prominent people', or ' opinion leaders'. It prefigures, as it were, a neo-corporatist mode of governance. This interpretation is supported by the described processes of 'multiple filtering' to which every step in the controlled process of transition is subjected. Engineering bureaus, staff, councils (with representatives of big business, state organisations, and expert systems), ministeries (*ibid.*: 21), the State Planning Office (*ibid.*: 23), etc. are consulted time and again, presumably to tune the different proposals. The role of expert systems is remarkably dominant in all this.

Conspicuous by its absence is the Parliament. This particular arena for mediation (with, admittedly, all the associated problems) is not mentioned once. The concept of 'negotiation'

[40] It is telling (and ironic) that programmes like TransForum are steered by actors from (and representing) these same dominant interests, agribusiness groups, banks, state agencies, expert systems, etc.

(and more generally, understanding transition as contested and simultaneously negotiated development and change) is also missing. In short: the state programmes for 'transition' are testing and polishing approaches that very well might reveal themselves to be as authoritarian (if not dictatorial) impositions[41] of a new order upon society[42].

Other important elements include the previously mentioned fact, that society as such represents the major problem. It is 'society' that is 'corroded' (*ibid.*: 26). The notion of 'society' refers here clearly to, say, ordinary people (or citizens) and/or to the 'micro-level'. The state, expert-systems, big business, and/or the interplay between the three and 'civil society' is not mentioned, not even once, as probably being a highly problematic aspect of today's social formations – that is as one of the main 'hindrances' to change (in terms of theory, the notion of a socio-technical regime is absent, even though the word is used a few times). 'Society' (as it functions in the narrative of Transform, InnovatieNetwerk and similar programmes) represents 'negative power' (in Dutch language 'hindermacht') (*ibid.*: 121). Hence it is to be kept 'at a distance' [*sic*] during the transition process (*ibid.*: 121).

16.5 Some final considerations

We live, admittedly, in very complex societies that are characterised by (and partly entrapped in) many interdependencies that often prevent timely responses to the many urgent problems we are facing. However, tackling such situations through highly biased problem identifications (by 'seeing like a state', as James Scott, 1999, argues) is far from helpful. Just as are the elaboration of 'solutions' that are simplistic for many reasons, among other things because they do not consider the many pitfalls and hindrances that might emerge when implemented. The subsequent 'schemes to improve the human condition' [*op. cit*] might be very damaging and, ironically, also very counterproductive, as they provoke their own contestation and resistance.

I will not develop a fully fledged comparison between the two contrasting approaches to transition discussed in the previous sections. Rather I prefer to point here to two important differences that have not fully emerged from the text so far. The first is the difference in vocabulary and tone. The tone in the InnovatieNetwerk approach is triumphant and pretentious (and this is even more so in the Transforum programme). The NFW proposals are, on the other hand, rather modest; they often reflect hesitation and insecurity. Nonetheless, the latter approach resulted in many effective changes as well as some major reversals. It also positively influenced (parts of) the knowledge infrastructure. The state programmes, on the

[41] 'From society itself there will not emerge any collaborative leadership', according to a 'director' of InnovatieNetwerk (Van Staveren, 2007: 67). Hence, it is to be composed *beyond* and independently from society. The then ensuing 'transition' is to be imposed, if not 'boosted'.

[42] Small steps and also a 'step-by-step' approach (as outlined in the previous section of this paper) represent, according to InnovatieNetwerk, 'a mental error' (Van Staveren, 2007: 87).

other hand, have hardly produced any effective changes, even when endowed with millions of euros.

Maybe this major (and disappointing) difference relates to a second one. In the transitional process occurring in the Northern Friesian Woodlands people know very well that their departure point is 'chaos'[43]. In a step-by-step fashion they create out of this a slowly unfolding, new 'order'[44]. The state programmes for transition, on the other hand, start with a two-levelled 'order'. With a starting point that is understood as the 'corroded order' and a 'spot on the horizon' that contains a well ordered new order as destination. Tragically, the 'boost' that is intended to interlink these two orders only produces chaos.

References

Cabello N. and Antonio, L., 2004. Politicas de Assentamento e Localidaded: os desafios da reconstituçao do trabalho rural no Brasil. PhD thesis. Wageningen University: Wageningen, the Netherlands.

Frouws, J. and Van der Ploeg, J.D., 1973. Over de landbouwvoorlichting: materiaal voor een kritiek op de voorlichtingskunde en de agrarische sociologie. Boerengroep: Wageningen, the Netherlands.

Goodman, D., 2004. Rural Europe redus? Reflections on alternative agro-food networks and paradigm change. Sociologia Ruralis 39: 17-38.

Hayami, Y. and Ruttan, V., 1985. Agricultural development: an international perspective, John Hopkins: Baltimore, USA.

Heyman, W., Hubregtse, M.H. and Van Ophem, J., 2002. Regional economic impact of non-standard activities on farms: method and application to the province of Zeeland in the Netherlands. In: J.D. Van der Ploeg, A. Long and J. Banks (Eds.) Living ountrysides: rural development processes in Europe – the state of the art. Elsevier: Doetinchem, the Netherlands.

Marsden, T.K., 2003. The condition of rural sustainability. Royal van Gorcum: Assen, the Netherlands.

Milone, P. and Ventura, F., 2008. La vita fuori della citta. CESAR: Assisi, Italy.

Morgan, K. and Sonnino, R., 2008. The school food revolution: public food and the challenge of sustainable development. Earthscan: London, UK.

Porter, M.E., 1985. Competitive advantage, creating and sustaining superior performance. The Free Press: New York, USA.

Reijs, J., 2007. Improving slurry by diet adjustments: a novelty to reduce N-losses from grassland based dairy farms. PhD Thesis. Wageningen University: Wageningen, the Netherlands.

Renting, H., Marsden, T. and Banks, J., 2003. Understanding alternative food networks: exploring the role of short food supply chains in rural development. Environment and Planning 35: 393-411.

Renting, H. Rossing, W.A.H., Groot, J.C.J., Van der Ploeg, J.D., Laurent, C., Perraud, D., Stobbelaar, D.J. and Van Ittersum, M.K., 2009. Exploring multifunctional agriculture. A review of conceptual approaches and prospects for an integrative transitional framework. Journal of Environmental Management, in press.

---

[43] This 'chaos' is not limited to the local level only. In their view it also embraces agrarian policy, the mechanics of the expert-systems, etc.

[44] This means a space where legally conditioned self-regulation is materialised.

Roep, D., van der Ploeg, J.D. and Wiskerke, H., 2003. Managing technical-institutional design processes: Some strategic lessons from environmental co-operatives in the Netherlands. NJAS 51: 195-216.

Rotmans, J., 2003. Transitiemanagement: sleutel voor een duurzame samenleving. Koninklijke van Gorcum: Assen, the Netherlands.

Scott, J.C., 1998. Seeing like a state: how certain schemes to improve the human condition have failed. Yale University Press: New Haven, USA.

Sonneveld, M., 2004. Impressions of interactions: land as dynamic result of co-production between man and nature. PhD thesis. Wageningen University: Wageningen the Netherlands.

Swagemakers, P., 2002. Verschil maken: novelty productie en de contouren van een streekcooperatie. CERES: Wageningen, the Netherlands.

Van der Ploeg, J.D., 2008. The new peasantries, struggles for autonomy and sustainability in an era of empire and globalization. Earthscan: London, UK.

Van der Ploeg, J.D. and Renting, H. (2004). 'Behind the redux': a rejoinder to David Goodman. Sociologia Ruralis 44 (2): 233-242.

Van der Ploeg, J.D., Long, A. and Banks, J., 2002. Living ountrysides: rural development processes in Europe – the state of the art. Elsevier: Doetinchem, the Netherlands.

Van der Ploeg, J.D., Verschuren, P., Verhoeven, F. and Pepels, J., 2006. Dealing with novelties: a grassland experiment reconsidered. Journal of Environmental Policy and Planning 8: 199-218.

Van der Ziel, T., 2008. Verzet en verlangen: het platteland in de greep van MKZ en stedeling. IJsselacademie: Kampen, the Netherlands.

Van Staveren, Arienne, 2007. Werken aan systeeminnovaties: hoe projectleiders bij InnovatieNetwerk hun ervaringen (inmiddels) zien. In: Grin, J. and A. Van Staveren (Eds.) Werken aan systeeminnovaties: lessen uit de praktijk van InnovatieNetwerk. Koninklijke van Gorcum: Assen the Netherlands, pp. 9-130.

Ventura, F. and Milone, P., 2004. Novelty as redefinition of farm boundaries. In: Wiskerke, H. and J.D. Van der Ploeg (Eds.) Seeds of transition: essays on novelty production, niches and regimes in agriculture. Royal van Gorcum: Assen, the Netherlands.

Ventura, F. and Milone, P., 2005. Traiettorie di Sviluppo: il sostegno a modelli di sviluppo endogeno. Dall'esperienze del distretto viti-vinicolo di Montefalco alla valorizzazione dell'area della Valnerina. CESAR: Assisi, Italy.

Verhoeven, F., Reys, J. and Van der Ploeg, J.D., 2003. Rebalancing soil-plant-animal interactions: towards reduction of nitrogen losses. NJAS 51: 147-164.

Wiskerke, H. and Van der Ploeg, J.D., 2004. Seeds of transition: essays on novelty production, niches and regimes in agriculture. Royal van Gorcum: Assen, the Netherlands.

Wiskerke, H. and Roep, D., 2007. Constructing a sustainable pork supply chain: a case of techno-institutional innovation. Journal of Enviromental Policy and Planning 9: 53-74.

Chapter 17

The relationship between description and prescription in transition research

Martijn Duineveld, Raoul Beunen, Kristof Van Assche, Roel During and Ronald van Ark

Abstract

In this chapter we will elaborate upon the relationship between the *analysis* and *representation* of long term societal change as well as the *recommendations* for actual practices that are based on this analysis and representation. We argue that the promise of transition management as a strategic science to solve socio-political problems in our society is partly based on unrealistic theories and empirical misrepresentations. This is primarily due to the fact the descriptions of transitions have not always been accurately constructed. Within transition management, the Foucauldian conceptualisation of power is still disregarded as an analytical tool for the production of 'realistic' representations of governance. The second reason is that there is, by definition, an inevitable gap between descriptions and prescriptions. We conclude that an amoral and realistic analysis of what is happening in actual practice is not only necessary for the (scientific) production of realistic representations of long term societal change but will also teach researchers and knowledge users to be realistic and therefore modest about their ability to 'manage' or direct their desired changes in society.

Keywords: transition management, social engineering, power, Foucault, governance

17.1 Introduction: the fictitious civil servant

The fictitious civil servant, Honoré wrestled with a huge problem. The ministry he worked for had assigned him the task to write a paper in which he was to clearly explain how the intensive livestock farming sector could make the transition towards sustainability. Pollution had to be reduced, animal wellbeing increased, the production had to be optimised and all this had to be realised without compromising the sector's competitive edge in the international market. Honoré, an intelligent man, understood only too well that this was a gigantic, if not impossible, task. He listed the criteria that would be needed to achieve this goal. At all levels of government, the political will to change would have to be created, regulations and laws would have to be adjusted, technical innovations would have to be stimulated ... and last but not least ... the conservative agricultural sector would have to completely revamp it's management style. As a modern functionary should, he turned to the computer to bring some order into the maelstrom of ideas that filled his head. He began to search the internet using terms like 'animal husbandry', 'sustainability' and 'innovation'. It wasn't long before he hit a site where similar problems were discussed, along with possible avenues to explore for the solutions to these problems. One word that he came across time and again was 'transition management'. He pursued this term further in his surf through cyberspace. The result of this brought a smile to his face. On the websites that he found he read, among other things, that transition management was the key to a sustainable society. He also read that it offered theoretical and practical frameworks to realise these ambitions and to legitimise them. It purported to offer an innovative perspective on steering social reforms. But its most important characteristic was that it appeared to be one of the most effective strategies for widespread quality improvement and the development of niche markets. Honoré was thoroughly chuffed and began to write.

17.2 The rise of strategic research

Changes in society have lead to the call for scientific research to be useful for purposes other than purely academic ones (Gibbons, 1994). Social and political relevance are considered to be just as important as scientific relevance. The rationale behind this is simple: scientific research needs funding. As a consequence of this researchers have to look for people and organisations who want to pay for their research and that they have to 'sell' their results. This is not necessarily a negative phenomenon. In fact, it has occurred for as long as research exists. As a result, the marketing of research and its uses has become increasingly important and professional in recent years. It is not only the quality of research that is important, but also its image.

To many people, policy research is considered to be of use if it helps to formulate more effective policies or if it strengthens the implementation of policies. Much of the present-day policy research is (co-)funded by governments. Policy research should therefore hold the promise of clear recommendations for effective policies. This forces scientists, chair groups and research institutes to explicitly present themselves as producers of strategic research and knowledge (Van Ark, 2005; Hoppe, 2002; Tress and Tress, 2003; Loos *et al.*, 2007; Spaapen *et al.*, 2007).

The popularity of strategic research, like transition management, can partly be explained by the promise it holds. For example, the promise that research on long-term societal changes can provide tools that help guide society towards sustainability. In practice, however, the usefulness of these recommendations is often limited. There are simply too many factors that influence the implementation of policies and therefore these policies will not always work out as expected (e.g. Pressman and Wildavsky, 1979). This insight is not new and has, for instance, led to more attention given to research perspectives in which the role of government is limited (Pierreand Peters, 2000). In this chapter we will deconstruct the promise that policy research, including transition research, often holds (cf. Fischer, 2000, 2003). Our criticism is aimed both at the way transition research is conducted and at the way practical recommendations are derived from it. We argue that there is a clear distinction between the analyses and the recommendations and that research does not tell us how to organise the world. We do not argue that policy research is useless. On the contrary! The value of policy research becomes clear after we have shown what research cannot do. Once we recognise its limitations we can focus on the strong points of research, which we will. We will put forward an alternative approach, by elaborating on the useful functions of policy research and how research should be conducted to fulfil these functions. This will bring us to some modest recommendations. Recommendations that take into account the fact that it is difficult and sometimes even impossible to define generally valid policy measures from policy studies.

In order to probe the relationship between analysis and research and the concrete conclusions for policy that are drawn from them, we will first describe how socio-political long term processes of societal change are analysed by Dutch transition managers and the recommendations they derive from that. Next, we will argue that these processes are unrealistically represented as more or less mechanical processes. Following that, we will explore to what extent realistic recommendations can be derived from process analysis. We will criticise the frequently held misconception that process analysis can be translated almost directly into specific strategies for policy. Finally, we will come to the most important conclusions and we will make some recommendations for anyone who prefers 'realism' to the 'idealism' of social engineering.

17.3 The Dutch transition management discourse

Transition management is an example of a successful type of strategic and applied policy and management research. Much research into long term societal changes and transition management is done at the request of ministries and institutions like LNV (Ministry of Agriculture, Nature and Food Quality), BuZa (Ministry of Foreign Affairs), V&W (Ministry of Transport, Public Works and Water Management) and VROM (Ministry of Housing, Spatial planning and the Environment). These and other organisations use the concept of transition management as a guideline for transition policy (Rotmans *et al.*, 2005). As described in other chapters of this book, all means that (could) lead to changes can be identified as (forms of) management and a transition as 'a structural social transformation that is the result of interacting and corroborating developments in the fields of economy, culture, technology, institutions, nature and the environment (...). (...) [They] are gradual transformations that take a long time, at least one generation (25-50 years)' (Rotmans *et al.*, 2005 [translated by the authors]).

17.3.1 Recommendations for problem solving

Guiding social transformations is not a goal in itself. Transition research carries the promise of contributing to the solutions to socio-political and environmental problems. According to experts on transitions these are problems of a special kind, because they are 'persistent problems that have been around for decades for which there are no cut and dried solutions

(...). These problems are persistent because they are deeply rooted in our social structures and institutions (...)' (Rotmans *et al.*, 2005 [translated by the authors]). In order to solve these so-called 'persistent problems' in, for instance, agriculture, water-management, transportation, education and healthcare, transition researchers produce recommendations and policy-measures in order to manage a transition. It is believed that transition management is able to offer a conceptual framework 'that enables one to come up with a specific mix of ways to steer things in the right direction.' (Rotmans *et al.*, 2005 [translated by the authors]) For this purpose it is, among other prescriptions posed in transition literature, that: (1) management at system level is essential; (2) newcomers should create a new regime; (3) a pluralistic approach is desirable; and (4) it is thought to be important for the actors involved within transitions to get to know each others perceptions of reality (Rotmans *et al.*, 2005). Even more specific are the recommendations for setting up a transition arena and developing transition coalitions and a transition agenda (Rotmans *et al.*, 2005). These and other recommendations make it clear that transition research is a good example of promising research in the effort to provide definite solutions for the problems governments are dealing with.

17.3.2 Describing transitions

The abovementioned and other recommendations and guidelines are deduced by transition researchers from theoretical and empirical based representations of long term societal changes. It is believed to be possible to determine from these analyses to what extent long term societal changes can be guided and the way this should be done. Among other things, these insights could teach us how a transition works and which factors and mechanisms play a part in it. Transitions would consist of different phases, each characterised by their own dynamics. Also, they'd be brought about by system changes at different scale levels. Transitions are viewed as processes involving several people and organisations, such as ordinary citizens, governments, businesses and social organisations (Rotmans *et al.*, 2000).

17.3.3 Analysing transitions

The descriptions of the way transitions function are derived from theoretical and empirical analyses of long term societal changes. These analyses are based, firstly, on existing theories and insights from, among others, public administration studies, sociology and political science. These include new kinds of governance theories, complexity theories, network theories and system theory. Secondly, they are derived from the analysis of existing societal change processes and long term societal change processes in the past (Rotmans *et al.*, 2000; Rotmans, 2003). Thirdly, transition researchers take part in societal change processes themselves and they base their analysis partly on their own experiences within these processes. This type of research is called (reflexive) action research (Termeer and Kessener, 2006: 30; cf. Zuber-Kerritt, 1991; Tress and Tress, 2003).

17.4 Deconstructing the promise

The previous paragraph shows that transition management is a good example of research that promises to provide definitive solutions for the problems governments are faced with. This largely explains its popularity. As we have already mentioned in paragraph 1, we have some serious doubts about these promises. In this paragraph we will explain our doubts. We criticise the way research is done as well as the way recommendations are deduced from this research. Before we can investigate to what extent it is possible to deduce recommendations from analyses of long term societal changes and other social and political processes, we first have to answer the question, how realistic are these analyses and descriptions.

17.4.1 Deconstructing transition analysis and representation

What is remarkable about the manner in which socio-political changes are represented within some transition studies, is the high level of abstraction (e.g. Rotmans, 2003) Transitions are, for example, represented as a set of factors or conditions that, if they all work together, will cause a desired change – as if they are the result of more or less mechanical, instrumental processes. We will argue that these abstract representations do not offer a realistic view of the factors and mechanisms that, in fact, influence social, administrative and political processes. We believe it to be more realistic to analyse these processes within the framework of power, in the manner described by Machiavelli (1988), Nietzsche (1977), Foucault (1988, 1998, 2001; Gils *et al.*, 2004) and, more recently, Flyvbjerg (1998, 2002). Because the context of power has remained under-exposed within transition research, just as it has been in disciplines such as planning (Van Assche, 2004) and public administration (Korsten and Hoppe, 2006), we will first define our use of the term 'power'.

In an interview, Foucault says that to him, 'power' is shorthand for the expression he generally uses: 'relations of power'. 'But there are readymade models: when one speaks of power, people immediately think of a political structure, a government, a dominant social class, the master and the slave, and so on. I am not thinking of this at all when I speak of relations of power. I mean that in human relationships, whether they involve verbal communication (...), or amorous, institutional, or economic relationships, power is always present: I mean a relationship in which one person tries to control the conduct of the other. So I am speaking of relations that exist at different levels, in different forms; these power relations are mobile, they can be modified, they are not fixed once and for all' (Foucault, 1997: 291-292). According to Foucault, power is not an external factor, but it is everywhere and it is exercised from different viewpoints and positions (Foucault, 1998: 93). Moreover, relations of power are always connected to a certain objective, they are intentional relations: 'There is no power that is exercised without a series of aims and objectives' (Foucault, 1998: 95). It has to be stressed at this point that the word 'power' doesn't hold a negative connotation, in contrast with the everyday use of it. Power is neither good, nor evil. It can be repressive as well as productive: power produces some discourses, realities, knowledge, values, subjects et cetera and makes others impossible, marginalises or subjugates them (Foucault, 1998: 81-102, cf. Foucault, 1994).

Flyvbjerg has extensively studied a planning process from the perspective of power, strongly influenced by Machiavelli, Nietzsche and Foucault. His book '*Rationality and power: democracy in practice*' is the result of detailed empirical research into planning practices in the city of Aalborg. Aalborg's local administration received an award for its innovative long-term transportation plans for the inner city. These plans were said to have been developed in an innovative manner, involving new concepts, new strategies and new partners. Transition managers would have called it a successful 'transition'. However, Flyvbjerg's analysis did not underline this success. In this book he exposes the power strategies the different actors, often with opposing interests, used to attain their objectives. One of those strategies was the selective use of (scientific) knowledge and the conscious concealing or marginalising of research that did not support their case.

Those who, like Flyvbjerg (or Foucault), analyse socio-political processes, including long term societal change processes, will understand that in the socio-political arena, many groups of people, organisations, parties and governments use various means and strategies to attain their ideal society and reinforce their claims (Duineveld, 2004, Roth *et al.*, 2006). Examples of these strategies are: lobbying, the formation of networks, coalitions and alliances, playing the media, the use of rhetoric, the selective use of the results of scientific research, making and implementing laws, formal rules and procedures and the formation and transformation of institutions.

In short, those who follow Machiavelli, Foucault and Flyvbjerg, in analysing political and social processes and practices as 'the continuation of war by other means' (Foucault, 2003), will gain insight into the factors, processes and mechanisms that instigate changes or ensure stagnations. These factors will partly be in accordance with the factors already deduced from theory and empirical research within transition research. Nevertheless, factors will come to light that are rarely or never mentioned in the descriptions of (transition) processes. They remain invisible because of the existing analysis methods, and perhaps also due to the fact that some factors are considered to be so immoral or undemocratic, that they have become a blind spot for the researchers (cf. Van Assche, 2004).

17.4.2 Recommendations for problem solving and the inevitable gap between description and prescription

The above must be read as a criticism on the way in which long-term societal changes are analysed and at the same time as a recommendation for another, more realistic way of analysing them. But even if the analysis within transition research could be more accurate, the question would still remain: can this knowledge, these descriptions, be used for prescriptive purposes? To answer this question we must first acknowledge that with the so called 'persistent problems' (environmental, political and social), aims and means are not necessarily a given but can be both the outset and the result of social interaction, political decision-making and conflict (Ringeling, 2002, Peters, 2005). Just like set goals and means, problems are constructed by people and are therefore always subjective (Ringeling, 2002). Some people, for instance, see nuclear energy as the answer to the energy problem. Even though the advocates of nuclear energy generally share the same problem analysis as the adversaries, they have totally different views on what means should be used to solve this problem. This implies that even in the utopian situation of knowing everything, this knowledge still would not tell us how to act. It would remain a choice which would mean different things to different people. This fact, however, seems to be hard to accept and many governments and researchers collaborate in their quest for control (Von Gunsteren, 1976, Scott, 1998). This phenomenon, called high modernism by Scott, is often referred to as a form of malleability thinking or social engineering. (We should add here that malleability is the translation we use for the Dutch word 'maakbaarheid', a term often used within the Dutch context. The term refers to the assumption that governmental and non-governmental actors can reach certain goals using guidelines and other directive means. The term is related to the more broadly and internationally used concept of social engineering.) Since malleability seems to be a persistent phenomenon (see e.g. Frissen, 1996) it is important to keep emphasising the difference between description and prescription.

Transition experts state that it is a misconception to presume that the implementation of the theory 'will lead to a deterministic collection of directing rules' (Rotmans *et al.*, 2005). Though recognising the fact that guidelines may not be deterministic, it doesn't hinder the transition researchers in producing concrete recommendations, guidelines, methods and techniques that are presumed to have real effects, and which can be used to attain certain objectives and solve certain problems. This presumption can be considered as a (new) kind of malleability thinking or social engineering. According to Terpstra, 'malleability refers to the practical or pragmatic question: what are the possibilities of effectively implementing a certain (political) decision in accordance with the therein-contained intention?' He believes it is 'the conscious use of the present means for attaining (...) set goals (...)' (Terpstra, 1995). He illustrates the idea of malleability with a simple but telling example: 'Suppose, I have to go to Amsterdam and I want to travel by train. I can make an accurate mental image of how to reach my goal. I know that if I buy a ticket and get on the right train, I will eventually arrive in Amsterdam. I can completely put this image into practice (...). What is important here is that there is a beautiful similarity between my mental image of the coming events (I call it: my policy plan) and the events that actually take place during the execution of this plan. (...) There is a connection between designing (my mental image of the trip and the accompanying reasoning), making (in this case: that which has been made, together with my own actions) *and* knowing (I know from experience that it usually works). There is a connection between my action, the result of it and the prior justification for it (...). This connection, as said, is the essence of the idea of malleability' (Terpstra, 1995)[45].

We will name three strongly interwoven arguments that can be made against the idea of malleability. The first we partly derived from Terpstra. According to him malleability is not always possible, because machines are rarely perfect or function as such. The railway-company he uses as an example 'is, as we all know, not a perfectly working machine. In reality, machines like the railway-system will frequently falter, for example, either because of a human error or weather conditions' (Terpstra, 1995). Consequently, the imperfections of the railway-company limit the possibilities for coming up with a strategy that will almost certainly succeed. This applies even more so for socio-political (transition) processes. After all, these do not work according to the mechanical principles that are characteristic for the railways. Still, some transition processes are presented as such. By representing socio-political (transition) processes as a system, wherein various processes, mechanisms and factors affect each other in more or less regular patterns, the suggestion is made that a transition can be represented as a complex 'machine'. This makes it relatively simple to suggest that long term societal changes can be steered or engineered to a certain extent. In other words: the 'representation of a transition as a machine' wrongly suggests that one only has to point out 'which buttons to push' to steer the transition process in the desired direction.

A second argument against the possibility of malleability is the almost insurmountable difference between the perspectives of those that analyse and those that take part in a (transition) process. Many interpretive anthropologists, constructivists and postmodernists teach us that the way people, including scientists, perceive and represent the world, has to been seen as a construct (Von Glaserfeld, 1995; Eco, 1993; Potter, 1996; Geertz, 1973; Branaman, 2001; Howarth, 2000). This means there is no direct connection between the world outside us and the way we perceive it. In principal, people's constructs (or worldviews, discourses, frames, configurations, perspectives) can constantly change and often differ between various groups of people. This implies that many analyses can not be converted into guidelines, because discrepancies (could) exist between analyses and between first and second order observations. Following a similar line of reasoning Maturana and Varela (1987) state that first order observations, the observations of people that act, are by definition different from second order observations, those of researchers observing other people's actions. This is why analyses and recommendations cannot be synchronised. They illustrate their argument with the following example, paraphrased by us.

> *'Two observers on the shore watch a submarine make it's way through a very dangerous reef without damaging the submarine nor the reef. As the crew comes ashore, observers compliment them on their helmsmanship. They ask the crew how they managed such a delicate operation. "How did you manage to avoid the reef?" The crew answers: "Reef? What reef!"'*

[45] All the quotes from Terpstra (1995) have been translated out of Dutch by the authors.

From their position, the environment in which they manoeuvred looked totally different. What they perceived through measuring instruments and monitors were numbers, graphs and other abstract representations of the surroundings. But no reef (Maturana and Varela, 1987). This simple example illustrates the theory that the way a system functions (the actions of people, the functioning of organisations as seen through a first order observation) and the analysis of the dynamics (second order observation) of this system in its environment should not be confused. The dynamics of the system do not, by definition, work with the same representation of the environment as the observer's one. No matter how good and solid the analyses of the observers on the shore are and no matter how detailed their insight into the relationship between the crew and their environment, one cannot expect them to be able to offer the crew concrete advice on how to steer their boat.

A third argument, partly intertwined with the second, against malleability is that there will be, by definition, a difference between the context which is analysed, and the context for which the recommendations are drawn. Socio-political processes are the result of power games that have a different outcome every time. This happens, amongst other things, because in every process, transition or social change, different questions play a part; different problem definitions dominate; there are different actors and balances of power; and different means are considered to be necessary. Therefore, it is impossible to deduce concrete plans, designs, instruments or guidelines from an analysis of the factors or mechanisms that influenced one process and impose them in a different situation. A much-heard recommendation, for instance, is that in order to stimulate changes and innovations, it is important to start a social learning process (cf. Gray, 1997). The aim is to get the actors involved in a transition to use more or less the same definitions of reality. The idea behind this recommendation is that transition processes can go wrong because people think and act from different perspectives, configurations or images of reality. Although this recommendation could turn out to be useful, this doesn't mean that one should, by definition, initiate social learning processes, or reframe people nor does it say how this should be done. Perhaps the desired objectives can be attained in a different context or situation by excluding those actors that have a different perspective, or by pressuring them, or merely give them the illusion of being involved and listened to.

The ideas of malleability that we disputed above, promulgate a misconception in transition management and other strategic policy research. This misconception more or less synchronises description and prescription, in other words synchronises the process analysis and the recommendations that come from it. For example: analyses from sociology, political sciences and public administration show that steering processes are no longer dominated by the sovereign position of governments (Pierre and Peters, 2000; Hajer and Wagenaar, 2003; Bevir, 2004). Rather, steering is the result of the working of networks in which various actors participate in hierarchic and non-hierarchic, horizontal relationships. However, these analyses don't automatically imply that governments or other actors that want a transition should set up networks in order to attain this objective. Perhaps some things should be still be directed hierarchically. Besides, it is still not sure that networks can be planned. The intentional or unintentional confusing or mixing-up of description and prescription is not only typical for transition management. Pierre and Peters (2000) declare that this is also the case with the concept of governance, which is currently popular within political sciences. This concept is used both in the descriptive and the prescriptive way and sometimes confused (Pierre and Peters, 2000; Bevir, 2004).

17.5 How to make policy research useful

17.5.1 Realistic analyses...

We propose that it is impossible to predict and verify the possible effects of (policy) scientific recommendations. Our recommendations will therefore be modest. Just like other (policy) researchers, we can't tell administrators what they should do in order to attain their objectives. We can, however, help the people that take part in a political, administrative and/or social process of transition act more realistically, by providing them insights into the reality and consequences of policy. Below, we will first give five recommendations for researchers that analyse socio-political (transition) processes. We will conclude with a description of the possibilities and impossibilities of deducing recommendations from realistic analyses.

A few recommendations can be made with regard to the manner in which realistic scientific research should be conducted. Firstly, we think 'power' has to be the central perspective for the starting point of the analysis. One needs to view socio-political (transition) processes as power relationship transformations. Furthermore, the organisational, institutional and disciplinary systems should be viewed as factors of power as well, which both enable and constrain the behaviour of actors. This implies that the formation of networks and coalitions, the construction and use of policy instruments *and* the compliance to rules should also be considered as factors of power. The functioning of these factors in practice relies on, amongst other things, the interpretation and use of these factors by the actors (Beunen and Van Ark, 2007).

Secondly, one shouldn't define the strategies and means that play a part in socio-political processes beforehand. It is recommended that the researcher should try to analyse each case without too many (theoretical) presumptions. Then, one is in a better position to deduce from the analysis those factors that influenced or played a role in a specific process. When this is done, one can investigate to what extent these factors correspond with analyses and descriptions of other processes. A conscious 'open-mindedness' at the start of the analysis process reduces the chance that one 'discovers' mainly those factors, strategies or mechanisms that confirm and/or match existing presumptions and theories. 'Open-mindedness' can prevent one from intentionally or unintentionally uncovering things that confirm the set principles (Strauss and Corbin, 1990).

Thirdly, one also will have to put aside ones (own) ambitions for the development of a certain transition during the research process (Van Ark, 2005). This also applies for presumptions on how strategies, laws, rules, organisations and such should ideally work. An analysis shouldn't be made starting from an idea of how a transition should be conducted, how planning should ideally work or how a political-administrative transition process should develop. It should be about analysing what really happens (see also Wissink, 2000; Van Ark, 2005).

In the fourth place, a realistic analysis demands an amoral position from the researcher. This means that for the duration of the research the researchers temporarily try to put aside their own moral frame and describe as realistically as possible what happens in the transition processes. Just to be clear: an amoral analysis doesn't imply that the researcher is immoral, nor that he intentionally overrides the moral rules of a community. He is amoral because the official codes of conduct and the desired ways of thinking cloud his view on the real power games and lead to false conclusions (Van Assche and Duineveld, 2004; cf. Machiavelli, 1988). An amoral analysis implies that the researcher also takes things into account that might be considered undesirable.

In the fifth place: the modernistic idea that theories on administrative, political and social processes can be perfect, has to be considered a fairytale. It is more realistic to use existing theories and theories that still have to be constructed in a very pragmatic way, customised for a specific problem, in the manner of Foucault and Rorty (Foucault, 1994: 250; Foucault, 1997: 172; Rorty, 1989; Malachowski, 1990) The pragmatic use of theories in this way implies that the researchers try, neither to pretend to construct *the* transition theory nor *the* guidelines. In each individual case, they look for useful theories from disciplines like philosophy, psychology, sociology, public administration and political sciences that can help them analyse and describe the transition.

17.5.2 ...and modest recommendations

This chapter should not be interpreted as a repudiation of the possibilities to formulate guidelines and recommendations. Social and political transition processes do take place and will continue to do so. Many factors contribute to them. The answer to the question whether social science can be a factor of importance is a resounding 'YES'. With the additional comment that research is able to fulfil various (sometimes unpredictable) functions and malleability itself has various gradations.

Between setting the original objectives and analysing the outcome of the research, a degree of uncertainty should be expected from socio-scientific research that aims to analyse a policy process, the functioning of an institution or transition. Whether or not the analysis succeeds, depends on the researchers, the methods of evaluation and the predisposition of the people and organisations whose functioning is under evaluation. If all these things are in order, it is likely that an evaluation will be properly executed. It is, however, difficult to use research to direct the actual use of the results of the analysis. This depends, amongst other things, on the 'users' of the research and the institutional conditions in which the research is used. There is as much chance that people and organisations adjust their behaviour as a result of the research, as there is that it disappears into the bottom drawer.

A higher degree of uncertainty between the original objectives and the outcome of a research is expected from the deducibility of recommendations and guidelines from a process analysis or evaluation. Especially where recommendations for the 'design' of long term, complex processes are concerned. Naturally, one can always construct a recommendation, but chances are slim that following the guidelines will lead to the desired objectives in the long run. After all, the context changes with every step of the process and this demands new recommendations that are adjusted to the new situation (Scott, 1998). The more complex a process becomes, the smaller the level of malleability will be.

It is impossible to deduce a definite set of useful tools, instruments and guidelines from process analysis. Similar to the theories and research methods, the use and functioning of specific means will depend on the particular circumstances and these cannot be predicted beforehand. It is preferable to view the results of studies into policy, transitions, political processes et cetera as a set of tools that can or might help steer processes in the desired direction, but these results don't do that by definition. One cannot predict what the choice of certain tools should be, nor their effect on a specific case, in advance. It *can* be expected, though, that recommendations on the means that are to be applied (tools, guidelines and such) that are based on a thorough (Foucauldian) analysis of the specific process, will be much more realistic and will have a bigger effect. The following analogy might clarify this recommendation: we think it's wrong to answer the question 'how do I cross a river?' (how can I influence a process?) with: 'by boat, because this has proven to be a very adequate way to cross a river in the past'. It is better to take the river as an object of research. From the analyses of the river it can be deduced that it would be better to avoid this river, or that a bridge would be a good and sustainable solution, or that the purchase of a boat would be a fine solution. Maybe it is even possible to swim across, which would be the cheapest solution (cf. Hopkins, 2001).

We started this chapter with the statement that the popularity of strategic research such as transition management can partly be explained by the promise it holds. We end it by adding that this promise can only be sustained if transition managers and transition researchers become more modest about their ability to 'manage' or direct the desired long-term changes in society: bearing in mind that modest intentions and expectations are often the basis for surprising outcomes.

References

Beunen, R. and Van Ark, R., 2007. De politieke dimensie van natura 2000. Hoe wil NL het EU-beleid uitvoeren? Landschap 24 (1): 13-20.

Bevir, M., 2004. Governance and interpretation: what are the implications of post-foundationalism? Public Administration 82 (3): 605-625.

Branaman, A., 2001. Self and society. Blackwell Publishers: Oxford, UK.

Duineveld, M., 2004. Bedankt voor de bloemen. Over machtige burgers, bloemen en bakstenen. In H. Venhuizen (Ed.) Geest en grond. Bureau Venhuizen: Rotterdam, the Netherlands, pp. XVI-XXVIII.

Eco, U., 1993. De grenzen van de interpretatie. Bert Bakker: Amsterdam, the Netherlands.

Fischer, F., 2000. Citizens, experts and the environment: The politics of local knowledge. Duke University Press: London, UK.

Fischer, F., 2003. Reframing public policy: discursive politics and deliberative practices. Oxford University Press: Oxford, UK.

Flyvbjerg, B., 1998. Rationality and power: democracy in practice. University of Chicago Press: Chicago, UK.

Flyvbjerg, B., 2002. Bringing power to planning research. One researcher's praxis story. Journal of Planning Education and Research 21: 353-366.

Foucault, M., 1988. De orde van het spreken. Boom: Meppel, the Netherlands.

Foucault, M., 1994. Power. Essential works of Foucault 1954-1984. Volume 3. The New Press: New York, USA.

Foucault, M., 1997. Ethics: subjectivity and truth; the essential works of Michael Foucault, 1954-1984. Allen Lane: London, UK.

Foucault, M., 1998. The will to knowledge. The history of sexuality: 1. Penguin Books: London, UK.

Foucault, M., 2001. Discipline, toezicht en straf. Historische Uitgeverij, Groningen, the Netherlands.

Foucault, M., 2003. 'Society must be defended' Lectures at the College de France, 1975-1976. Picador: New York, USA.

Frissen, P.H.A., 1996. De virtuele staat. Politiek, bestuur, technologie: een postmodern verhaal. Academic Service: Schoonhoven, the Netherlands.

Geertz, C., 1973. The Interpretation of cultures: selected essays. Basic Books, New York, USA.

Gibbons, M., 1994. The new production of knowledge: the dynamics of science and research in contemporary societies. Sage: London, USA.

Gils, W., Hoeks, H., Kate, L. and Van Rooden, A. (Eds.), 2004. Foucault. Breekbare vrijheid. Teksten en interviews. Uitgeverij Boom: Amsterdam, the Netherlands.

Gray, B., 1997. Framing and reframing of intractable environmental disputes. In: R. Lewicki, R. Bies & S.B. (Eds.) Research on negotiation in organizations. Vol. 6. JAI Press: London, UK, pp. 163-188.

Hajer, M. and Wagenaar, H. (Eds.),. 2003. Deliberative policy analysis: understanding governance in the network society. Cambridge University Press: Cambridge, UK.

Hopkins, L.D., 2001. Urban development: the logic of making plans. Island Press: Washington, USA.

Hoppe, R., 2002. Van flipperkast naar grensverkeer. Veranderende visies op de relatie tussen wetenschap en beleid. AWT Achtergrondstudie nr. 25: Adviesraad voor het Wetenschaps- en Technologiebeleid.

Howarth, D., 2000. Discourse. Open University Press: Buckingham, UK.

Korsten, A. and Hoppe, R., 2006. Van beleidswetenschap naar kennissamenleving. Voortgang, vooruitgang en achteruitgang in de beleidswetenschap. Beleidswetenschap 4.

Loos, A., Spaapen, J., Wamelink, F. and Drooge, L., 2007. Zichtbaar maken van maatschappelijke relevantie van kennis. De sci-Quest methode in praktijk. NWO: The Hague, the Netherlands.

Machiavelli, N., 1988. The prince. Cambridge University Press: Cambridge, UK.

Malachowski, A.R., 1990. Reading Rorty: critical responses to philosophy and the mirror of nature (and beyond). Blackwell: Oxford, UK.

Maturana, H.R. and Varela, F.J., 1987. The tree of knowledge. The biological roots of human understanding. Shambhala Publications: Boston, USA.

Nietzsche, F., 1977. Morgenrood. De Arbeiderspers: Amsterdam, the Netherlands.

Peters, B.G., 2005. The problem of policy problems. Journal of Comparative Policy Analysis: Research and Practice 7(4): 349-370.

Pierre, J. and Peters, B.G., 2000. Governance, politics, and the state. Macmillan: Basingstoke, UK.

Potter, J., 1996. Representing reality: discourse, rhetoric and social construction. Sage: London, UK.

Pressman, J.L. and Wildavsky, A.B., 1979. Implementation: how great expectations in Washington are dashed in Oakland: or, Why it's amazing that Federal programs work at all, this being a saga of the Economic Development Administration as told by two sympathetic observers who seek to build morals on a foundation of ruined hopes (2nd ed.). University of California Press: Berkeley, USA.

Ringeling, A., 2002. An instrument is not a tool. Paper presented at the Paper presented at Conference on Policy Instruments, McGill School of Law, Montreal, Ontario, Canada.

Rorty, R., 1989. Contingency, irony, and solidarity. Cambridge University Press: Cambridge, UK.

Roth, D., Warner, J. and Winnubst, M., 2006. Noodverband tegen hoog water: waterkennis, beleid en politiek rond noodoverloopgebieden. Wageningen UR: Wageningen, the Netherlands.

Rotmans, J., 2003. Transitiemanagement: sleutel voor een duurzame samenleving. Van Gorcum: Assen, the Netherlands.

Rotmans, J., Kemp, R., Van Asselt, M., Geels, F., Verbong, G. and Molendijk, K. (2000). De casus van een emissiearme energievoorziening. International Centre for Integrative Studies, Universiteit Maastricht: Maastricht, the Netherlands.

Rotmans, J., Loorbach, D. and Van der Brugge, R., 2005. Transitiemanagement en duurzame ontwikkeling; Co-evolutionaire sturing in het licht van complexiteit. Beleidswetenschap 19(2): 3-23.

Scott, J.C., 1998. Seeing like a state: how certain schemes to improve the human condition have failed. Yale University Press: New Haven, USA.

Spaapen, J., Dijstelbloem, H. and Wamelink, F., 2007. Evaluating research in context: a method for comprehensive assessment. Consultative Committee of Sector Councils for Research and Development (COS): The Hague, the Netherlands.

Strauss, A.L. and Corbin, J.M., 1990. Basics of qualitative research: grounded theory procedures and techniques. Sage: London, UK.

Termeer, C.J.A.M. and Kessener, B., 2006. Vitaliseren van gestagneerde organiseerprocessen. Management en Organisatie 2: 26-38.

Terpstra, M., 1995. Ontwerpen en weten, maken en verkeren. Argumenten voor en tegen maakbaarheid in beleid en wetenschap. Best/Damon: Nijmegen, the Netherlands.

Tress, B. and Tress, G., 2003. Interdisciplinary and transdisciplinary landscape studies: potential and limitations. Delta Program: Wageningen, the Netherlands.

Van Ark, R.G.H., 2005. Planning, contract en commitment: naar een relationeel perspectief op gebiedscontracten in de ruimtelijke planning. Eburon: Delft, the Netherlands.

Van Assche, K., 2004. Signs in time. An interpretive account of urban planning and design, the people and their histories. Wageningen University: Wageningen, the Netherlands.

Van Assche, K., and Duineveld, M., 2004. Groot archeologiepark Leidsche Rijn: over de rol van cultuurhistorie in planning en ontwerp. Wageningen Universiteit: Wageningen, the Netherlands.Wissink, B., 2000. Ontworpen en ontstaan: een praktijktheoretische analyse van het debat over het provinciale omgevingsbeleid. Sdu Uitgevers: The Hague, the Netherlands.

Van Gunsteren, H.R.,. 1976. The quest for control: a critique of the rational-central-rule approach in public affairs. Wiley: London, UK.

Von Glasersfeld, E., 1995. Radical constructivism: a way of knowing and learning. Falmer Press: London, UK.

Zuber-Kerritt, O., 1991. Action research for change and development. Avebury: Aldershot, UK.

Chapter 18

Transitions in history

Pim Kooij

Abstract

In this article we explore the historical background to transitions. We will look at what developmental characteristics determine the label 'transition', and go back into time to see what transitions occurred in the past couple of hundred years. Transitions can occur in one of several socio-economic categories. We will look at each of these categories in turn to see what transitions have occurred within them. Finally, we will see how these categorical transitions have interacted with one another in the past using an analysis of changes that have occurred in one village in Groningen from 1770 to 1914.

Keywords: categorical transitions, revolution, the Enlightenment, globalisation, interaction of transitions

18.1 Transitions in historical science

Historical researchers define transitions as important processes of change in time, which originated in one societal domain but affected all the other domains. Transitions span a long time, in most cases more than a century, are irreversible and involve the complete substitution of one situation by another. Transformations on the other hand refer simply to development processes. In this way you could say that transformations are a subset of transitions.

The first time the word transition can be found in historical textbooks was in describing demographic transition in the 1950s. In fact this is the only process of change which received the designation transition at that time. Having said that, there were at the same time many processes of change in all societal domains, and looking back, it is possible to discern more transitions: the economic transition, the political transition, the cultural transition, the religious transition (which was sometimes incorporated into the cultural transition), and the social transition. We can also define a spatial transition which is now called globalisation.

The origins of all these transitions mentioned above, are placed by historians primarily in the nineteenth century. The start of most transitions, however, can be traced back as far as the eighteenth century, while most transitions are considered to have been completed in the first half of the twentieth century.

In the next paragraphs we will look more closely at the individual transitions while in the last paragraph the interaction of these transitions will be analysed.

18.2 The demographic transition

The demographic transition shows a change from high mortality rates and high birth rates into low ones, from 30 to 50 per thousand inhabitants to 10 per thousand or even less. This process was characterised by the following three stages:

- *Stage 0:* Population growth was followed, due to lack of food, by subsistence crises, with increased mortality, which on their turn were followed by growth because of a rise of marriages and births. The birth rate fluctuated between 30‰ and 50‰ while the death rate did the same. These two trends kept each other in balance and as a result there was only minimal population growth.
- *Stage 1:* From 1750 the death rate started to fall, while the birth rate initially remained the same. Later on there was even a slight rise of the birth rate because people tended to marry at an earlier age. A stronger population growth was the result.
- *Stage 2:* In the middle of the nineteenth century the birth rate also started to fall which mitigated the population growth.

This process took place in western Europe and parts of the United States, the first industrial nations. In fact the first formulation of the demographic transition was based on a analysis of the British demographic figures. The term was launched by the American demographer Frank W. Notestein (Notestein, 1945) but the principle had already been described by the American demographer Warren Thompson (Thompson, 1929), the French demographer Adolphe Landry (Landry, 1934), and the British sociologist A.M. Carr-Saunders (Carr-Saunders, 1936).

After the formulation of the demographic transition, two debates emerged. The first one focused on the causes of the fall of the death and the birth rate. And the second one pivoted around the question how synchronic the transitions were in the different countries.

The fall of the death rate was attributed by some discussants to growing medical knowledge. The invention of the smallpox vaccination would have had an early positive effect. Others however argue that the economic growth which was induced by industrialisation was the main cause. This meant that most people could afford better food which resulted in a better resistance against diseases. With larger incomes and more tax for the government, more public facilities for hygiene, like waterworks and sewer systems were developed as well as public health services and hospitals. In general there is more support for this latter view (Schuurman, 1991).

The most dramatic fall in the birth rate was the consequence of the introduction of birth control. Already in pre-modern times there were practices that limited family size, like marriage at a later age, *coitus interruptus*, abstinence from sexual intercourse and prolongation of the period of breast feeding which reduced fertility. Even condoms made of sheep intestines were already available. But widespread birth control was only possible when it became economically attractive and morally accepted (Lesthaeghe and Wilson, 1986). This happened in the industrial period. It was also the period in which, in some regions, the anti-birth control influence of the Catholic church diminished. For the Netherlands this process has been revealed in an analysis within in a regional setting by Engelen (1987), Hillebrand (1991) and Delger (2003). Only recently Schoonheim discovered that birth control among Roman Catholics in poor, mainly agrarian communities was only accepted after 1960 (Schoonheim, 2005).

The motivations for people to start practicing birth control included their desire to give their children a good education and to finance this they had to limit their number. Moreover thanks to the fall of infant and child mortality the chance that these children would reach adulthood, became greater. The introduction of social security also meant that children were no longer the parents' only financial security for their old age.

From the beginning the demographic transition has been described as a global process with forerunners and followers (Reher, 2004). In forerunners, like Great Britain and France, the transition had already started in the eighteenth century while in others the process has just started recently (Table 18.1). A variation in the transition characteristics with some of the followers has been that a decline in mortality took place thanks to the exportation of medical insights and medical care from developed countries, while motivation for and acceptance of birth control were still lacking, which caused fast population growth. In other countries, especially in Africa, HIV has prevented the death rate from falling.

Table 18.1. The span of the demographic transition (sources Van der Woude, 1985; Chesnais, 1986; Livi-Bacci, 2007: 102).

Country	Period	Duration (years)
Great Britain	1750-1960	210
France	1785-1970	185
Sweden	1810-1960	150
The Netherlands	1850-1960	110
Germany	1876-1965	90
Italy	1876-1965	90
Russia	1896-1965	70
Taiwan	1920-1990	70
Mexico	1920-2000	80
China	1930-2000	70

In all countries in Table 18.1 the fall of the death rate marked the starting point of the demographic transition. This fall was linked with economic modernisation. Among the forerunners the acceptance of birth control took much more time than among the followers. In fact all these leading countries showed a minimum level in the 1960s. In most countries there was a still slight rise in the birth rate after that decade. Thanks to economic growth, a growing number of couples could afford more children and children even became a status symbol. Of course China with its draconian one child policy remained an exception. Within each country there were strong regional differences, especially in countries where the demographic transition started early. In the Netherlands for instance, the mortality in the towns in the west of the country remained high. The rapid expansion and the related growth of industry created very unhealthy living conditions, especially with respect to housing. In the last decades of the nineteenth century, however, the death rate started to fall quickly thanks to better public health conditions, especially the separation of the supply and drainage of water. In the southern parts of the country mortality remained relatively high until 1925, especially infant mortality. This was caused by a preference for bottle feeding instead of the more hygienic breast feeding.

The upward shift in the birth rate after the 1960s did not last. The introduction of the contraceptive pill and the second wave in the emancipation of women, which in the Netherlands was accompanied by the slogan *Baas in eigen buik* [Boss in my own womb], meant a new impulse for family size limitation. As a result the birth rate fell below reproduction level. Some demographers, like Ron Lesthaeghe and Dirk van de Kaa propose that this development should be defined as a second demographic transition (Van de Kaa, 2003). Apart from a further fall of the death and birth rates this transition was characterised by a shift from the 'bourgeois family model', a solid institution, to the 'individualistic family model', which was characterised by growing changes in relationships, because of divorces and the rise of new forms of cohabitation, a growing acceptance of abortion, and clear discussions between the partners about whether or not to have children. The number of childless partnerships grew, but at the same time the demand for advanced medical techniques to treat infertility increased as well.

In my view this last development can be better considered as the outcome of the demographic transition. The same is the case with the rise of migration, which in many highly modernised countries serves as a means to repair the population deficit. In any case this notion of demographic transition has never been considered a true scientific theory. It has been seen rather as a generalisation of demographic development during the last two centuries. Even the time frame is not fixed, because the fluctuations in the birth and death rates differed in their timing from country to country, from region to region, and even between town and country. The precise causes are not totally clear, although the categories in which they fit are the same everywhere.

In a recently published, very popular demographic textbook, a distinction has been made between the mortality transition, the fertility transition and the migration transition in order to get a clearer view of those phenomena (Weeks, 2005). This works well, but in this analysis, the interaction between these transitions is neglected. This is a problem with many descriptions of the demographic transition. In fact, this transition was a complicated interrelation of developments in terms of causes and results of birth and death rates, in which migration formed just one element.

In spite of many national and regional differences the demographic transition as a general phenomenon still stands. In a growing number of countries the process of a transition from high to low death rates and birth rates has already been completed. Other countries are still in that on-going process.

18.3 The economic transition

The essence of the economic transition is the substitution of muscle, wind and water power by energy forces derived from fossilised sources: steam, gas, electricity. In fact this change started in the Dutch Republic in the seventeenth century, when the exploitation of peat and its use in industry became one of the main contributors to the Golden Age (De Zeeuw, 1978). The accepted origin of the economic transition, however, is placed in the mid-eighteenth century in Great Britain, where, around 1750, coal became a major source of energy for industry.

Initially this shift was called the industrial revolution. This term was already used in the nineteenth century, among others by Friedrich Engels (Engels, 1845) but was given a firm base by Arnold Toynbee, Paul Mantoux and T.S. Ashton (Toynbee, 1884; Mantoux, 1906; Ashton, 1948). In their opinion, England changed in a very short time from a rural into an industrial nation characterised by steam and steel, where chimneys dominated the landscape and railways brought modern times to all corners of the country. This all took place between 1760 and 1830. 1760 was taken as a starting point because at that time major innovations took place in the textile and iron industries. Around 1830 a point of no return was reached. To cap it all off, political reforms took place in 1832, which gave the new industrial cities voting powers for parliament at the expense of the so called rotten boroughs, tiny settlements which had been left behind by the changes, and had never outgrown the Middle Ages or Early Modern times.

Later research, however, made it clear that the term Industrial Revolution was a misnomer (Cameron, 1989). In the first place there was no sudden change. The industrialisation process started slowly and spanned almost 100 years before it became omnipresent. Secondly it started only in two branches, the textile industry and the iron industry, and the textile industry initially used water-power, not steam engines (the Arkwright water frame). In the third place modern industry started only in a few regions (Shropshire, the Midlands, parts of Scotland).

These three features show that the industrialisation process is a lot less revolutionary than the term suggests. Moreover the start can be situated even further back in time. The use of coal had begun already around 1600 because of a lack of wood in the London region. At that time wood was used for building houses, for shipbuilding, for heating and cooking, and in industry it was primarily utilised in the production of charcoal which was used for purifying iron ore. The exploitation of coal lead to the development of steam-engines to remove water from the mines. The steam engines by Savery and Newcomen were introduced already in about 1663 and 1708 respectively. James Watt did not invent the steam engine in 1776, but succeeded in turning the pumping movement into a rotating one, which enabled the development of the factory system and of the railway.

In light of this, it is better to speak of a industrial transition than of a industrial revolution. Mechanisation also affected the other sectors. In agriculture, for example, iron ploughs and other iron engines were introduced, like the reaper. Locomobiles were used on the land and some agricultural activities like threshing and the production of butter, were transferred from the farm to the factory. In the service sector the introduction of steam-power in the form of railways and steamships meant a major impulse for mobility and a new stage in globalisation. Moreover the development of the factory system implied the genesis of the office, with typewriters and other machinery, while the related rise of banking and insurance companies was an extra impulse in that direction. Because this industrial change affected all three economic sectors it is more accurate to speak of an economic transition than an industrial one.

The start of this transition has been traced back to around 1730, when the first industrial invention (flying shuttle) took place in the textile industry while in agriculture new techniques developed and became widespread, like crop rotation, which created an agricultural surplus which in turn could feed an additional labour force outside of agriculture itself. The end of the transition is sometimes taken to be 1851, the year that the first World Exhibition took place in London. This showcase of new technology highlighted the performance of British industry and made clear how it had taken the lead worldwide. The main exhibition building, the Crystal Palace, was a symphony of art and technology. Its glass and iron structure impressed visitors from all over the world.

There are some accounts of the contribution of the Netherlands to the world fair. These are not very flattering. The Dutch stand was hidden behind the Austrian one and despite of the display of a magnet (Baudet, 1967) had minimal attraction value. In fact economic modernisation of the Netherlands was not primarily industrial. It was agriculture which took the lead. Industrialisation in textiles and machinery initially took place in the southern part of the Kingdom of the Netherlands, a political entity which was created in 1815 at the Congress of Vienna, and which included both the Netherlands and Belgium of today. When Belgium (the industrial part of the low countries) revolted against the Kingdom of the Netherlands in 1830, the mainly agricultural Netherlands had to make up for it's developmental backwardness.

This proved to be rather difficult because around 1830 Germany was also in the full throes of becoming an industrial nation (mainly in the Ruhr valley, and the area around Berlin). The city of Maastricht profited from its location between the two industrialising nations but in other parts of the country economic growth on an industrial basis didn't really get underway until 1860 (Van Zanden and Van Riel, 2000).

A lot of comparative research on the origins of the industrialisation process was done in the 1960s. A major stimulus for this kind of research was given by W.W. Rostow, who defined five stages of economic growth, which were applicable in every country but where the timing of the developments varied (Rostow, 1960, 1963). Some developed countries had already reached the last stage 'the age of high mass consumption', while others only were in the initial stage 'the traditional society'. The crucial stage in his model was the 'take off' in which the economy modernised rapidly, in the first place by a rise in the investment rate. In many countries scholars tried to fix this take off in time. Rostow himself determined 1840 as the starting point for de United States. For the Netherlands 1895 was mentioned, which later on proved to be too late (De Jonge, 1968).

Rostow's book was a politically correct 'cold war study'. Communist economies were depicted as unproductive deviations from the general pattern. Other researchers like Angus Maddison extended this comparative research in a much more sophisticated way. By means of precise comparison of economic growth patterns in different countries they tried to establish which factors especially favoured economic growth (Maddison, 2006).

Rostow's 'stages' have become obsolete now but there are other phases of development that have been established that are still considered realistic. These are called business cycles, especially the ones formulated in 1926 by the Russian scientist Nicolai Kondratieff, which span about 40-60 years. Periods of growth which were followed by periods of depression, especially the Great Depression of 1875-1895 and the world depression of 1929-1940 are well known. In 1939 Joseph Schumpeter, an Austrian economist who had moved to the USA, proposed that depressions could be overcome by *Neue Kombinationen* (new combinations); innovations which were an impulse for economic growth. By looking back into the past, he observed that the development of the railway for instance played an important role in the economic recovery around 1848, while the introduction of electricity marked the end of the Great Depression. Extrapolating on his ideas, one can see that the 1930s economic crisis came to an end with the introduction of the mass production of cars, but also by the war preparations. The last major economic crisis, the oil crisis of the 1970s was ended by the introduction of the (personal) computer, among other things.

Taking a global perspective, the economic transition seems to be the transition that has taken the longest to run its course. It already began in Europe in the eighteenth century but in other parts of the world it started only very recently as shown in the cases of Brazil, India, and China, where the economies are only now booming.

18.4 The political transition

The political transition involved the change from oligarchy to democracy, from a situation in which power is confined to a select group into a situation in which every citizen is *de jure* allowed to vote and to be elected. This process was accompanied by the formation of the nation state, in which the national assembly became the legislative and political centre, with jurisdiction over the towns and provinces, which during the *Ancien Régime* had themselves been the centres of power. The first forms of democracy appeared in ancient times, although in most cases many groups – women, slaves – were excluded. The actual origin of the political transition addressed in this chapter, started in 1775 in the United States with the revolt against Britain. The Declaration of Independence, which was promulgated one year later, stated that 'all men are created equal'. The Americans founded a democratic republic, while in Europe the monarchy was at its zenith. The Dutch Republic and the Republic of Venice had preceded the American version, but their influence had waned by this time. The American *res publica* with its charter rooted in the Enlightenment epitomised the modernity of this political system.

The ideas of the American revolution were a source of inspiration for the French revolution of 1789. France was an absolute monarchy. The parliament, which let itself be spoon fed by the king, had not met since 1614. Since then a lot had changed. The third estate, the bourgeoisie, consisting of merchants, lawyers, entrepreneurs, teachers and other people who paid the most taxes, inspired by philosophers like Voltaire and Montesquieu, demanded more influence, at the expense of the other estates, the nobility and the clergy.

At this time, the French King was forced to call on the Estates General (the parliament) to raise taxes. He had financial problems because he'd lost some wars and his treasury was also drained by his financial aid to the Americans in their fight against Great Britain, France's arch-enemy. This was followed by a series of events which repeated themselves later on in many other countries. Riots like the one that preceded that storming of the Bastille (France's national prison) created a precarious situation in which the third estate, the bourgeoisie, tried to seize power. A declaration of the rights of men with the slogans *liberté, égalité, fraternité* (freedom, equality, brotherhood) was launched, armed farmers came to Paris to help the third estate, a National Convention was created and the monarchy became constitutional.

Then the revolution radicalised. The possessions of the church were confiscated in order to get rid of the growing state deficits and a growing number of noblemen lost their heads at the guillotine. The king and queen underwent the same fate in 1793. Finally, a general, Napoleon Bonaparte, took over power. He started out as a consul but soon declared himself to be emperor with absolute power. After his Waterloo the French monarchy was restored. The new king was Louis XVIII, the brother of the murdered Louis XVI. (The dauphin who should have become the seventeenth Louis had, in the meantime, died in prison). Bonaparte's brothers, who had been appointed kings in several countries, had to leave.

At first, it looked as if the old situation had returned. But there were some significant changes. France remained a constitutional monarchy. In all west and north European countries a parliament was created, most of the time consisting of two chambers. The First Chamber, which was modelled after the House of Lords in Great Britain consisted of noblemen appointed by the kings. It controlled the Second Chamber (the House of Commons) which was comprised of elected members. It wasn't universal franchise, however, because the right to vote depended mostly on the ownership of property. One had to pay a substantial amount of taxes to obtain the right to vote.

In the decades after the fall of Napoleon there were many movements aimed at restoring power to different groups on a local regional and national level. There were some years in which this aim became so clearly manifested that one could speak of revolution years. The first step was taken in 1830. In that year Belgium declared itself independent of the Kingdom of the Netherlands, which had been created by the conference of Vienna in 1815 where peace treaties and new boundaries in Europe were arranged. The Kingdom was meant to function as a buffer between Great Britain and France, but it did not work out because of religious conflicts (catholic versus protestant) and economic ones (industrial versus agrarian).

In France the autocratic king Charles X, successor of his brother Louis XVIII, was replaced by Louis Philippe from the house of Orleans, who was more in favour of a substantial representation of the bourgeoisie. The Chamber of Peers ceased to be hereditary and the number of electors in the Chamber of Deputies was doubled to 200,000, all possessors of substantial real estate, which meant that now one out of thirty adult French males was allowed to vote.

In Germany, at that moment still a patchwork of kingdoms, duchies, counties and free towns, the power of Prussia was growing. One of the reasons was its acquisition of Rhineland in 1815 which included the industrialising Ruhr area. The Prussian kings were not in favour of a constitution and forbade political meetings. This fostered the rise of harmless looking associations like sport clubs, reading clubs and dining clubs, which were in fact acting as fronts for associations that stimulated political discussion. Moreover, the liberals who at that time were the alternative for the conservatives made big gains in the elections in the 1840s (Altena and Van Lente, 2006: 148). In Great Britain the first preparations for the Reform Bill which was promulgated in 1832, were already being made in 1830. This bill, among others things broadened the franchise and reassigned 143 parliamentary seats from old 'rotten' boroughs to new industrial towns.

1848 was the real year of revolution. In that year the *Communist manifesto,* written by Karl Marx and Friedrich Engels was published. In their view a revolution which would bring a society which was characterised by complete equality was inevitable. Most revolutionary groups had less far reaching aims. They wanted to extend the influence of the middle classes and labourers in politics, by enlarging the group of electors.

Once again, the revolution started in France. In February the king was exiled and France became a republic again. Napoleon, son of the former king of Holland, Louis Napoleon, brother of the emperor Napoleon, was elected president. In the Netherlands king William II was forced to accept a constitution which turned the Netherlands into a constitutional monarchy and moved political responsibility to the ministers. There was a slight extension of the right to vote by broadening the franchise, while the First Chamber ceased to be appointed by the King. In Great Britain the Chartist movement again organised petitions signed by millions of people for a people's charter, which demanded the right to vote for all men. Labourers believed that they would be able to do something about the bad working conditions once they were able to enter parliament. This movement failed, however, and was followed by the foundation of labour unions which in the long run would prove more successful.

In the German states and in the Habsburg monarchy, which apart from Austria and Hungary also encompassed a great part of the Balkan, serfdom was abolished. In Frankfurt a parliament consisting of representatives from all the different German states met in order to make a blueprint for the unification of Germany. This initiative failed mainly due to the opposition of the Prussian monarchy. Prussia in fact only got its constitution in 1850. One third of the second chamber was elected by big tax payers, one third by people paying medium taxes, and one third by the numerous group of small payers. At the time, this system was rather progressive. In the Habsburg monarchy, in spite of the actions of a large number of radical groups, nothing significant happened apart from the abolition of serfdom.

The revolution did not last long in France. In 1850, universal suffrage, which was introduced in 1848, was abolished. The lower classes, which might have harboured socialist sympathies, were excluded. One year later the whole Assembly was sent home. From that moment, Napoleon governed as a dictator and in 1852 he declared himself emperor.

The period up until 1870, was characterised by nationalism, a movement which had been manifest for decades, but which now, apart from nation building in existing states and the creation of some new ones like Greece (1821) and Belgium, led to the dissolution of some empires and the creation of new ones.

The Crimean war (1854-1856) in which Great Britain and France together successfully assisted the Turks in their fight against Russia – which wanted expansion to the south – resulted in the creation of Rumania. In 1859 the unification of Italy was completed, with the exception of Venice and the papal state Rome, which joined in 1866 and 1870 respectively. In Germany Bismarck was creating a North German League under Prussian dominance and the Habsburg monarchy in 1867 was turned into a Dual Monarchy, to keep Hungary within Austria's sphere of influence. Both countries became constitutional parliamentary states, however, with complicated voting systems which favoured big landowners. In the United States, the American Civil War (1861-1865) initially formed a serious threat to the first constitutional state, but in the end the constitution was amended to establish that every inhabitant of every state should be not only a citizen of its own state but also of the United States. The war ended with the abolition of slavery in 1865.

The German unification was effected in 1871, after their victory in the French-German war, which was provoked by Napoleon to inhibit growing Prussian dominance. The Prussian king became emperor of Germany. At the same time there was universal (male) suffrage for the Reichstag, but this did not mean that much in practice because Germany was in fact a federation of monarchies (Bavaria, Würtenberg, etc.) with their own constitutions. The appearance of the German emperor heralded the disappearance of the French one and France became a republic once again. The Third Republic, with its own parliament, was unfortunately ineffectual because of competing political parties.

The last quarter of the nineteenth century was characterised by the growth of the socialist movement. One branch preached the proletarian revolution, but a growing number of people were in favour of universal suffrage. Industrialisation which was now widespread, was accompanied by a growing number of unions and associations which promoted real parliamentary democracy. But the higher social strata were afraid of the growing influence of the masses and events like the revolt of the Paris commune in 1871 reinforced these fears. During this revolt, labourers in fact took over the city for a short time, but the uprising was bloodily suppressed by the people in power. In some countries, the government tried to marginalise socialism by implementing some sort of social security. Germany under chancellor Bismarck is the most striking example of this ploy.

In the end however pressure from below resulted in the creation of universal male suffrage. Belgium started in 1892 and the Netherlands were relatively late in 1917. The extension to general suffrage for women didn't take so long. In many countries women had done men's jobs during the First World War, and as a result demanded equal rights. Moreover the Russian revolution caused other countries to take measures to prevent this event happening in their country. Soon after the First World War, suffrage for women was effectuated in most western countries. In the Netherlands in 1919.

We may conclude that the political transition ended around 1920 when universal suffrage became widespread while at the same time the process of creating the nation state received a new impulse by the treaty of Versailles (1918). This treaty meant the dissolution of multi-nation states like the Habsburg empire and the Ottoman Empire (Turkey). But in addition it is useful to point out, that the setback formed by the *inter bellum* in Russia, Germany, Italy and Spain showed that the process of nation building could also coincide with anti-democratic movements.

18.5 The cultural transition

The cultural transition can be defined as the dissemination of the ideas of the Enlightenment into society. This meant a change from a worldview in which societies were completely dominated by religion into a view in which man himself had the capacity and responsibility for his own development and that personal and social progress could be achieved by means of science and education. According to most scientists this new worldview came into being in the second half of the eighteenth century, the 'age of reason', especially in France. Recently, however, Jonathan Israel pointed out that the Enlightenment had already manifested itself in the seventeenth century in the Dutch Republic, especially in Spinoza's writings (Israel, 2006). In any case the term was used by some eighteenth century intellectuals, while its roots go back to the scientific 'revolution' of the sixteenth and seventeenth centuries.

To understand the cultural transition it would be better to look at the impact of the Enlightenment than at its roots. In most West European countries the ideas of the Enlightenment spread via salons, coffee houses, learned societies, books and pamphlets, etc. They promulgated general suffrage, the innovation of education, freedom of speech and thought, and a critical eye on religion. But first and foremost, a faith in progress was created. This progress not only was seen in technical terms, in the form of a significant series of inventions and innovations, with the steam engine as the catalyst, but also in terms of social standards and learning, which would create a more civilised, educated society.

The Enlightenment is related to the writings of a number of philosophers and intellectuals like Voltaire and Locke. But the most characteristic book proved to be the Encyclopaedia by Diderot and d'Alembert. In 17 volumes the whole corpus of human knowledge was displayed. It set out all known human achievement and created an intellectual climate which supported the idea of unstoppable progress. Knowledge is power became the adage.

In the religious sphere, the Enlightenment caused a change to a more sober, no-nonsense religion. The belief in miracles and other superstitious elements were dropped because these could not be explained in a scientific way. In politics, the absolute monarchy was challenged. There was less support for kings, noblemen and clerics who proclaimed that their position was ordained by God. People no longer believed in the worldview that supported structural inequality.

The conviction that not everything was arranged in heaven and that on the contrary people could take their lives into their own hands, played an important role in the American Declaration of independence (1776) and in the French revolution (1789). Here is a clear link between the political and the cultural transition. But one could say that all transitions are linked by the central theme of this cultural transition: the growing belief that people had the right to decide their own fate and, therefore, needed to obtain knowledge to be able to make the right decisions.

In fact this vision could be found in every aspect of society. But one can say that the most important impact of the Enlightenment in the cultural sphere was the growing significance attached to education – the spread of knowledge – and the decreased interest in religion.

Almost immediately after the start of the revolution, leading politicians in France started to substitute church-run schools by free state schools. They also tried to create a national curriculum which was to include everything a civilised citizen ought to know. This did not work out very well, because Napoleon had other priorities, but in the course of the nineteenth century, more efforts were made. In other countries, like the Netherlands, the same development took place. National governments subsidised only their own state schools. The schools which were owned and run by the churches were financially handicapped. The provision of subsidies for confessional schools in the nineteenth century became one of the hottest political items of the time.

In the preceding paragraphs culture is presented as a set of human achievements, both in the sphere of artefacts (machines, books) as in the intellectual sphere (science). But even when a more limited definition of culture is taken – the Arts – the influence of the Enlightenment is clear. The Arts were seen as a necessary attribute of the civilised citizen. Orchestras ceased to be the monopoly of kings and members of the nobility. Mozart operas were also performed for the middle classes and paintings were not only made to order, but also for the market. Not only that, every city started to build its own theatres, while museums were built as an alternative for private collections.

As has been pointed out before, the Enlightenment influenced thinking and behaviour of people in Western Europe and America in a way that is still felt today, and which spread to other parts of the world. Sometimes the inspiration proved too radical. Many of Descartes' ideas became universally accepted, especially his idea that reasoning and investigation were the foundation of knowledge rather than the dogmas of the church (*cogito, ergo sum*). But his assertion that animals were only automatons, without feelings or senses, did not last. Ideas about the equality of men and women which were already plighted for in 1673 by the French priest François Poulain de la Barre in his treatise *De l'égalité des deux sexes* (Stuurman, 2004) and which were promulgated by the famous *salonnières*, (Goodman, 2004) were not widely accepted.

The ideas of the Enlightenment also evoked counter reactions. Soon after the French Revolution, when Napoleon had been defeated, a form of restoration took place, not only in politics but also in the domain of culture. The Romantic movement stressed the importance of feelings not as an alternative but as an extension to reason (Altena and Van Lente, 2006: 114). The world was no longer seen as a machine, but as a living, dreaming, feeling creature. Both the Enlightenment and Romanticism had a utilitarian vision of nature, but whereas the Enlightenment pointed to nature primarily as a provider of food and raw materials, Romanticism also saw nature as a source of beauty, health and joy.

In the nineteenth century world of religion, the downplaying of Gods almightiness meant, for a growing number of people, a turn to agnosticism or at least latitudinarianism. At the same time, however, this threat to mainstream religion caused other people to turn more fundamental. In the Netherlands this led to two secession movements, the first in 1834 (*Afscheiding*) the second in 1896 (*Doleantie*) which created orthodox protestant churches, which merged later on. In Great Britain the rise of Methodism points to the same process and in the Roman Catholic religion the rise of ultramontanism is a comparable reaction.

In my view these reactions are also part of the cultural transition, which makes this transition more dialectical than the other ones. It is not easy to mark the end of the cultural transition but one could say that at the end of the nineteenth century, most ideas of the Enlightenment had become such common property that no justification was needed anymore. Even the cultural pessimism of the 1930s or postmodernism of the 1990s could not change that.

18.6 The social transition

The social transition marks the change from a society in which social inequality was based on birth and law, into a society in which people are *de jure* equal but differ *de facto* because of their social, economic and political position in combination with their level of education. This social transition was firmly linked to the political and cultural transition and was also induced by the Enlightenment which promoted equality, especially in the context of the French Revolution.

Before the French revolution, society consisted of estates; social classes based on birth and specific prerogatives. In France there were three estates, the first estate: the nobility, the second estate: the clergy, and the third estate *'tiers état'*: the bourgeoisie. This third estate, consisting of traders, industrialists, lawyers, doctors, teachers and shopkeepers amongst others, as well as labourers, was discriminated against by the other ones. At the beginning of the French revolution, the third estate proclaimed the declaration of the rights of men, which put forward the idea that all men are equal. The American constitution, which was launched some years before, had done the same.

As we have seen in the paragraphs on the political transition, this equality, which was also introduced in countries conquered by the French, did not mean that all people got the same rights. It took almost a century before general suffrage was effectuated. But inequality in the judicial system was abolished, and when there was a suspicion of class injustice, this caused heavy commotion, as in the Dreyfus affair. By the nineteenth century, some leading sociologists had put this new equality into a model. One could even call this a theory. The most influential of these theorists were Karl Marx and Max Weber.

According to Marx, every historical period was characterised by the presence of competing classes, having opposite economic interests and, because of this, social and political conflict as well. In modern industrial society these classes polarised into two main opponents: the bourgeoisie and the proletariat (Marx and Engels, 1848). The central category in Marxian inequality was the possession of the means of production. This was an economic criterion which also fixed social and political inequality.

Max Weber, however, put forward a three-dimensional theory of inequality. He pointed to an economic criterion: the position in the labour market; a social criterion: way of life, status and prestige; and a political criterion: power. This led to a threefold inequality consisting of classes, estates, and parties. Weber was not very clear about the relation of these three to one another but in one of his last works, on inequality in India, he stated that in periods of economic progress and stability the social elements were dominant, while in periods of fast change or economic depression the economic elements were more prominent. In books in the Weber tradition social class mostly refers to status and prestige as well as lifestyle. And since the Weber theory was not as revolutionary as the Marxist one, it got more support, especially in the USA (Duijvendak and Kooij, 1992: Chapter 3).

Within the context of Weber's theory, the social transition can also be defined as a shift from ascribed status, to achieved status. Education became the main vehicle to obtain status. Especially the middle classes used the secondary schools to improve their position. On the whole, the nineteenth century showed examples of success of this strategy. In the 1860s, for instance, the officers in the German army ceased to belong exclusively to the nobility. People from the highest ranks in society worked more and more together with successful businessmen in public service, and sometimes these social classes even merged by marriage.

In the twentieth century new theories of social inequality were formulated because Weber's criteria were insufficient to explain the complicated, flexible twentieth century society. But even today influential theories like the occupational status theories of the structural functionalists, and the social/cultural capital theory of Pierre Bourdieu pay tribute to Weber's theory.

These new theories did not, in fact, play a role in the social transition, because this transition was already completed at the beginning of the twentieth century, at the same time that the political and cultural transition came to an end. Together they effectuated that the interaction of people changed fundamentally. People learned to exploit their own possibilities and could expect to be rewarded for these efforts, by money or by status. Of course obstacles remained but they were no longer inherent and could be removed.

Sometimes the social transition is not indicated as a transition of its own but incorporated in the other transitions (see below for an example of this). This is because all transitions are the result of human acting, and this acting does not need to be indicated separately. On the other hand, distinguishing a social transition underlines how demographic change, economic change, intellectual change, and political change affected inter-human relations in a fundamental way.

18.7 The spatial transition: the first stage of globalisation

The spatial transition refers to world unification, the creation of a global economic system which was created in the nineteenth century, driven by modern imperialism and the destruction of distance by trains and steamboats. It is better not to name this movement globalisation, because in most studies the globalisation process is placed in the twentieth century. Scholars like D. Held c.s., J.R. and W.H. Mac Neill, and A. Schuurman however made clear that globalisation started long before that time (Held *et al.*, 1999; MacNeill and MacNeill, 2003; Schuurman, 2001 and 2007). In fact twentieth century globalisation only consisted of an intensification and condensing of contacts and networks on a global scale which, for the greater part, had taken shape in the nineteenth century.

The roots of the global transition go rather far back in time. One could start with the spread of people from Africa to other continents in pre-historic times. But it is more common to begin with the explorations of the Portuguese along the west coast of Africa in the course of the fifteenth century and the discovery of America by Columbus in 1492. From that time some countries in Western Europe – Portugal, Spain, England, the Dutch Republic, France – started to create colonies. Large parts of the world were incorporated into trade movements which already had the shape of a network:

- Europe provided Africa with arms and textiles in exchange for slaves.
- These slaves were exported to America.
- America provided Europe with silver, gold, tobacco, sugar.
- Silver and gold were shipped from Europe to Asia.
- Silk, spices and porcelain came back from Asia to Europe.

Sometimes this system is called trade capitalism. The intercontinental trade only formed a very small part of total trade movements, but it was very striking and profitable.

A second stage in the process of global unification emerged in the second half of the eighteenth century. In a number of colonial wars, for instance the seven years war (1756-1763), the British succeeded in demarcating a global sphere of influence, comprised of, amongst others, North America and India. They did so at the expense of the French and the Dutch. Perhaps we should place the start of the spatial transition in this period because these demarcations also played an important role in the nineteenth century, not only because they expressed the division of the world between West European nations, but also because they induced the losers to try again.

The scramble for Africa meant an opportunity for revenge for France. It conquered a vast area in the north west, while newcomers like Belgium and Germany also took their share and Portugal kept its old positions. Once again, Great Britain emerged as the big winner with possessions from Cairo to the Cape. Spain had already lost all its colonies in South America at that time, which together with the Philippines came under the influence of the USA. France established a number of colonies in East Asia, while un-colonised China and Japan made their entrance into the world trade system.

This new colonisation was not performed by isolated ships with a limited crew as in the past. Steamships and railroads transported armies all over the world and the introduction of the electric telegraph, around 1840, and the telephone, around 1880, enabled fast communication. Now a real global economic system was established, which sometimes is called industrial capitalism, because the colonies provided the western European industries with raw materials and their labourers with food. At the same time, these global transformations were closely linked with nation building in western Europe, which was a characteristic of the political transition.

In the context of nation building another development became obvious which also could be seen as a part of the spatial transition. This was the national unification, the placement of regions, provinces and cities into a national framework (Knippenberg and De Pater, 1988). One could define this as a globalisation process in miniature but in fact it was part of a broader movement. Thanks to national unification, nations could play a role on the international scene. We have already seen that this unification movement had political, cultural and economic dimensions, but also a spatial one. The infrastructure, especially the railways but also highroads and canals, proved an important instrument in this process of national unification.

From the beginning this unification process evoked counter forces. The central governments kept a keen eye on peripheral regions and cities, which sometimes tried to stress their relative independence. To maintain their control central governments ensured there was a strong police presence in these areas. To underline the simultaneous emphasis on regional and local roots and identity the orientation on a unifying world, called globalisation, the word glocalisation was formulated (Robertson, 1995). It refers to the last decades of the twentieth century but glocalisation seems to have been a nineteenth century movement as well.

18.8 The interdependence of the transitions

It has been stressed many times in the previous paragraphs that the transitions are interrelated. An example of this interrelation is shown in a study on a Dutch village, Hoogkerk, near the city of Groningen. In the context of the project 'Integrated History', which covers the period 1770-1914, an analysis has been made of the effects of the big nineteenth century transitions on regional and local communities. This small scale of one village has been chosen to get a clear view of the interactions between the various transitions. This resulted in the following matrix (Table 18.2).

As you can see in this matrix the effects of the individual transitions on the societal domains are given. In this analysis no social transition was defined, but the effects of the other

Table 18.2. Transitions at a local level (sources: Kooij and Sleebe, 1991; Kooij, 1993).

	Demographic transition	Economic transition	Political transition	Cultural transition	Religious transition
Demographic domain	Birth rate → Mortality rate↓ Age of marriage ↓	Labour migration	Codification opinions on morality	Family planning Education	Differentiation opinions on marriage and family
Economic domain	Enlarging labour market	Industrialisation Changing occupational structure Enlarging of the market Economic growth	Free market Taxation Sharpening labour relations	Disappearance moral economy	Corporatism versus individualism
Political domain	Development social infrastructure	Development economic infrastructure	Patriotism Liberalism Socialism	Discipline Issuing of rules	Cleavage
Cultural domain	Development social infrastructure	Commercialisation	Formulation group codes	Disappearance Popular culture	Differentiation of social control
Religious domain	Acceptance birth control	Commercialisation religious care	Dis-establishment church from state	Latitudinarianism	Secularisation versus orthodoxy
Social domain	Shifting denominations	Increasing social distance	Sharpening social relations	Increasing class consciousness	Pluralism

transitions on the social domain are included. The cultural transition is split into a 'secular' and a religious part because on a local scale religion proved too important to put it into a general cultural context. Because the scale of one village was chosen, it was not practical to include an explicit spatial transition in the analysis, but of course some spatial elements like the extension of the market and political unification do make part of the analysis.

There is no room here to explain the whole matrix. What it shows is that the transitions affected all societal domains and by doing so they influenced each other. By focusing on a local level, some transitions manifested themselves as miniatures of the general transitions mentioned before. This was the case with the demographic transition, although this also implied a local variation because the birth rate did not fall that much. The two cultural transitions at the local level also reflect the general trends. But in the case of the political transition this was different. Here, it was obvious that political power was situated elsewhere, so in this case it was clear that the effects were more important than the transition itself. In the case of the economic transition a combination of effects of developments elsewhere and local developments in line with the transition could be found.

A local study like the one on Hoogkerk, shows the extent to which the nineteenth century transitions were interrelated, how they reached every corner of society and how even at this local level, Hoogkerk in its turn, participated in the formation of these transitions.

References

Altena, B. and Van Lente, D., 2006. Vrijheid & rede. Geschiedenis van westerse samenlevingen 1750-1989. Verloren: Hilversum, the Netherlands.

Ashton, T.S., 1948. The industrial revolution 1760-1830. Oxford University Press: London/Oxford, UK.

Baudet, H., 1967. De dadels van Hassan en de start van de Nederlandse industrialiteit. In: Bedrijf en samenleving. Aangeboden aan prof. dr. I.J. Brugmans. Samsom: Alphen aan den Rijn, Netherlands, pp. 1-17.

Cameron, R., 1989. A concise economic history of the world. From paleolithic times to the present. Oxford University Press: Oxford, UK.

Carr-Saunders, A.M., 1936. World population: past growth and present trends. Clarendon Press: Oxford, UK.

Chesnais, J.C., 1986. La transition démographique. Presses Universitaires de France: Paris, France.

De Jonge, J.A., 1968. De industrialisatie in Nederland tussen 1850 en 1914. SUN: Nijmegen, the Netherlands.

De Zeeuw, J.W., 1978. Peat and the Dutch Golden Age. The historical meaning of energy-attainability. A.A.G. Bijdragen 21: 3-33.

Delger, H.E., 2003. Nuptiality and fertility. An investigation into local variations in demographic behaviour in rural Netherlands about 1800. Verloren: Hilversum, the Netherlands.

Duijvendak, M.G. and Kooij, P., 1992. Sociale geschiedenis. Theorie en thema's. Van Gorcum: Assen, the Netherlands.

Engelen, Th.L.M., 1987. Fertiliteit, arbeid, mentaliteit. De vruchtbaarheidsdaling in Nederlands-Limburg, 1850-1960. Van Gorcum: Assen, the Netherlands.

Engels, F., 1845. Die Lage der arbeitenden Klasse in England. Leipzig, Germany.

Goodman, D., 1992. The republic of letters. A cultural history of the French enlightenment. Cornell University Press: Ithaca, USA.

Held, D., McGrew, A., Goldblatt, D. and Perraton, J., 1999. Global transformations. Politics, economics and culture. Polity Press: Cambridge, UK.

Hillebrand, J.H.A., 1991. Van motivatie tot acceptatie. Een onderzoek naar de daling van de vruchtbaarheid in de provincies Utrecht en Groningen, 1879-1960. PhD: Nijmegen, the Netherlands.

Israel, J., 2006. Enlightenment contested: philosophy, modernity and the emancipation of man 1670-1752. Oxford University Press: Oxford, UK.

Knippenberg, H. and De Pater, D., 1988. De eenwording van Nederland. SUN: Nijmegen, the Netherlands.

Kooij, P. and Sleebe, V., 1991. A small village in a changing world. Integral history at a local level. Economic and social history in the Netherlands 3: 19-37.

Kooij, P., 1993. Hoogkerk integraal. In: Kooij, P. (Ed.) Dorp naast een stad. Hoogkerk 1770-1914. Van Gorcum: Assen, the Netherlands, pp. 339-357.

Lesthaeghe, R. and Wilson, C., 1986. Modes of production, secularization, and the pace of the fertility decline in Western Europe, 1870-1930. In: Coale, A.J. and S. Cotts Watkins, (Eds.) The decline of fertility in Europe. The revised proceedings of a conference on the Princeton, European Fertility Project. Princeton University Press: Princeton, USA, pp. 261-293.

Landry, A., 1934. La revolution démographique. Librairie du Recueil Sirey: Paris, France.

Livi-Bacci, 2007. A concise history of world population. Fourth edition. Blackwell: Oxford, UK.

MacNeill, J.R. and MacNeill, W.H., 2003. The human web: a bird's-eye view of world history. W.W. Norton & Company: New York, USA.

Maddison, A., 2006. The world economy. OECD: Paris, France.

Marx, K. and Engels, F., 1848. Communist Manifesto. London, UK.

Notestein, F.W., 1945. Population, the long view. In: Schultz, T.W. (ed.), Food for the world. University of Chicago Press: Chicago, pp. 37-57.

Mantoux, P., 1906. La révolution industriѐlle au XVIIIe siècle. Essai sur le commencement de la grande industrie moderne en Angleterre. Société Nouvelle Librairie et d'édition: Paris, France.

Reher, D.S., 2004. The demographic transition revisited as a global process. Population, Space and Place 10: 19-41.

Robertson, R., 1995. Glocalization: time-space and homogeneity-heterogeneity. In: Featherstone, M., Lash, S. and Robertson, R. (Eds.) Global modernities. Sage: London, UK, pp. 25-44.

Rostow, W.W., 1960. The stages of economic growth. A non-communist manifesto. Cambridge University Press: Cambridge, UK.

Rostow, W.W. (ed.), 1963. The economics of take-off into sustained growth. Macmillan: London, UK.

Schoonheim, M., 2005. Mixing ovaries and rosaries. Catholic religion and reproduction in the Netherlands, 1870-1970. Aksant: Amsterdam, the Netherlands.

Schuurman, A.J., 1991. Historische demografie: bevolkings- en gezinsgeschiedenis. Walburg Pers: Zutphen, the Netherlands.

Schuurman, A., 2001. Globalisering en geschiedenis. Tijdschrift voor Sociale Geschiedenis 27: 385-411.

Schuurman, A., 2007. Globalisering, geschiedenis en ruimte. Tijdschrift voor Sociale en Economische Geschiedenis 4 (3): 15-36.

Stuurman, S., 2004. François Poulain de la Barre and the invention of modern equality. Harvard University Press: Cambridge Mass., USA.

Thompson, W.S., 1929. Population. American Journal of Sociology 34: 959-975.

Toynbee, A., 1884. Lectures on the industrial revolution. Rivingtons: London, UK.

Van de Kaa, D.J., 2003. Second demographic transition. In: Encyclopedia of population. Macmillan Reference USA: New York, USA.

Van der Woude, A.M., 1985. Bevolking en gezin in Nederland. In: Van Holthoon, F.L. (Ed.), De Nederlandse samenleving sinds 1815. Wording en samenhang. Van Gorcum: Assen/Maastricht, the Netherlands, pp. 19-71.

Van Zanden, J.L. and Van Riel, A., 2000. Nederland 1780-1914: staat, instituties en economische ontwikkeling. Balans: Amsterdam, the Netherlands.

Weeks, J.R., 2005. Population. An introduction to concepts and issues. Ninth edition. Thomson Wadsworth: Belmont, USA.

Chapter 19

Kondratieff, Williamson and transitions in agriculture

Krijn J. Poppe[46]

Abstract

Transitions are long term deep structural changes in society. The economic theory that comes closest to this concept is that of the Kondratieff Business Cycles. These describe the five industrial revolutions western societies have experienced in the last 250 years. Each revolutionary wave is characterised by 4 phases. This paper compares the concept of transition with that of industrial waves and explores the effects of these five waves on agriculture. The analysis suggests that in the coming years ICT might have a major impact on the organisation and structure of agriculture, comparable to the deployment of the tractor and pesticides in the 1950s.

Keywords: post-productivism, industrial revolutions, business cycles

19.1 Introduction

Agriculture is changing. Over the last 20 years there has been an extensive debate as to what extent productivist agriculture is being replaced by post-productivist agriculture. One could also say *has to be* replaced, as the debate is sometimes more normative than objective. Productivist is equivalent to Fordism and mass production. As Wilson (2007) showed in his recent overview of the literature, it is less clear what post-productivist stands for, but Table 19.1 gives the main characteristics of the two systems (see also Marsden, 2003). By applying transition theory Wilson (2007) deconstructed the productivist/post-productivist notion. And by replacing post-productivism by non-productivism he redefined multifunctional agriculture as a (Deleuzian) transitional pathway between productivist agriculture and non-productivist agriculture.

These discussions, productivist/post-productivist as well as the conceptualisation of multifunctionality as a transitional pathway, fully concentrate on agriculture, although they recognise the equivalence between productivist agriculture and Fordism. In this paper, a more radical view is explored: that agriculture is not a sector on its own, but for the last 250 years the major developments in the sector originate in the rest of the economy. That implies

[46] An earlier version of this paper was published in Dutch as Poppe (2008). This version has been revised including suggestions by Prof. Dr. Guido Van Huylenbroeck as discussant in a session organised at ILVO, Merelbeke, Belgium, organised by Prof. Dr. Ludwig Lauwers and by the group discussion at a Round Table at the 3rd EAAE Forum on System Dynamics and Innovation in the Food Industry in Igls, Austria, February 2009.

Table 19.1. Dimensions of productivism and post-productivism (based on Wilson, 2007).

	Productivism	Post-productivism
Farm production	Increased mechanisation Decline in labour inputs Increased use chemical inputs Environmental impacts Intensification of land use Specialisation Concentration Surplus production Agricultural treadmill	Reduced intensity of farming Extensification Physical inputs replaced with knowledge Diversification. Pluri-activity Move from production to consumption of countryside Less emphasis on self-sufficiency
Food chain	Fordist regime Atlanticist food order, USA oriented Industrialisation Commercialisation	Post-fordist regime: non-standardised demand Changing consumer behaviour Free market liberalisation, free trade
Governance and policy	Corporate relationship farm lobby and agriculture ministry Agricultural policy community small but powerful Strong financial state support Protectionism and price guarantees Security of property/land use rights	Agricultural policy community widened Changing power structures in agricultural lobby Counter-urbanisation Reduced financial state support, decoupling New forms of rural governance Increased regulation of agricultural practices Increasing planning regulations for agriculture Loss of property rights
Ideology	Central hegemonic position of agriculture in society Agricultural exceptionalism Belief in farmers as best protectors of countryside Rural defined in terms of agriculture	Loss of ideological and economic sense of security Changing attitudes of the public towards agriculture, agriculture as a villain Rural defined independently of agriculture Contested country sides

that transitional pathways for agriculture and the rural area should be identified using an analysis of de structural changes in society at large.

Over the last 250 years, Western society has experienced a number of industrial revolutions (see Chapter 18 for a critical discussion if these revolutions were revolutions or transitions). Each introduced new technologies and products: railways, steel, cars, plastics, and computers. Each industrial revolution also had an important impact on agriculture. In the 19th century, new forms of efficient transport, railways and steam boats, brought cheap cereals from America to Europe. These imports had significant ramifications for agricultural policies as a result. The industrial revolution in the 20th century based on oil, cars and the chemical industry introduced the tractor and pesticides that substituted labour. This too had a marked impact on agriculture.

Such dramatic changes can be labelled as transitions. Transitions are defined as fundamental changes in the ideas on the organisation of society, including production and consumption, often due to changes in the relative scarcity of international goods. These fundamental changes include the perception of stakeholders on their own role and contribution, and the emergence of new stakeholders. Developments in agriculture are determined by trends in society at large. For this reason it is useful to understand these trends and their effects on agriculture, and especially those that arise from the industrial revolutions. That is exactly the objective of this paper.

In the next section we describe the five recent industrial revolutions and argue that these industrial revolutions are clear examples of transitions. Section 19.3 extends the description by showing the effects of these transitions on agriculture. It leads to the conclusion that in several industrial revolutions agriculture was heavily influenced in a way that couldn't have been predicted at the start of the revolution. Section 19.4 analyses the revolutions/ transitions by applying the Kondratieff wave theory[47]. Using the recent Schumpeterian description by Perez (2002; see also Freeman and Perez, 1988), it is concluded that in the last and current wave, agriculture is relatively far from the epicentre of the new technology, and that it is only in the second period of the wave that agriculture will be transformed deeply. Section 19.5 takes up Perez' point that institutional changes are needed to reap the benefits of the new technology in the second period of the wave. Here we supplement Perez' analysis with Williamson's framework for institutional change (Williamson, 2000). Section 19.6 speculates on the trends in agriculture in the coming 25 years, based on the transition

[47] There is room for a more in-depth investigation into the relationship between industrial revolutions and Kondratieff economic business cycles. Kondratieff himself used three periods per cycle but nowadays most scholars agree on the four (spring-summer-autumn-winter) periods and on the Schumpeter-Freeman-Perez paradigm explained in the text. The winter period in the Kondratieff Business Cycle characterised by a depression can probably be linked to the crash halfway through the industrial wave. This timing issue does not influence the conclusion of this paper. Thanks to Guido Van Huylenbroeck (Gent University) for pointing this out.

theory and empirical evidence in the previous sections of the paper. We conclude with some suggestions for further research.

19.2 Industrial revolutions as the mothers of transition

History suggests that Western society has experienced five industrial revolutions over the last 250 years, the last of which is still in progress. Based on Perez (2002), Table 19.2 summarises these revolutions. The first one was based on water power, canals and cotton. It moved the textile industry from the rural area to the manufacturing hall in the city. It set in a relative decline of agriculture in the economy, but the resulting agricultural inventions ensured that the European population could double within a century.

The second revolution was based on steam and brought forth the railways. The third one brought steel, electricity and heavy engineering. These developments were highlighted in the World expos at the Crystal Palace in London and the Eiffel Tower in Paris. It also heralded the advent of steel steamships and cheaper transport. The last century started with the fourth industrial revolution: that of the car, oil (with plastics from the chemical industry), and mass production. At the beginning of the 21st century we are now in the middle of the fifth industrial revolution where information and communication technology is the main driver of change. The cheap air transport of the previous revolution and the cheap communication technology of the current one go hand in hand in the very powerful trend towards globalisation (Friedman, 2005).

In all industrial revolutions there was one driver of change that became extremely cheap: hydro power, coal and steam, steel and electricity, oil and calculation power or data transport. This cheap input drastically changed relative prices, and made new products and new production methods feasible. Henry Ford's model T is one classic example of this, the microprocessor that doubles its capacity each 18 months (Moore's law), is another.

For every industrial revolution a break-through moment can be defined. Perez' choices are in Table 19.1. These are the moments where a technology leaves the experimental stage (or niche) and becomes main stream. The T-Ford (the breakthrough product of the 4th industrial revolution) was not the first car. For several years cars had been built as a toy for the rich or for special purposes. But the T-Ford was a mass-product based on the conveyor belt assembly line (copied from a slaughterhouse); the product became widely available and infrastructure was created for the car (motorways) and as a result of the car (suburbs). In a similar way Intel's chip in 1971 had its predecessors in transistors and computers before the computer was put on a chip. In her book Perez (2002) lists the fundamental changes that characterise each of the industrial revolutions. The current ICT revolution for instance leads to information-intensive processes, de-centralised integration and networks, and a commercial valuation of knowledge as capital. It is these fundamental changes, as well as the emergence of new stakeholders from the new industries that causes industrial revolutions to be classified as transitions.

Table 19.2. Five technological revolutions (Carlota Perez, 2002).

Name and technology	Breakthrough and location	Some fundamental changes
1. The industrial revolution	1771 Arkwright's mill in Cromford (UK)	Factory production Mechanisation Productivity measurement Water as power source and transport mode
2. Steam and railways	1829 The Rocket steam engine railway Liverpool - Manchester (UK)	Economies of agglomeration: cities, national markets Scale as progress Standard parts/machine-made machines Energy where needed via steam Interdependent movement
3. Steel, electricity and heavy engineering	1875 The Carnegie Bessemer steel plant in Pittsburg (USA)	Giant steel structures Economies of scale and vertical integration Science as a productive force Cost accounting for control Worldwide networks and empires
4. Oil, automobile and mass production	1908 T-Ford in Detroit (USA)	Mass production and mass markets Horizontal integration Energy-intensity and synthetic Centralisation Metropolis as centre National powers and confrontations
5. Information and Telecommunication	1971 Intel first micro-processor in Santa Clara - Cal. (USA)	Information intensity Decentralised integration and networks Knowledge as capital and value added Heterogeneity, segmentation Economies of scope Globalisation Instant contact and action

19.3 Industrial revolutions and agriculture

This section extends the description of the industrial revolutions by showing the effects of these transitions on agriculture. An important effect of the first wave was the textile industry's shift from the farm to manufacturing halls in the towns. It was also accompanied by a large increase in the population. The second wave brought North America into the global economy because of cheaper transport in the form of railways. In the USA land was cheap and labour expensive. The induced innovation theory (Hayami and Ruttan, 1970) explains that this lead to innovation in labour-saving technologies. The invention of the grain-reaper by Cyrus McCormick is a famous example. The third wave brought a new dimension in cheap transport; this time in the form of steel ships driven by steam. As a result Europe was flooded with cheap American (and Russian) grain. It led to mass emigration from the countryside on the European continent, and to radical changes in agricultural policy. These changes ranged from protectionist policies in e.g. France and Germany (that also wanted to protect its infant industry from the third revolution) and more liberal ones (with quality standards, education and extension as core elements) in open agricultural and trade economies like the Netherlands and Denmark. In the United Kingdom, they developed extremely liberal ones to feed the industrial cities.

The deployment of the fourth wave took some time, partly due to the two World Wars (that are not completely exogenous to the industrial revolutions). Agriculture was also severely damaged by the economic crisis in the 1930s, leading to more protectionist policies. The main effects of this revolution in Europe, however, took place in the 1950s. In a relatively short period, tractors replaced horses and manpower on the farm. Tractors were a typical product of the car-century, although it was technologically influenced by the development of the military tank in the First World War. Pesticides were another example, elaborating on the earlier introduction of chemical fertilisers in this revolution. These developments were encouraged by the (common) agricultural policies of the 1950s and 1960s to reduce the food shortages of the war period, to create export and to enhance living conditions in the rural area. The countryside was also heavily influenced by the arrival of the car, roads, electricity (a product of the third revolution, now mass marketed!) and telephones.

At the end of each industrial revolution the negative effects show up. Environmental problems in agriculture (pesticides, nitrogen, climate gas emissions) are some examples from the last wave; the mass migration out of remote rural areas (like mountain areas) is another one. This overview of the effects of the industrial revolutions on agriculture leads to the conclusion that agriculture was heavily influenced more than once in a way that could not be predicted at the start of the industrial revolution. This shows that agriculture was part of the transition. The overview also suggests that in the first three industrial revolutions West-European agriculture, being a larger part of the economy (and geographically closer to the epicentres of the innovations), was influenced much more quickly than in the 4th and 5th revolution. However this needs further detailed analysis.

19.4 Kondratieff cycles and industrial waves revisited

Kondratieff (1935), a Russian economist, was the first to provide an economic interpretation of these long waves or long business cycles. Schumpeter (1939) added to the theory of business waves in the book with the same name. In 1983, Jaap Van Duijn revisited the theory.

Recently Carlota Perez (2002) published a detailed analysis that inspired a large part of this paper. She argues that each industrial revolution or Kondratieff Business Cycle takes about 50 to 60 years and is characterised by two periods: the Installation period and the Deployment period (Figure 19.1). These two periods are divided by a financial crisis: the revolutionary year 1848, the Wall Street crisis of 1929, the NASDAQ-crash of March 2000 (as Perez [2002] suggests) or the Credit Crisis of 2008 (as we might suggest today). Both periods each have two phases. The installation-period starts with the Irruption phase where new technologies are fuelled by financial capital that can no longer be invested profitably in the old industries of the previous wave. This is the time of financial 'promoters' and 'business angels', to use some names from quite different industrial revolutions. This is so successful that financial capital becomes more prominent than production capital (two terms originally introduced by Schumpeter). This is the Frenzy period. Market values bear no relationship to real values, stock markets explode and interpretations are offered that we are in a 'new economy' where old economic rules are not valid anymore.

This is the time of venture funds, hedge funds and other new financial instruments (financial innovation) and also a time for unfettered capitalism with a large, unequal distribution of income. Some individuals become extremely rich (Carnegie, Vanderbilt, Buffett, Gates), others live in circumstances as documented by Dickens in 19th century London and that we now see in the third world. It is also a time marked by dubious practices and even fraud; the South Sea bubble, Barings, Enron, and Parmalait being examples of this. Bubbles burst and after the crash there is a sense of urgency for institutional changes. Roles of the private and

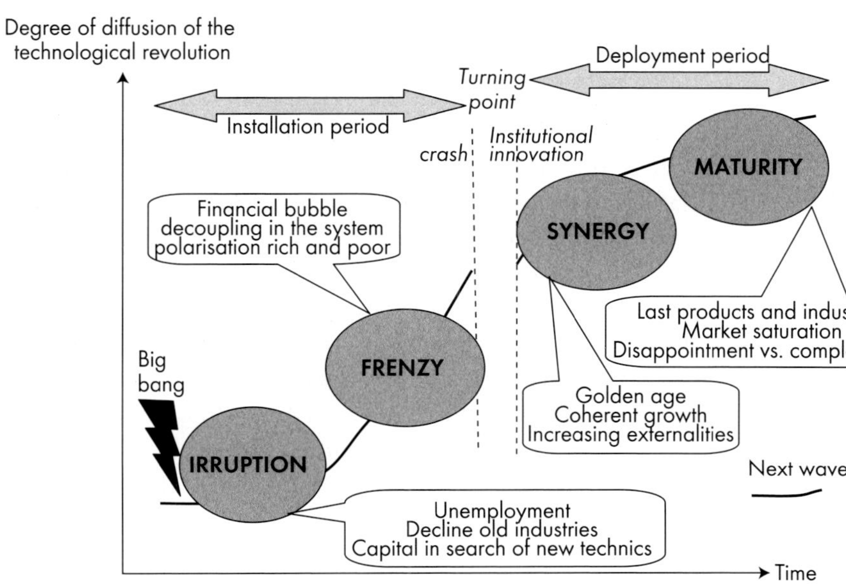

Figure 19.1. The different phases of a wave in an industrial revolution (based on Perez, 2002).

public sector are redefined. Where techno-economic innovations in the installation period are realised relatively easily, socio-institutional frameworks are more rigid and adapt less easily.

The deployment period after the crash and the institutional changes also have two phases. It starts with Synergy: a golden age like the 1950s and 1960s where production capital leads the way. The new technologies are rolled out to sectors and regions that were not yet much affected by them. Businesses finance themselves mainly via internal cash flows. Some firms merge to cope with declining prices that are the result of the new technologies. It is in this period that work processes are changed in many (old) sectors to incorporate the new technology. The first cars were built as horse coaches, in this later deployment period SUVs, caravans and campers are built to revolutionise recreation. The synergy phase is followed by the Maturity phase: more and more externalities of the new technology become a problem and profitable investment opportunities for the technology, using the available cash flows, decrease markedly. The industrial revolution has run its course. Young people and artists show their dissatisfaction with the dominant paradigm. Financial capital starts looking for new ideas which are brewing in the nation's 'garages'. In short, this is the theory of the industrial wave cycle as interpreted and expanded recently by Perez (2002). Application of this theory to the current industrial revolution suggests that the last cycle started in 1971 with Intel's announcement of the microprocessor and that the NASDAQ March 2000 crash was the end of the installation period (Perez 2002). Or the current (2008) credit crisis – future historians will decide which. That implies that we are now in a period of institutional change and that the 'golden age' is ahead of us.

If we relate this to the situation in agriculture, we can assume that major changes are still to come. We suggested in the previous section that in the last and current wave, agriculture is relatively far away from the epicentre of the new technology, and that it is only in the second period of the wave that agriculture will be transformed significantly. This suggests that in the next 15 years we can expect a change in agriculture due to the deployment of ICT, comparable to the introduction of the tractor and pesticides in the 1950s in Northern and Western European agriculture. Detailed registration, precision farming, detailed product information (tracing and tracking), unmanned tractors or small robots, mobile robot milking parlours that replace immigrant labour, links between health, ICT and food, our imagination is probably too limited to predict the real development. The main relevance of this for transition theory is that this transitional pathway is very different from the transitional pathway of multi-functionality in Wilson (2007) or the post-productivist views. In both the analysis of those pathways and from a transition theory perspective, the most interesting question is perhaps: what is happening in institutional reframing in the agricultural sector?

19.5 Williamson and institutional economics

Institutional changes are needed to make the new technologies fully profitable, especially in older industries. Williamson (2000) provides a framework to understand concepts of institutional change, based on new institutional economics that complement Kondratieff, Schumpeter and Perez. Williamson showed (Table 19.3) that business and consumer decisions based on costs and benefits are influenced by transaction costs and property-rights. These are based in social theory and change only slowly: Williamson indicates that property rights are stable for a period of between 10 to 100 years. The average is by coincidence (or not?) 55 years, the average size of the Kondratieff Business Cycle.

Property rights design the institutional environment in which people live and work. If the government monopolises broadcasting technology, you cannot set up your own commercial television company or broadcast a commercial. If there is no intellectual property right defined on GMO seeds, you cannot commercially develop them. Property rights and technology determines the transaction costs that people and businesses face in exchanging goods and services. Low transaction costs lead to market based solutions: it is attractive to outsource activities to specialists. High transaction costs leads to organisations where the activity is done internally, to prevent even higher transaction costs in a market situation.

Williamson's scheme makes clear why institutional change is much more difficult to achieve than techno-economical change. Social norms and property rights are much more rigid. That has the positive effect that it reduces uncertainty in the society and makes investments possible. But it makes change in a transition more difficult.

19.6 The current and future transitions in agriculture

The theory in the previous sections leads to a number of observations on the current transition in agriculture. First of all, policy makers and, therefore, researchers and farmers are very much concerned with the problems (externalities) of the 4th industrial wave, that was based on oil, cars and mass production. It lead to a high level of welfare, but also to unprecedented

Table 19.3. Levels of change in institutional economics (based on Williamson, 2000).

Level	Core elements
1. Social theory (stable for 100-1000 years)	Embeddedness: informal structures, customs, traditions, norms, religion
2. Economics of property (stable for10-100 years)	Institutional environment: formal rules of the game – especially property (judiciary, bureaucracy, policing)
3. Transaction cost economics (stable for 1-10 years)	Governance: playing the game – especially organisation forms like markets, contracts, command and control, vertical integration, etc.
4. Neo-classical economics, contract and agency theory	Resource allocation (prices, quantities, information, incentive alignment)

pollution. Policy makers are trying to repair those holes in the system caused by ill-defined property rights (e.g. the right to a clean environment, a noise-free environment, dark nights in the neighbourhood of greenhouses, clean water, healthy and sufficient fish) This is often labelled as a move to 'sustainable' agriculture.

Secondly, it can be observed that in this transition, policy makers are hindered by globalisation. Globalisation (Friedman, 2005) is a very robust phenomenon, as it is based on the combined forces of the 4th (cheap air transport) and 5th (ICT and cheap data transport) wave. This brings governments into a very competitive environment where people and businesses become nomadic and move easily to another jurisdiction if government policies are not in line with their own strategy.

In the third place, the effects of the 5th industrial wave (ICT) have not yet influenced agriculture significantly and this is certainly true for agricultural institutions. Of course computers, RFID tags and even robots have entered the production process, but they have not yet really changed the working methods, nor the organisation of production. The analysis above suggests that in the coming years ICT might have a major impact on the organisation and structure of agriculture, comparable to the deployment of the tractor and pesticides in the 1950s. How this might effect the organisation of the food chain has been described by e.g. Boehlje (1999) and Menard *et al.* (2005). Boehlje (1999) uses institutional economics to describe how spot markets are replaced by vertical ownership, if the programmability of the production process and the asset specialisation increases, and the contribution of business partners in the chain become more interdependent. Menard and Valeschini (2005) show the importance of hybrid organisational forms that minimise transaction costs.

The trend towards programmability of the primary production process, based on ICT and biotechnology, also threatens the existence of the family farm in its traditional form. One of the explanations (Allen and Lueck, 2002) for the existence of family farms in Northern and Western Europe, and not plantations (like in the third world for palm oil or tea production) is that family farms are competitive due to the fact family members have a relatively high level of education and self management, are willing to work at a relatively low cost (especially in overtime), take up risks that are non-tradable, and that their hourly performance can not be measured easily (agency theory). ICT might change these factors considerably: new types of knowledge are needed, labour becomes scarce due to the demographic transition, risk has become tradable, and the production process becomes very observable.

A last observation concerns the 6th industrial wave. Perez (2002) suggests that biotechnology could be a very good candidate for becoming the driving force of the next wave. That would imply that somewhere around 2020/2025 a breakthrough is to be expected, comparable with the T-Ford and Intel's microprocessor. Perhaps it will come in the field of life sciences, maybe it will be in relation to farming and pharmacology and it has yet to be seen if the location will be the USA, China or India. Bearing in mind the GMO and other ethical debates it is not very likely that Europe will be the epicentre of that wave.

19.7 Conclusion and suggestions for further research

This paper shows that researchers interested in agriculture and the rural areas should not restrict their analyses too much to their own subject of interest. Discussions on transitional pathways from productivist agriculture to post- or non-productivist agriculture and the character of multifunctional agriculture (Wilson, 2007) are hampered by the fact that in the analysis agriculture is isolated from the rest of society. The fact that productivist agriculture is a form of Fordism, should be the basis for a reflection of the background of Fordism, the development of this '-ism' (as in the history of industrial revolutions and Kondratieff Business Cycle theory) and explore the effects of this development on agriculture and the rural area.

This paper suggests a number of topics for further research. One of them is to set up foresight studies on the role of ICT on the organisation of the food chain. There is also more work to be done in historical economics whereby further analysis of the adoption of technology by agriculture in previous waves could lead to more insight as to how agriculture might be affected by this in the current one. This could help explain why agriculture has been closer to the epicentre of the new technologies in earlier industrial revolutions than in the last ones. A third issue relates to how ICT might be used to solve the problems caused by the previous industrial revolution and what this might mean for institutional change today.

References

Allen, D.W. and D. Leuck, 2002. The nature of the farm. MIT Press: Cambridge, USA.

Boehlje, M., 1999. Structural changes in the agricultural industries: how do we measure, analyze and understand them? American Journal of Agricultural Economics 81 (5): 1028-1041.

Freeman, C. and Perez, C., 1988. Structural crises of adjustment, business cycles and investment behaviour. In: Dosi, G., C. Freeman, R. Nelson and L. Soete (Eds.) Technical change and economic theory. Pinter Publishers: London, UK, pp. 38-66.

Friedman, T.L., 2005 The world is flat. Farrar, Straus and Giroux: New York, USA.

Hayami, Y and Ruttan, V.W., 1970. Factor prices and technical change in agricultural development: The United States and Japan, 1880-1960. The Journal of Political Economy 78 (5): 1115-1141.

Kondratieff, N.D., 1935. The long waves in economic life. The Review of Economics and Statistics 17 (6): 105-115.

Marsden, T., 2003. The condition of rural sustainability. Van Gorcum: Assen, the Netherlands.

Ménard, C. and Valeschini, E., 2005. Institutions for governing agriculture and rural areas. European Review of Agricultural Economics 32: 421-440.

Perez, C., 2002. Technological revolutions and financial capital. Edgar Elgar: Cheltenham, UK.

Poppe, K.J., 2008. Transities en Kondratieff. Technische en Kwantitatieve Analyse 2008 (4): 36-39.

Schumpeter, J.A., 1939. Business cycles: a theoretical, historical, and statistical analysis of the capitalist process. McGraw-Hill Book Company: New York, USA.

Van Duijn, J.J., 1983. The long wave in economic life. Allen & Unwin: London, UK.

Williamson, O.E., 2000. The new institutional economics: taking stock, looking ahead. Journal of Economic Literature 38: 595-613.

Wilson, G.A., 2007. Multifunctional agriculture – a transition theory perspective. CABI: Wallingford, UK.

Chapter 20

Where are we now? Where do we go from here?

Krijn J. Poppe, Catherine J.A.M Termeer and Maja Slingerland

20.1 Introduction

The greatest current challenge to the agricultural sector is for it to become sustainable in all three of the dimensions profit, people and planet. This challenge is even more urgent in highly urbanised countries like the Netherlands, where high land prices, rising consumer concerns for issues like animal welfare and the highly negative environmental externalities of the agricultural industry combine with new demands from the city for recreation, health care and local food products. These are some of the developments in our society that are forcing agriculture to change. The government, farmers, the agri-food industry and the retail sector struggle to meet this challenge. In the Netherlands, the government has called for a 'transition towards sustainable agriculture' and it is investing in this programme with its research and education policy. Similar trends have been observed in other countries.

The chapters of this book present the expertise that has been accrued from at least five years of Dutch research in this area. Our aim was to collate the results of our experiments, to learn from them, to confront them with existing theory, and to share them with a larger audience in order to foster learning about transition. Given the leading position of the Netherlands in global agriculture, in a highly urbanised setting, and its leading position in the study of transition theory this raises a number of interesting questions to reflect upon before we draw this book to a close.

In the next section we ask what have we learned from our experiments on the state of Dutch agriculture and its transition. Do we have a better understanding of the challenges and are these being addressed properly? Following on from this reflection on the goals for the research reported in this book, we reflect on the research theories and methods used in the papers in this book (paragraph 20.3). This raises the question as to what makes the Wageningen UR approach special, if anything.

Many of the experiments presented in this volume have not been conducted behind the researcher's desk, but are based on repeated interaction between the scientist and the environment she or he is studying. Often this involves participatory research, sometimes theory-inspired action research. In paragraph 20.6 we reflect on the role of the scientist in such experiments. Inevitably at the end of book like this, we list a number of topics for future research. Our hope is that this proves to be a stimulant for further scientific forays into this exciting new research domain.

20.2 Dutch agriculture

The Dutch agricultural transition is the result of the dichotomy between the interests and concerns of Dutch society and the divergent needs of its international agri-food industry (Figure 20.1). This industry, an important part of the Dutch economy, has traditionally tried to improve its competitiveness by further industrialisation, economies of scale and innovation for export markets. But the general public is critical about the environmental performance of agriculture and the food industry, animal welfare, the loss of landscape, the use of modern technology and is actively concerned about food safety and health issues. Globalisation has lead to a renewed interest in regional issues. All these things have lead to a quest for sustainability, although this is an ongoing, complex, multi-dimensional concept (Chapter 2 by Dewulf *et al.*).

Many of the papers in this volume have reported on experiments that are directly or indirectly linked to the clashes between the general public's concerns (dominated by those of city dwellers) and the needs of the (export) industry. These concerns address the loss of landscape (Chapter 4 by Salverda *et al.*), animal disease (Chapter 14 by Termeer and Van der Peet), animal welfare (Chapter 12 by Bos and Groot-Koerkamp), innovation (Chapter 10 by Fortuin and Omta) and animal production problems (Chapter 7 by Wielinga and Geerling-Eiff). In all these areas we now have a better understanding of the issues at stake than five years ago.

There appears to be both a micro and a macro approach for bridging this clash between town and country and for reconnecting both sides. The micro approach stresses innovation in the food industry and on the farm, as well as methods to support this innovation process. Many authors in this volume report on examples in which this road has been taken, some for the productivist agribusiness farms (e.g. Chapter 7 by Wielinga and Geerling-Eiff; Chapter 6 by

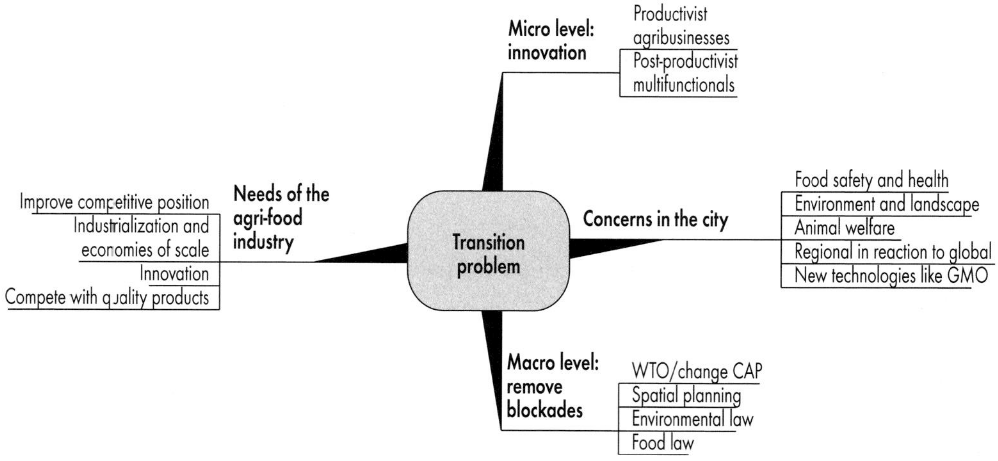

Figure 20.1. The transition problem for Dutch agriculture.

Vogelezang *et al.*; Chapter 8 by Van Mierlo and Arkesteijn), others for the post-productivist multifunctional farms (Chapter 4 by Salverda *et al.*). To quote Van Dam *et al.* (Chapter 5): transition starts with people. Either by self organisation (Chapter 5 by Van Dam *et al.*; Chapter 4 by Salverda *et al.*; Chapter 10 by Fortuin and Omta), or as in a situation where the government organises a learning process (Chapter 8 by Van Mierlo and Arkesteijn; Chapter 7 by Wielinga and Geerling-Eiff).

The second approach to reconnect industry and city stresses the macro aspect: in this view transitions are very much based on macro trends. Some authors argue that it is impossible to have systems in the making (Kooij in Chapter 18 is probably the most outspoken author that defends such a view in this volume). Others support the idea that the government can subsidise, organise learning trajectories (Chapter 11 by Wijnands and Vogelezang; Chapter 7 by Wielinga and Geerling-Eiff) or help to plan them (Chapter 12 by Bos and Groot-Koerkamp; Chapter 13 by Visser *et al.*). The multi-level perspective on transitions as stated by Geels (see Dewulf *et al.*, Figure 2.1 in Chapter 2) suggests that government can also play a role by removing blockades in regulations in an effort to bring niche-innovations up to the regime level, but this is not stressed much in the contributions here. Along this line Woodhill (Chapter 15) argues that transition management (and development in the South) is largely about changing institutional arrangements of the past that, given the problems of today, now make no sense. At the present time, intervention strategies like subsidies and innovation support in kind (in a netwerk approach) seem to be the favoured tools used by the Dutch government and Wageningen UR researchers. This is consistent with the choice for experiments at the niche level.

The emphasis on the micro-approach and experiments with learning and innovation is also understandable in the framework of path dependency. In the past, the Ministry of Agriculture and the Common Agricultural Policy was a dominant factor in the industry's decision making. A transition to a sustainable agriculture affects the Ministry too, and demands less governmental dominance. This can only be made clear by ensuring that institutions do not take the lead in the transition itself and send off the signal that the responsibility for innovation lies with the industry. As a former Minister, Cees Veerman, said: the Ministry's role should move from 'caring for' towards 'caring by'. This implies that the government needs to assume a facilitating role via research and innovation programs for experiments, not (yet) a macro approach by steering developments from the landscape/regime level. However in a next phase of the transition process it could be important to strengthen the link between the niche experiments and the regime level by learning from the experiments exactly which institutional changes are necessary for the transition to take place.

We end this section with the one major question that the reader will have on his mind after finishing this book: *does it work*? Is the transition process towards a more sustainable Dutch agriculture on track? It is not the purpose of this book to address that question, and good scientific methods to effectively monitor the transition process are possibly lacking (Ten

Pierick *et al.*, 2005; 2009; see Syscope (2008) for a more positive view). The question how to monitor depends also on the perspective of the evaluator on transitions: are small irreversible steps, tipping points or S-curves the phenomena to look for?

With this in mind, the different chapters suggest that there is still a long way to go. The fact that the Dutch Ministry of Agriculture has put the transition to a sustainable agriculture high on the political agenda is an important step, but it is normal for transitions to span 20 to 30 years or more. That makes it only logical that none of the authors reports outright success. There might also be some bias in the experiments reported here, as only the particularly interesting cases were selected for discussion in this book; cases in which a more in-depth scientific analysis seemed necessary for progress. These are not necessarily the biggest success stories.

Statistical analysis of income developments (De Bont *et al.*, 2008), structural change (Bruchem *et al.*, 2008) and environmental performance (Boone *et al.*, 2007) also suggest steady improvements but no big breakthroughs. A reason for this might be that not many of the transition experiments are really aiming to create a totally different agriculture in the short run. Dutch agriculture has traditionally based its competitiveness on a cost price reducing commodity model (Figure 20.2): profits are used to invest in process innovation which increases production (increased capacity) that, in turn, makes price reductions possible, which create extra demand by improving competitiveness and maintaining profits. The consumers' concerns in the Dutch cities suggest that either the production systems in this model should be changed to reduce negative externalities or that the base commodity model should be replaced by a value-added model (Figure 20.2). If profits were invested in product innovations that could be sold with a higher profit margin or cater for new types of demand, prices could be increased and profits maintained without increasing total production (capacity) with its negative external side effects. Such a value-added model is more in line with post-productivist agriculture, but a transition to this model is difficult and takes time.

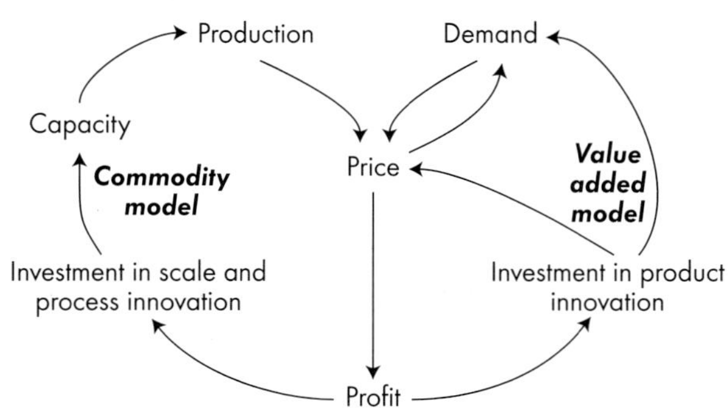

Figure 20.2. Commodity model and value added model.

This competition between two types of business models (one based on competition driving prices down, the other on heterogeneity in the consumer market) raises more questions. Are our experiments mere window dressing? Are they by-passing the issues? Do they offer fundamental change? Poppe *et al.* (2009) argue that government initiatives and those of the sector designed to break away from the commodity model by promoting chain management and taskforces on organic farming and multifunctional farming, are still mainly by-passes that do not (yet) change nor challenge the base model of Dutch agriculture. These experiments at the regime level create niche-innovations, e.g. in regional products or 'green care', that have as an interesting side effect that they improve the image of the total sector. Their main *raison d'etre* in the transition is, however, the potential to change the regime level in the years to come.

Generally speaking the experiments reported in this volume suggest that innovation is being adopted and that there are some successes with system innovations (e.g. Chapter 7 by Wijnands and Geerling-Eiff; Chapter 8 by Van Mierlo and Arkesteijn; Chapter 12 by Bos and Groot-Koerkamp). In line with what might be expected from transition theory and the changed role of the Ministry, innovations crossing agriculture's system by linking it to other sectors (like chemistry) or important innovations in institutions at the regime level have not been reported here.

20.3 Transition theories

The research and experiments to support the transition to a sustainable agriculture presented in the previous chapters have built on concepts from a variety of disciplines, as reported by Dewulf *et al.* (Chapter 2). They suggest the use of multiple theories to understand and steer transitions. Gamma sciences like economics, sociology, management science and communication studies are represented in several papers. Ordinary economics theory with its emphasis on scarce resources with costs and benefits is absent. Although a reflection on the transition aspects of the price boom (and bust) in 2008 might be interesting, the economics discipline is more useful in transition studies with the new institutional economics that features in the paper written by Salverda *et al.* (Chapter 4). Sociology with its emphasis on norms and values and social relationships is more in vogue in transition studies. It is interesting to see that several authors base their work on prominent and popular contributions in the discipline like Foucault's ideas on power (Chapter 17 by Duineveld *et al.*), Luhmann on social system theory (During *et al.* in Chapter 9) and Castells on the network society (Van Dam *et al.* in Chapter 5). Management science is, of course, present with innovation theory (Fortuin *et al*). It is not a surprise that learning theories (originating from communication or management science) are also popular (Chapter 15 by Woodhill; Chapter 6 by Vogelezang *et al.*; Chapter 7 by Wielinga and Geerling-Eiff, although this team is also inspired by ecology).

There are also several papers that have chosen their methodology from the beta sciences and related design theories: the DEED framework (Chapter 13 by Visser *et al.*) and Reflexive Interactive Design (Bos *et al*). Technology development with forecasting and backcasting

(Chapter 11 by Wijnands and Vogelezang) is another. The change theories discussed by Dewulf *et al.* (Chapter 2) have surprisingly been only cited in few of the papers. Sometimes the assumptions from a change theory are implicit: for instance the small steps approach advocated by Van der Ploeg (Chapter 16) is not that different from continuous change. Organisational change/development is reflected in the emphasis that several authors give to learning, but otherwise it doesn't play a very important role here as the papers seldom deal with one specific organisation or organisational issues as such. The paper by Termeer and Van der Peet (Chapter 14) is an exception. Social learning, network governance and multi-actor collaboration resonate in several papers. Adaptive governance and policy agenda setting theory have not been used in the papers presented in this volume, and that reflects the Wageningen UR approach in general. Power issues and political aspects do not get much attention in the transition research. A recent overview of a Dutch research program on the transition to sustainable energy reached a similar conclusion (Van der Hoeven, 2009). It is not considered part of the researchers' brief (see Duijneveld *et al.* in Chapter 17) and it is not likely to be applied in active research to change a situation. Woodhill (this volume) states that we should not be naïve about the influence of power. It could be useful to look to a political economy approach as e.g. used in the 'competing claims' approach (Giller *et al.,* 2008).

This book provides multiple theoretical perspectives on understanding and steering transitions. Most of the papers focus on the application of theory within their transition experiment. In addition to that some of the authors reflect on transition theory or transition management itself. Examples are the paper by Dewulf *et al.* (Chapter 2) that describes the theory in relation to other theories. The paper by Van der Ploeg (Chapter 16) questions the validity of the S-curve as a suitable model for transition management and proposes instead an incremental, path dependent innovation theory. Woodhill shows how methods in North and the South are similar and could strengthen each other. He also argues for institutional innovation and stakeholder engagement. Poppe questions the productivist/post-productivist paradigm in his paper, favouring instead the theory of industrial waves. Learning networks have been discussed using different methodologies in three chapters (Chapter 6 by Vogelezang *et al.*; Chapter 7 by Wielinga and Geerling-Eiff; Chapter 8 by Van Mierlo and Arkesteijn) This suggests that discussing a single experiment from multiple methodological perspectives might be an interesting way forward to enhance the understanding of different stakeholders' perspectives in the transition and to further develop and test theories.

20.4 The characteristics of transition research at Wageningen UR

Taking a broad view, it can be said that the transition research that is undertaken at Wageningen UR has its own special character. We do not claim that we are, or that we are becoming a particular 'school of thought' but it is interesting to look more closely at the nature of this research and how it is influenced by our special area of expertise in the field of agriculture, food chains and peri-urban areas.

Many of the papers reflect a participatory or action research approach in applying or improving a certain theory. In these cases they address an emerging problem and the research has an immediate objective, to do something concrete or alter behaviour in the field. The research reported in this volume has developed as a response to practical problems and needs, not as a way of testing hypotheses derived from theory. On the other hand, a relevant theory is often sought out to assist in dealing with the problem, as is shown in the papers by Termeer and Van der Peet (Chapter 14) or Salverda *et al.* (Chapter 4). From the professional conversations and the written history of some of the papers we know that there is also quite a lot of experimentation by researchers with different theoretical approaches, and there is a common search for new approaches. The result is that the knowledge gained can contribute to the evaluation of theories, refining them and it can also define new theoretical challenges. This all makes the research interdisciplinary, multi-disciplinary and even transdisciplinary. The problem-oriented, something-has-to-be-done attitude in this type of research seems to be linked to the engineering approach that is central to Wageningen UR's background. Actors play an important role too, in this engineering approach. Actors think and act holistic, taking many aspects of a problem into consideration, and that means that the actor approach is often multi-disciplinary by nature.

This engineering approach also leads to a certain result orientation or goal orientation, which can not always be guaranteed to achieve something in terms of improved sustainability or change in agriculture. At the very least it can facilitate processes for (social) learning and deliver a new learning system. The importance of learning, multi-actor collaboration and network governance in the Wageningen UR approach is not in itself remarkable. The transition to sustainable farming involves 80,000 farmers to start with, not to mention agri-business, farm workers, policy makers, consumers, etc. Many of those involved have their own typical unsustainable practices. That makes it difficult to bring even a small number of stakeholders to a central table and broker a deal, in the way it was done in the famous Dutch energy transition from coal to gas in the 1960s (see Correljé and Verbong in Elzen *et al.*, 2004). Instead of changing the policy agenda or institutional arrangements, the Wageningen UR transition approach focuses on learning and innovation. Human capital is central in this approach: not only the drawing table but also the kitchen table is a locus of change. The next challenge is to connect the kitchen table with the board room of the big players that dominate the regime: retailers, food multinationals, top level politicians, infrastructure development companies.

In paragraph 20.2 we have already suggested that this choice to emphasise learning and innovation is not only consistent with transition theory, but that it has a logical consequence for the role of the Ministry of Agriculture. It must refrain from attempting to orchestrate the transition within the industry from above, and assume a more facilitative role to achieve that end from within the ranks. Bearing in mind the history of the sector, with its own Ministry and the Common Agricultural Policy (CAP), we have reached a setting in which the policy makers support a sector which is in transition from a protected CAP environment to the more hostile competitive market. This new facilitative role for the Ministry has consequences for the objectives of the Ministry's main national policy instrument – the knowledge system. A new task for that system had to evolve, following in the wake of the traditional hierarchical model of research and innovation.

This emphasis on learning and innovation is strongly influenced by the character of the sector. Dutch agriculture is characterised by many small farms. In the light of technological progress most of them are essentially too small to be viable. They are, therefore, motivated to explore alternatives to ensure their survival. These farms are characterised by low profitability, which is typical for farming over the last centuries. Low profitability, defined as being a low reward for the investment of capital, means also a low risk profile on investments. Agriculture is in that sense different from ICT (where small start-ups could attract quite some risk capital as rewards can be high) or big corporations with deep pockets (like Nokia to develop mobile phones or Toyota to develop the Prius). Small steps that are not too risky seem to be a characteristic for innovation behaviour in agriculture.

Is this a problem? The experiences of the authors revealed in these papers, as well as some of the innovation and transition theory (e.g the contribution of Geels in Elzen *et al.*, 2004) suggest that it is not necessarily so. Van der Ploeg (Chapter 16) makes it clear in his contribution to this book: small steps work and are characteristic of transitions in agriculture. Niche management is more important than the S-curve in rolling out. System innovation is a much more societal process than simply bringing an idea from the lab to the consumer's table. Small changes can become the tipping point for larger institutional changes. And once something works, Dutch farmers are fast adopters, following a very steep learning curve. In the 1970s, the total area of silage maize cultivation in the Netherlands grew after successful farmers' trials in just a few years from 0 to 200,000 ha, at the expense of rye. Biological pest control and growing on substrate in glasshouses (invented in Denmark) were adopted very quickly in the Netherlands. In the early 1960s a similar quick adoption of mechanisation took place, resulting in a large relocation of farm workers to the cities. A similar process is probably typical for other large groups like consumers, city dwellers, employees in the agribusiness, etc.

There is another reason why this should not be a problem. The largest transitions in society do not originate in the agricultural sector or at Wageningen UR. The contributions of Kooij (Chapter 18) and Poppe (Chapter 19) in this book show that the big demographic and economic transitions matter more, and perhaps matter most, but are beyond the control of the agricultural sector – including its big institutions like the Ministry of Agriculture or Wageningen UR. Anticipation and adaptation are relevant but take place by translating the big societal trends into smaller innovations and testing what works in practice.

And last but not least it is not problematic because the relationship between description and prescription in transition research is still a problematic issue. As long as power issues are neglected, researchers (as Duinveld *et al.* argue in Chapter 17) should be modest about their ability to manage their desired changes in society. The role of these research programs by Wageningen UR seems to be to make those small steps towards sustainability, each of which contribute piece by piece to the giant step that is needed for mankind (to paraphrase the astronaut, Neil Armstrong).

20.5 Monitoring and evaluation of the transition[48]

The Wageningen UR approach, which is doing experiments that contribute to transitions by small steps, raises the questions of monitoring and evaluation (M&E). M&E seems to be especially important in these situations as a guide to public investment in research. M&E is not easy in cases where it is in the first instance even difficult to detect the change (as Termeer and Van der Peet show in Chapter 14). There is often a time lag between change and outcome in terms of improved sustainability, especially when measured at a more aggregated sector level (Poppe, 2008). This suggests that there is a micro and macro view for evaluating the results of a transition (see Termeer and Van der Peet in Chapter 14).

Against this background and in line with the emphasis on learning (see above) it is logical that monitoring and evaluation in the Wageningen UR approach should use social learning. It should be noted however that there are quite different views on M&E in the literature.

In transitions, knowledge creation is not seen as a linear top down process but as a complex process with many iterations. Gibbons (1999) labelled this as the change from Mode 1 to Mode 2 science (Table 20.1). It is a Triple Helix approach (Leydesdorff and Etskowitz, 2003) in which three independent institutional structures (government, business and science) interact from time to time with each other, steered rather autonomously by their own development. This framework for analysis stresses the importance of the dynamics of networks and alliances between institutions, instead of the 'how' and 'where' of creation of knowledge (extra-mural over intra-mural). This is also reflected in the trend from second to third generation universities (Chapter 3 by Rabbinge and Slingerland).

Monitoring and evaluation tools are not very well developed for this kind of post-modern framework. Some measure people, profit and/or planet indicators (Boone *et al.*, 2007; Ross, 2003). Based on the work by Rotmans and others (2001), Ten Pierick *et al.* (2006) made some suggestions for indicators. The concept of Learning Histories, developed at MIT by Kleiner and Roth (1997) seems to work as an M&E tool as it provides context-specific information that makes results more transferable to other contexts. Others work on 'Reflexive Process Monitoring' (see Bos and Groot-Koerkamp in Chapter 12). Many of these monitoring tools come from a social learning environment (Wals, 2007). Arkesteijn and Van Mierlo (Syscope, 2008) identified three different views on monitoring and evaluation:

[48] This section is based on Poppe, 2008.

Table 20.1. Mode 1 and Mode 2 science, according to Gibbons (1999).

Mode 1	Mode 2
Academic	Oriented towards application
Discipline-oriented	Transdisciplinary
Homogeneous	Heterogeneous
Linear and stable	Non-linear and volatile
Academic quality control	Quality management on a broader set of criteria
Accountable to peers	Accountable to society

- *Result-oriented methods*: accountability and control are the objective and results are measured. LogFrames and Theory of Change are some of the well-known methods.
- *Constructive methods*: mutual learning and management processes are the objective and the learning process is evaluated. Learning histories, responsive M&E and most significant change are some of the methods.
- *Reflexive methods*: they address not only mutual learning but also the quality of the learning process. It is a constructive method in which also norms and values are evaluated to reach a holistic, integrative system approach.

Applying theories by Weick (2000), and a hypothesis formulated by Beer and Nohria (2000), Termeer (2006) suggests that, from a public administration/public management perspective, the programmes or instruments that managers apply do not matter that much, as long as they contribute to the basic conditions for creating meaning or relevance that are essential for learning, adapting and changing in a turbulent world. These basic conditions are achieved by:
- motivating people to keep moving and experimenting to make unknown possibilities known (vitalising);
- creating a general direction to evaluate experiments;
- promoting a process of adapting to local situations (updates) by precise attention to developments, context and meaningful details;
- facilitating open interactions in which trust, reliability and self respect can grow in such a way that people can appraise the situation and developments as objectively as possible.

Mode-2 has not yet solved the representativity and legitimacy issue: which stakeholders should be taken on board, how to organise accountability to society? To prevent problems, the stakeholders in a project should from time to time reflect on the external effects of their work and see if some stakeholders are missing. In relation to the experiments carried out in the papers published here, it should be noted that transition experiments are purposely often set up to involve innovative, forward thinking farmers and in areas where change is likely or needed (so called transition points, often areas with inconvenient truths). This is in line with transition theory but raises issues about how to upscale sustainable practices into the mainstream.

20.6 The role of the scientist

The papers in this book make it clear that some researchers in agriculture have left the isolation of their technical laboratory and experimental fields. The real world, with all its different actors, have become his or her working environment. Research is sometimes carried out in the analytical, observatory way of the social scientist, but often the research methods are participatory or even action oriented: the scientist becomes part of the change team, using engineering-style design capacities. In this way, the Mode 2 research model replaces Mode 1. The researchers involved in the experiments that are published in this book clearly work in a third generation university model (Rabbinge and Slingerland in Chapter 3). This leads to several questions concerning the role of the scientist. In some of the papers, the researcher is the traditional observant analyst looking at things from the sideline. This role seems to hold true for Van Dam *et al.* (Chapter 5) when observing self organisation and Fortuin *et al* in researching innovation processes in the multinational food business. Duineveld *et al.* (Chapter 17) explicitly discuss the ethical issues facing researchers. They defend the objective role of the researcher. In their view the researcher's task is deconstruction and he or she should take an amoral position. Other authors are in a much more participatory or action research mode. The papers that report on the learning processes in farmers' networks (e.g. Chapter 6 by Vogelezang *et al.*; Chapter 8 by Van Mierlo and Arkesteijn) are clear examples. This is perhaps not strictly true for the research teams that wrote these papers as 'trainers of the trainers'. Their job is to set up learning processes and coach their colleagues. But the colleagues they coach are clearly working in a participatory or action research mode.

This raises the question about the true role of these researchers. In the professional conversations that took place during the writing of this book, the authors recognised the importance of this issue, but to date there has been little reflection on the subject. From the consultancy profession we know that consultants can assume several roles: they can act as a plumber (fixing the holes), as an employee (helping to do the work), as an architect (designing the end result), as a film director (coaching different persons in their roles), as a guru (showing the principles for the way forward), as a therapist (counselling stakeholders with problems) or in several other roles. In these transition projects, it is not as clear which role(s) the researchers should take up without turning into a consultant, an extension worker or a teacher. In the event that it would become necessary for a researcher to fulfil (one of) the last three roles, this would have a marked effect on the role and organisation of the third generation university concept outlined by Rabbinge and Slingereland (Chapter 3). The researchers in these transition projects are often active as organiser of the learning and innovation process and act as a knowledge broker. This clearly requires special, T-shaped skills, enabling them to be able to communicate across disciplines and between research and practice. It is not as clear whether researchers are also able to successfully provide perspectives for changing behaviour for farmers or their co-scientists working with farmers. Innovators in the field are not interested in theory or explanations but would like to have workable strategies for their business.

Wielinga and Geerling-Eiff (Chapter 7) have reflected on this role for the researcher from the perspective of the requirements for achieving a process of transition. They emphasise the need for a free actor. Stakeholders in self organisation do not appear to be in a position to organise this for themselves nor to appoint someone to fulfil this free actor function. The authors have not yet been able to establish who should pay for this service, but at the present time the free actor appears to be a 'public good', paid for out of public funds.

The issue of quality assurance in research also needs to be addressed. Once the role of the researcher is clarified, there should be guidelines as to how the experiences won in projects can be generalised to move the scientific frontier forward. How can we do a better job in validating our results and make them more generic? Standard methods originating from social sciences (and perhaps especially from sociology and anthropology, but also case-based business research) could be useful. Given the large number of farmers that could take part in research, it would be interesting to find out to what extent control experiments could be set up to validate innovations in methodology.

20.7 What next?

Traditionally books like this end with the 'more research is needed' section, and we are certainly no exception. In our work and in writing this book we have increased our understanding of transition towards a sustainable agriculture and food chains in peri-urban areas, but although we have attained a higher level of understanding, we have a long way to go. As a result of the work published in this book, along with input received during the conference we organised in 2008 (www.agricultureintransition.eu), we can isolate the following areas as being important for our agenda for future research:

- *Extend the work on learning processes and innovation.* The work presented in this book places the learning processes from experiments central to innovation and transition towards a sustainable agriculture and food chains in peri-urban areas. Several reasons have been given to explain this micro approach to bridge the gap between the general public's expectations and the needs of the industry.
- *A closer alignment of the practical work carried out in projects with academic work.* If this could be achieved in areas of organisational change, multi-actor collaboration, network governance and social learning, it could be advantageous for reaching our goal. It is empirically obvious from these transition projects that there is not just one route towards sustainable agriculture but there are many possible routes. Entrepreneurial skills, co-creating knowledge, alignment of stakeholders, consumer feedback and upscaling seem to be key issues along these routes. Recent discussions on peri-urban agriculture suggest

that such issues are perhaps more easily solved by linking them to the metropolis than to (inter)national policy.

- *Develop better methods for monitoring and evaluation of transition.* In line with the previous point and the suggestions in paragraph 20.5 above, these methods should support social learning. This learning process should indicate the bottlenecks in the transition to sustainable agriculture: lack of technical know-how, conservatism, outdated institutions, the power of some stakeholders to block innovations, etc. Social learning process should include the results of the more traditional monitoring of outputs and outcome of agriculture to see the effects of the transition on sustainability so far.

Developments in these three areas could help to improve the sustainability of agriculture and food chains. As mentioned earlier, we believe that they will help in small ways, which is characteristic for the sector in which we are working. It is unlikely that this will lead to a sudden, big change that creates a very different type of agriculture as advocated by Michael Pollan (2008) and other food writers. The policy agenda setting theory suggests, in the words of Frank Baumgartner (Baumgartner and Jones, 2009), that nobody can control that type of change, but that you can prepare for the crisis. In this respect it seems that transition *management* has a limited potential for agriculture (Woodhill in Chapter 15). To prepare for these options (that might eventuate even if you don't approve of them, or have good reason to doubt their suitability) it seems to us that much more work should be done on issues like policy agenda setting, power relations, institutional arrangements and linking experiments with learning at the regime level. Foresight studies on trends outside of agriculture (including very diverse areas such as technology, geo-politics and art) could also be useful in this respect. Some of this work is probably more easily placed on the academic agenda than commissioned by an agriculture ministry.

The research and education system is both an important part of the agri-food complex and part of the transition. Because of this, we have a special interest in how the research system should be changed within the transition to sustainable agriculture and food chains. Several European countries have made changes to their systems over the last few years and it makes sense to monitor and evaluate these developments. More reflection on the role of researchers along the lines discussed in paragraph 20.6 above is also important.

All these issues could be addressed with more international cooperation on this topic than we have seen so far. In his recent book on transition in agriculture Wilson (2007) started his discussion on transitions in agriculture noting that transition in agriculture seems to be mainly an English debate, although he questioned if it was only relevant for England. In the Netherlands we organised the conference Transitions in Agriculture (www.transitionsinagriculture.eu) with a similar idea that we should make our experiences internationally available. The conference itself also concluded that there is room for more international collaboration, using EU projects or joint programming efforts to organise this. We hope that this book contributes towards making that vision a reality.

References

Baumgartner, F. and Jones, B., 2009. Agendas and instability in American politics (2nd edition). Chicago University Press: Chicago, USA.

Beer, M. and Nohria, N., 2000. Breaking the code of change. Harvard Business School: Boston, USA.

Boone, K., De Bont, C., Van Calker, K.J., Van der Knijff, A. and Leneman, H., 2007. Duurzame landbouw in beeld. Resultaten van de Nederlandse land- en tuinbouw op het gebied van people, planet en profit. LEI Wageningen UR: The Hague, the Netherlands.

De Bont, C.J.A.M., Van Everdingen, W.H. and Van der Knijff, A., 2008. Actuele ontwikkeling van bedrijfsresultaten en inkomens in 2008. LEI Wageningen UR: The Hague, the Netherlands.

Elzen, B., Geels, F.W. and Green, K. (Eds.), 2004. System innovation and the transition to sustainability. Edgar Elgar: Cheltenham, UK.

Gibbons, M., 1999. Science's new social contract with society. Nature 102: 11-18.

Giller, K.E., Leeuwis, C., Andersson, J.A., Andriesse, W., Brouwer, A., Frost, P., Hebinck, P., Heitkönig, I., Van Ittersum, M.K., Koning, N., Ruben, R., Slingerland, M., Udo, H., Veldkamp, T., Van de Vijver, C., Van Wijk, M.T. and Windmeijer, P., 2008. Competing claims on natural resources: what role for science? Ecology and Society 13(2): 34.

Kleiner, A. and Roth, G., 1997. How to make experience your company's best teacher. Harvard Business Review 75: 172-177.

Leydesdorff, L. and Etskowitz, H., 2003. Can the public be considered as a fourth helix in university-industry-government relationships? Report of the 4th Triple Helix conference. Science and Public Policy 30: 55-61.

Pollan, M., 2008. Farmer in chief. New York Times, October 9, 2008.

Poppe, K.J., 2008. Economic assessment of Dutch agricultural research. Report 2008-032. LEI: The Hague, the Netherlands.

Poppe, K.J., De Bont, K., Luttik, P., Pleijte, M., Schepers, H., Vogelzang, T. and De Vries, H., 2009. Toekomstverkenning van het kennissysteem en de belangenbehartiging in de agrosector. LEI, The Hague, the Netherlands.

Ross, J.P.M., Van den Born, G.J., Drissen, E., Faber, A., Farla, J.C.M., Nagelhout, D., Van Overbeeke, P., Weltevrede, W.R., Van Wijk, J.J. and Wilting, H.C., 2003. Methodiek voor de evaluatie van een transitie. Casus: transitie duurzame landbouw en voedingsketen. RIVM rapport 550011001/2003. RIVM: Bilthoven, the Netherlands. Available at: http://www.rivm.nl/bibliotheek/rapporten/550011001.pdf.

Rotmans, J., Kemp, R. and Van Asselt, M., 2001. Transition management in public policy – more evolution than revolution. Foresight 3: 15-31.

Syscope, 2008. Monitoring & evaluatie. Special Syscope 17.

Ten Pierick, E., Goddijn S. and Meeusen, M., 2006. Naar een gereedschapskist voor transitiemonitoring. LEI: The Hague, the Netherlands.

Ten Pierick E. and Van Mil, E., 2009. Multi-level perspectives. LEI Wageningen UR: The Hague, the Netherlands.

Termeer, C.J.A.M., 2006. Vital differences: on public leadership and societal innovation. Wageningen University: Wageningen, the Netherlands.

Van Bruchem, C., Silvis, H.J., Berkhout, P., Van Bommel, K., De Bont, K., Van Everdingen, W., De Kleijn, T. and Pronk, B., 2008. Agrarische structuur, trends en beleid: Ontwikkelingen in Nederland vanaf 1950. LEI Wageningen UR: The Hague, the Netherlands.

Van der Hoeven, D., 2009. Nieuw licht. NWO: The Hague, the Netherlands.

Wals, A.E.J. (Ed)., 2007. Social learning towards a sustainable world. Wageningen Academic Publishers, Wageningen, the Netherlands.

Weick, K., 2000. Emergent change as a universal in organisations. In: M. Beer and N. Nohria (Eds.) Breaking the code of change. Harvard Business School Press: Boston, MA, USA, pp. 223-241.

Wilson, G.A., 2007. Multifunctional Agriculture – a transition theory perspective. CABI: Wallingford, UK.

Contributors

N. Aarts, Prof. Dr.
Noelle Aarts is Professor of Strategic Communication at the University of Amsterdam and an Associate Professor at Wageningen University, where she also obtained her doctorate. Building on her background in Cultural Anthropology she studies inter-human processes and communication for creating space for change. She has published work on several topics such as negotiating conservation policies, dealing with ambivalence concerning farm animal welfare, and self-organisation and network-building for innovation.
E-mail: noelle.aarts@wur.nl

M.C.M. Arkesteijn, MSc.
Marlèn C.M. Arkesteijn (MSc. Rural Sociology, Wageningen University 1990) has specialised over the last 15 years in the monitoring and evaluation of (innovation) projects and programs in development cooperation, agriculture, and conservation. She not only uses written reports and reflection in her work, but wherever feasible she uses video and images for monitoring and evaluation purposes. Currently she is involved as monitor in an innovation network in integrated crop protection in maize production for Communication and Innovation Science, Wageningen UR.
E-mail: marlen.arkesteijn@wur.nl

R. Beunen, MSc.
Raoul Beunen is Assistant Professor in the Land Use Planning Group at Wageningen University. He is currently finishing his PhD thesis on the implementation of Natura 2000. His research interests include grassroots planning and implementation studies, with a particular focus on conservation and leisure management.
E-mail: raoul.beunen@wur.nl

A.P. Bos, Dr.
Bram Bos was trained as a biologist (Free University Amsterdam, 1993) and holds an MSc in Science and Technology Dynamics (University of Amsterdam 1993). He obtained his PhD in philosophy in 2004 at the Free University Amsterdam with a thesis on 'The role of living organisms in technological systems'. Since 2002 he has been involved as a senior researcher in several interdisciplinary innovation projects for sustainable animal husbandry at the Animal Sciences Group, Wageningen UR. His research interests lie in the field of technology studies, philosophy of technology and policy studies on technological innovation processes.
E-mail: bram.bos@wur.nl

G. Breeman, Dr.
Gerard Breeman graduated in public administration in 1996 and earned his PhD with a thesis on policy and trust in 2006 (Leiden University). He was Assistant Professor at Leiden University from 2004 to 2006, when he was appointed to the position of Assistant Professor at

Wageningen University. In 2009 he also became a fellow at the Montesquieu Institute in The Hague. His current research interests are the politics of attention, political agenda-setting, and governance and trust.
E-mail: Gerard.breeman@wur.nl

A. Dewulf, Dr.

Art Dewulf (1975) studied Organisational Psychology (Master's degree in 1999) and Social and Cultural Anthropology (Master's degree in 1998) at the University of Leuven, Belgium. He received his PhD. in organisational psychology in February 2006 at the same university, with a dissertation entitled 'Issue framing in multi-actor contexts' under the supervision of Prof. Dr. René Bouwen. After a year of interim lectureship and post-doc research at the University of Leuven, he started working as an Assistant Professor in the Public Administration and Policy Group at Wageningen University in October 2007. Art has been involved in inter-university cooperation with the University of Cuenca (Ecuador) in the fields of participatory technology development and land and water management, and in EU-sponsored projects on participation in river basin management and adaptive water management. His main research interests are collaborative governance, multi-actor conflict and negotiation, issue framing, natural resources management, transition management and adaptive governance.
E-mail: art.dewulf@wur.nl

M. Duineveld, Dr. Ir.

Martijn Duineveld is Assistant Professor in the Socio-spatial Analysis Group at Wageningen University. After obtaining his Bachelor's degree at the International Agricultural School Larenstein he went to Wageningen University to study spatial planning. After his graduation in 2002 he became a PhD student at the Socio-spatial Analysis group. He took his PhD degree in 2006 with a thesis on the politics of heritage. His current focus is on three interrelated lines of research: the multiple meanings of space and place, power strategies and the relationships between politics, policy and (scientific) knowledge. These lines are used to study the mechanisms of meaning and power production in the fields of spatial planning, landscape architecture and tourism and heritage.
E-mail: martijn.duineveld@wur.nl

R. During, MSc.

Roel During is Assistant Professor of the Belvedere Chair in Spatial Planning and Cultural Heritage and senior researcher in spatial planning and cultural heritage at Alterra, both part of Wageningen UR. He was educated as a landscape ecologist at the University of Utrecht (1988). He has lead many spatial planning projects and he currently leads an international project on cultural differences influencing interregional cooperation in multidisciplinary partnerships. At the same time, he is doing PhD research on the relevance and role of heritage in Europeanisation.
E-mail: roel.during@wur.nl

J. Eshuis, Dr.

Jasper Eshuis graduated from Wageningen University, where he also completed his PhD research on new institutional arrangements for conservation by farmers. His current research at Erasmus University of Rotterdam, Department Public Administration, is on governance in policy processes, with a special emphasis on public branding.

E-mail: eshuis@fsw.eur.nl

F.T.J.M. Fortuin, Dr.

Frances Fortuin graduated in biology at Groningen University in 1978. After a management career she took her PhD degree at Wageningen University in 2006 with a thesis on the management of innovation. She has worked as a researcher at the department of Business Administration at Wageningen University since 2007 and from 2008 she has combined this with the function of senior project manager for the Innovation Expertise Centre of the Food Valley Organisation. She is the author of many scientific articles and papers on innovation management, based on research in technology-based as well as in agri-food firms. Her current research interest encompasses innovation in regional innovation systems and in chains and networks in the agri-food industry.

E-mail: frances.fortuin@wur.nl

F.A. Geerling-Eiff, MSc.

Florentien A. Geerling-Eiff graduated in Communication Science and an additional minor study in Public Administration from the University of Amsterdam in 2000. She started working as a trainee at the Agricultural Economics Research Institute LEI Wageningen UR during her studies and has been working at LEI in social economic research ever since. She has worked as a researcher and project manager in several projects relating to knowledge application, knowledge management, communication, behaviour and innovation studies in the green domain. Since the end of 2007 she has been the manager of a cluster of policy supporting research projects regarding knowledge application and knowledge management within the green domain (BO-09 Cluster Knowledge).

E-mail: floor.geerling-eiff@wur.nl

P.W.G. Groot-Koerkamp, Prof. Dr. Ir.

Peter W.G. Groot-Koerkamp graduated in agricultural engineering at Wageningen University in 1990 and went on to take his PhD there in 1998. He has wide experience in the field of livestock husbandry and has held various positions at the Institute of Agricultural and Environmental Engineering and the Institute of Animal Husbandry, now the Animal Sciences Group. Since 2005 he has worked as a part time Professor in Bio-systems Engineering at Wageningen University. His focus is on technology development for sustainable livestock production, with special interest in environment, animal welfare and health. He combines this with the implementation of innovation theory in research projects.

E-mail: peter.grootkoerkamp@wur.nl

J.E. Jansma, MSc.

Jan Eelco Jansma has been active in several positions as a scientist within Applied Plant Research (Wageningen UR) since graduating in crop protection at Wageningen University in 1992. His main research interest is system innovation in agriculture, especially relating to town-country relations and sustainability. Jan Eelco is the founder of the *Agromere* research programme on Urban Agriculture in the city of Almere.
E-mail: janeelco.jansma@wur.nl

P. Kooij, Prof. Dr.

Pim Kooij graduated in 1970 at Groningen University in Economic and Social History and started as a lecturer at the Faculty of Economics. In 1986 he became a doctor in Economic Sciences (specialisation Urban History) from Groningen and was appointed as Professor in Regional History in 1988. Since 1998 he has combined chairs in Economic and Social History at Groningen University (Faculty of Arts) and Agrarian History at Wageningen University. His current research centres on town-country relations and the history of the green domain (mental as well as physical). He is chief editor of *Historia Agriculturae.*
E-mail: pim.kooij@wur.nl

J. Kruit, MSc.

Jeroen Kruit trained as a landscape architect (Wageningen University, 1997). He has been active as a researcher at the Alterra Research Institute for the Green World, Wageningen UR, for the past 10 years. His research and advisory functions focus on the locality (place) and its struggle with governmental spatial planning. Socio-cultural dimensions (understanding people's motives) in the development and conservation of landscape aesthetics are of pivotal importance in his research-work.
E-mail: Jeroen.kruit@wur.nl

J.R. Mulder

John Mulder is a researcher in Spatial Planning and Cultural Heritage at Alterra-Research Institute for the Green World, Landscape Centre Wageningen UR.
E-mail: john.mulder@wur.nl

S.W.F. Omta, Prof. Dr.

Onno Omta graduated in biochemistry from the University of Groningen in 1978 and, after a management career, he went on to obtain his PhD degree there in 1995, with a thesis on the management of innovation. In 2000 he was appointed as chaired Professor in Business Administration at Wageningen University. He is the Editor-in-Chief of the '*The Journal on Chain and Network Science*' and author of many scientific articles on innovation management. His current research interest encompasses innovation in chains and networks in the agri-food industry.
E-mail: onno.omta@wur.nl

K.J. Poppe, MSc.
Krijn J. Poppe was the manager of the project set up to publish this book. Trained as an economist (Erasmus University Rotterdam, 1981), he has been active in several positions at the Agricultural Economics Research Institute LEI Wageningen UR. His current research interest is in foresight studies with scenario analysis and system dynamics, especially within agricultural research and innovation systems. He has wide experience in policy research and is internationally active in EU research projects and is currently Secretary General of the European Association of Agricultural Economists. Within Wageningen UR, he manages a research program on transitions, institutions, governance and policy.
E-mail: krijn.poppe@wur.nl

R. Rabbinge, Prof. Dr. Ir.
Rudy Rabbinge was trained as a phytopathologist and received his PhD in Biological Control in 1976. He is a former Professor in Theoretical Production Ecology, in Crop Ecology and in Farming Systems and is currently Professor in Sustainable Development & Food Security at Wageningen University. He co-chaired the UN Panel InterAcademy Council on Food Security for Africa and the Millennium Assessment. He is currently actively involved as either a member or chairman, or on the Board of Directors of numerous organisations including Tropenbos International, Earth and Life Sciences Council of the Royal Netherlands Academy of Arts and Sciences (KNAW), Royal Institute for the Tropics (KIT), Alliance for a Green Revolution in Africa (AGRA) and the Science Council of the Consultative Group on International Agricultural Research (CGIAR). Until recently he was also an elected representative in the Dutch Senate.
E-mail: rudy.rabbinge@wur.nl

I.E. Salverda, MSc.
Irini E. Salverda was trained as a spatial planner (Catholic University of Nijmegen, 1999). She has been active as a researcher at Alterra-Research Institute for the Green World, Wageningen UR, for the past 10 years. Her current research interest is in the relation between self-organisation of citizens in the rural and urban landscape and governmental spatial planning. Socio-cultural dimensions (meaning, experience, bonding, language, etc.) of the landscape are of great importance in her research work.
E-mail: irini.salverda@wur.nl

H.B. Schoorlemmer, Ir.
Herman Schoorlemmer studied agricultural economics at Wageningen University. He is senior researcher in farm economics and management at Applied Plant Research, Wageningen UR. His current interest relates to the improvement of entrepreneurial skills of farmers with the help of interactive research approaches. He is internationally active in EU research projects about entrepreneurship and farm management.
E-mail: herman.schoorlemmer@wur.nl

L.H.G. Slangen, Dr.
Louis Slangen is Associate Professor of Agricultural Economics and Rural Policy at Wageningen University. He is teaching, doing research and supervising MSc and PhD theses in the field of resource economics, economics of land-use, new institutional economics and economic organisation theory. His publication record consists of 35 articles in journals subject to peer review, 15 chapters in books, and 10 books. He has extensive experience in many different types of research projects for various Dutch Ministries and for the European Union.
E-mail: louis.slangen@wur.nl

M.A. Slingerland, PhD.
Maja Slingerland is employed by Wageningen University Plant Production Systems Group and spends half of her time with the Sustainable Development and Food Security Group. It was with them that she took part in the publication of this book. Trained in farming systems in Wageningen University (MSc 1989; PhD thesis 2000) she lived in West Africa for 10 years and worked for FAO, SNV and the Wageningen University Outreach Centre in programmes on natural resource management. Her research focus is on the relationship between technology development and institutions. Over time she has specialised in research and scientific coordination of interdisciplinary research programmes, each with a multitude of partners. Currently she coordinates research programmes on Competing Claims on Natural Resources in Southern Africa, in Brazil and in the Netherlands and she is part of the research management team for the strategic Wageningen University programme Scaling and Governance. All programmes have strong linkages with policy makers and in each of them large numbers of PhD and MSc researchers are being trained.
E-mail: maja.slingerland@wur.nl

C.J.A.M. Termeer, Prof. Dr. Ir.
Catrien Termeer graduated in land use planning in 1986 at Wageningen University. She earned her PhD in 1993 with a thesis on 'Processes of change in pig husbandry networks' at Erasmus University Rotterdam. Since then she has worked at the strategic policy bureau of the ministry of Agriculture, Nature and Food and for a private consultancy firm. In 2005 she was appointed as Professor and Chair of the new Public Administration and Policy Group at Wageningen University. Her main fields of interests are change management; public leadership; innovations in public-private alliances and new modes of governance.
E-mail: katrien.termeer@wur.nl

K. Van Assche, Prof. Dr.
Kristof Van Assche is currently Associate Professor in the Planning and Community Development Program at Minnesota State University, St Cloud State. Before that, he taught at Wageningen University. He is interested in the cultural, political, institutional embeddings of spatial planning and design, topics on which he has published widely. For a number of years, he has been working on transitional institutions in Eastern Europe and the Caucasus.
E-mail: kvanassche@stcloudstate.edu

R.I. van Dam, MSc.

Rosalie van Dam studied Public Administration (University of Nijmegen, 2001). Since then she has been working as a researcher in Public Administration and Sociology at Alterra Wageningen UR. She has been active in different national and international projects involving, among other issues, the analysis and development of methods and strategies regarding spatial/heritage planning and policy processes; the role of culture in international cooperation and planning processes; power structures between involved actors; citizens' involvement; development and involvement of (local) communities; and social, spatial and administrative trends and their consequences. She is currently working on a PhD thesis dealing with self organisation (citizens' initiating and organising their living surroundings and heritage) and spatial and social bonding processes.
E-mail: rosalie.vandam@wur.nl

G.F.V. van der Peet, MSc

Geert F.V. van der Peet is senior researcher at the Animal Sciences Group of Wageningen UR. After his study in Animal Production at Wageningen University (1984) he had several jobs in education, research, consultancy and with the government. At the moment he works within the Animal Sciences Group at Wageningen UR on sustainable agricultural systems and more specifically on projects in the public/private arena such as innovation agendas, pig and cattle husbandry or the governmental agenda for animal health.
E-mail: geert.vanderpeet@wur.nl

J.D. van der Ploeg, Prof. Dr. Ir.

Jan Douwe van der Ploeg is Professor of Transition Studies at Wageningen University. He was Professor in Rural Sociology until 2004. Among other things he has supervised a large, NWO funded, comparative study on novelty production and transition processes, which was carried out in South Africa, Italy and the Netherlands. He is currently engaged in an international study on rural development processes in China, Brazil and the European Union.
E-mail: jandouwe.vanderploeg@wur.nl

B.C. van Mierlo, PhD.

Barbara C. van Mierlo works as a sociologist in the field of 'management of innovations, environment and ecology' in the Communication Science Group at Wageningen University. Her main interest is in processes of (system) innovation towards sustainable development by means of innovation projects. In her work she focuses on network building, debating, negotiating and learning in relation to change of practices and institutional change. She also focuses on the role of communication in environmental policy. She is Manager of the project *Reflexive Monitoring in Action*, which is conducted in cooperation with the Athena Institute, at the Free University of Amsterdam.
E-mail: barbara.vanmierlo@wur.nl

A.J. Visser, PhD.

Andries Visser graduated in eco-physiology in 1991 (University of Utrecht) and holds a PhD in ecology (1997, Free University, Amsterdam). He is coordinator of the research group Rural Innovations of the Applied Plant Research Group (Wageningen UR) and is also research manager of the research program on multifunctional farming systems (Ministry of Agriculture, Nature and Food quality). His current research interest is in multifunctional farming systems and sustainable urban-rural relationships.
E-mail: andries.visser@wur.nl

J.V.M. Vogelezang, Dr. Ir.

José Vogelezang was educated in horticultural sciences at the University of Wageningen. She earned her PhD in the field of energy saving strategies (Wageningen UR, 1993) and has been active in several positions at the Plant Sciences Group Wageningen UR as researcher and manager. Her current field of research is into methodological aspects of transition and innovation, and its implementation in national and regional innovation programmes and new institutional arrangements such as public/private partnerships. She now manages a cluster of system innovation programmes, financed by the Dutch Ministry of Agriculture, Nature and Food Quality (LNV) whose purpose is to facilitate the transition to a more sustainable agriculture. She is also a member of several regional platforms for innovation in the agricultural sector.
E-mail: jose.vogelezang@wur.nl

A.E.J. Wals, Prof. Dr. Ir.

Arjen E.J. Wals, (PhD, University of Michigan, Ann Arbor – 1991) holds a UNESCO Chair and is Professor in Social Learning and Sustainable Development within the Education and Competence Studies Group of Wageningen University in the Netherlands. He has (co) published and (co)edited over 150 articles, chapters, books and professional publications on topics such as: action research and community problem-solving, whole school approaches to sustainability, biodiversity education, and sustainability in higher education.
E-mail: arjen.wals@wur.nl

T.J. Weijschedé, Ir.

Titus Weijschedé is doing research on cultural heritage, related on spatial planning at Alterra Wageningen UR. His main interest is investigating the role heritage can have in the construction of new social partnerships between spatial planning and cultural activities. His work has enabled him to facilitate the connection between research and the social partners. He has specific experience with finding new uses for military heritage sites at the European level. His current research interest revolves around how the adaptation of immigrants can be positively influenced by preserving cultural heritage.
E-mail: titus.weijschede@wur.nl

R.A. Werkman, Dr.

Renate Werkman is Assistant Professor of the Public Administration and Policy Group at Wageningen University. She specialises in change in organisations and institutions. Her focus is on deadlocking patterns caused by actors in change processes, and how these patterns can be resolved. Her main field of research is agriculture. She has analysed deadlocking interaction patterns in the public debate about the future of agriculture, the rural area and the Common Agricultural Policy and the debate about the socialisation of agricultural policy in the Netherlands.

E-mail: renate.werkman@wur.nl

H.E. Wielinga, Dr. Ir.

Eelke Wielinga is a researcher connected in a part time capacity to the Agricultural Economics Research Institute LEI Wageningen UR, investigating the dynamics of networks in innovation and land development. He is working in collaboration on language and tools for networkers who mobilise actors around an initiative. In his PhD thesis 'Networks as Living Tissue' (2001) he developed an organic perspective for understanding the role of knowledge, leadership and government in Dutch agriculture since 1945. He studied Human Nutrition and Extension Science at Wageningen University (MSc in 1981), and worked in Bénin and the Philippines. As officer of the Ministry of Agriculture, Nature Management and Fisheries he has held several positions related to extension and knowledge policies in the Netherlands.

E-mail: eelke.wielinga@wur.nl

F.G. Wijnands, Ir.

Frank Wijnands trained at the University of Wageningen as an agronomist (1984) and began his professional career in farming systems research in arable farming and field production of vegetables. Together with pilot farm-based networks in plant production he is involved with pioneering work on the on-site development and testing of more sustainable farming systems such as integrated and organic systems in close cooperation between farmers, research and extension services. As a research programme leader he became involved in system innovation and transition projects in 2002. He currently coordinates the government sponsored research into organic agriculture in the Netherlands. He also manages 'Future for Farming', a large network based research project in the framework of the national agreement on crop protection. It aims to achieve a marked increase in the application of more sustainable crop protection measures in mainstream agricultural practice.

E-mail: frank.wijnands@wur.nl

A.J. Woodhill, PhD.

Jim Woodhill is Programme Director at Wageningen International. He has over 20 years experience in sustainable development across the fields of agriculture, food systems, natural resources and economic development. His PhD in political economics examined tensions between local level process of empowerment and larger scale processes of globalisation. He has worked on these issues at the international level and in Europe, Australia, Africa and Asia.

Jim has a special interest in applying systems thinking and participatory learning approaches to complex development issues. He has extensive experience as a process facilitator and in programme evaluation. Part of his work with Wageningen International involves supporting development practitioners and organisations to strengthen capacities for facilitating multi-stakeholder processes.
E-mail: jim.woodhill@wur.nl

A.N. Van der Zande, Prof. Dr.

André van der Zande holds the Belvedere Chair in Spatial Planning and Cultural Heritage at Wageningen University. He has been trained as a biologist (Leiden, University, 1976; PhD in 1980). He has worked for more than 25 years in the field of nature management and spatial planning with various management responsibilities. Since 2006, he has been appointed as Secretary General for the Ministry of Agriculture, Food Quality and Nature. Nearly all his professional work involves taking an international view on public policy making.
E-mail: andre.vanderzande@wur.nl

Index

G

H

I

N

O

P

R

S